STUDENT'S SOLUTIONS MANUAL
DEANA J. RICHMOND

MATHEMATICS FOR BUSINESS
SEVENTH EDITION

STANLEY A. SALZMAN
American River College

CHARLES D. MILLER

GARY CLENDENEN
University of Texas at Tyler

Addison
Wesley

Boston San Francisco New York
London Toronto Sydney Tokyo Singapore Madrid
Mexico City Munich Paris Cape Town Hong Kong Montreal

D1444430

ISBN   0-321-06921-8

1 2 3 4 5 6 7 8 9 10 VG 04 03 02 01 00

PREFACE

This manual provides complete solutions for many of the exercises in *Mathematics for Business,* Seventh Edition, by Stanley A. Salzman, Charles D. Miller, and Gary Clendenen. Solutions are provided for the odd–numbered section-level exercises (including those contained in the appendices) and for all of the Chapter Review exercises, Summary exercises, and Cumulative Review exercises.

This manual should be used as an aid as you learn to master your course work. Try to solve the exercises on your own before you refer to the solutions in this manual. Then, if you have difficulty, study these solutions. The solutions have been written so that they are consistent with the methods and format used in the textbook examples.

The following people have made valuable contributions to the production of the *Student's Solutions Manual:* Linda Buchanan and Cynthia Chang, accuracy checkers, and Judy Martinez, typist.

CONTENTS

Part 1 Basic Mathematics

Part 2 Basic Business Applications

CHAPTER 4 BANKING SERVICES

CHAPTER 5 PAYROLL

CHAPTER 6 TAXES

CHAPTER 7 RISK MANAGEMENT

Part 3 Mathematics of Retailing

CHAPTER 8 MATHEMATICS OF BUYING

CHAPTER 9 MARKUP

CHAPTER 10 MARKDOWN AND INVENTORY CONTROL

Part 4 Mathematics of Finance

CHAPTER 11 SIMPLE INTEREST

CHAPTER 12 NOTES AND BANK DISCOUNT

CHAPTER 13 COMPOUND INTEREST

CHAPTER 14 ANNUITIES AND SINKING FUNDS

CHAPTER 15 BUSINESS AND CONSUMER LOANS

Part 5 Advanced Accounting

CHAPTER 16 DEPRECIATION

CHAPTER 17 FINANCIAL STATEMENTS AND RATIOS

CHAPTER 18 SECURITIES AND DISTRIBUTION OF PROFIT AND OVERHEAD

CHAPTER 19 BUSINESS STATISTICS

PART 5 CUMULATIVE REVIEW CHAPTERS 16–19

APPENDIX A CALCULATOR BASICS

APPENDIX B THE METRIC SYSTEM

Chapter 1

Problem Solving and Operations with Fractions

1.1 Problem Solving

1. $80 + 75 + 135 + 40 + 52 = 382$

Beth rode 382 miles.

3. $200 - 70 = 130$

There were 130 more crimes in 1997.

5. $75 \times 16 = 1200$

The distance is 1200 miles.

7. $2425 - 582 + 634 = 2477$

The car will weigh 2477 pounds.

9. $24,000,000 - 7000 = 23,993,000$

There are 23,993,000 small and midsize business.

11. $15,293 \times \$25,000 = \$382,325,000$

The total cost of the buyouts was \$382,325,000.

13. $\$90 - \$65 = \$25$
$\$25 \times 5 = \$125$

The amount saved is \$125.

15. $(6 \times \$1256) + (15 \times \$895) = 20,961$

The total cost is \$20,961.

17. $1250 - (30 \times 25) = 500$ seats in the balcony
$500 \div 25 = 20$

There must be 20 seats in each row.

19. $4.4 \times 8 = 35.2$

35.2 hours would be needed.

21. $38 \div 0.58 = 65.5$

There are 65.5 million shares.

23. $221 \div 8.359 = 26$

26 coins can be purchased.

25. $124 \times \$0.29 = \$35.96$

The amount paid is \$35.96 million
(or \$35,960,000).

27. (a) $37 \times 4.3 = 159.1$

The manager worked 159.1 hours each month.

(b) $\$2365 \div 159.1 = \$14.86$

The manager earned \$14.86 per hour.

29. (a) $29,800 - 21,700 = 8100$

The difference in the salaries is \$8100.

(b) $44,523 - 27,038 = 17,485$

The salary of a person with a bachelor's degree is \$17,485 higher than that of a high school graduate.

1.2 Addition and Subtraction of Fractions

1. $1\frac{3}{8} = \frac{(8 \times 1) + 3}{8} = \frac{11}{8}$

3. $4\frac{1}{4} = \frac{(4 \times 4) + 1}{4} = \frac{17}{4}$

5. $22\frac{7}{8} = \frac{(8 \times 22) + 7}{8} = \frac{183}{8}$

7. $12\frac{5}{8} = \frac{(8 \times 12) + 5}{8} = \frac{101}{8}$

9. $\frac{8}{16} = \frac{8 \div 8}{16 \div 8} = \frac{1}{2}$

11. $\frac{40}{75} = \frac{40 \div 5}{75 \div 5} = \frac{8}{15}$

13. $\frac{25}{40} = \frac{25 \div 5}{40 \div 5} = \frac{5}{8}$

15. $\frac{120}{150} = \frac{120 \div 30}{150 \div 30} = \frac{4}{5}$

17. $\frac{132}{144} = \frac{132 \div 12}{144 \div 12} = \frac{11}{12}$

19. $\frac{96}{180} = \frac{96 \div 12}{180 \div 12} = \frac{8}{15}$

21. $2\overline{)7}$ with quotient 3, $\frac{6}{1}$ $\qquad \frac{7}{2} = 3\frac{1}{2}$

$$\begin{array}{r} 3 \\ 2\overline{)7} \\ \underline{6} \\ 1 \end{array} \qquad \frac{7}{2} = 3\frac{1}{2}$$

23. $$\begin{array}{r} 3 \\ 20\overline{)76} \\ \underline{60} \\ 16 \end{array} \qquad \frac{76}{20} = 3\frac{16}{20} = 3\frac{4}{5}$$

25. $$\begin{array}{r} 1 \\ 11\overline{)14} \\ \underline{11} \\ 3 \end{array} \qquad \frac{14}{11} = 1\frac{3}{11}$$

27. $$\begin{array}{r} 1 \\ 15\overline{)21} \\ \underline{15} \\ 6 \end{array} \qquad \frac{21}{15} = 1\frac{6}{15} = 1\frac{2}{5}$$

29. $$\begin{array}{r} 1 \\ 64\overline{)124} \\ \underline{64} \\ 60 \end{array} \qquad \frac{124}{64} = 1\frac{60}{64} = 1\frac{15}{16}$$

31. $$\begin{array}{r} 2 \\ 32\overline{)81} \\ \underline{64} \\ 17 \end{array} \qquad \frac{81}{32} = 2\frac{17}{32}$$

33. Answers will vary.

35. $\dfrac{2}{5} + \dfrac{1}{5} = \dfrac{2+1}{5} = \dfrac{3}{5}$

37. $\dfrac{7}{10} + \dfrac{3}{20} = \dfrac{14}{20} + \dfrac{3}{20} = \dfrac{14+3}{20} = \dfrac{17}{20}$

39. $\dfrac{7}{12} + \dfrac{8}{15} = \dfrac{35}{60} + \dfrac{32}{60} = \dfrac{35+32}{60}$

$$= \dfrac{67}{60} = 1\dfrac{7}{60}$$

41. $\dfrac{9}{11} + \dfrac{1}{22} = \dfrac{18}{22} + \dfrac{1}{22} = \dfrac{18+1}{22} = \dfrac{19}{22}$

43. $\dfrac{3}{4} + \dfrac{5}{9} + \dfrac{1}{3} = \dfrac{27}{36} + \dfrac{20}{36} + \dfrac{12}{36}$

$$= \dfrac{27+20+12}{36} = \dfrac{59}{36} = 1\dfrac{23}{36}$$

45. $\dfrac{5}{6} + \dfrac{3}{4} + \dfrac{5}{8} = \dfrac{20}{24} + \dfrac{18}{24} + \dfrac{15}{24}$

$$= \dfrac{20+18+15}{24} = \dfrac{53}{24} = 2\dfrac{5}{24}$$

47. $$\begin{array}{r} 82\frac{3}{5} \\ + \ 15\frac{1}{5} \\ \hline 97\frac{4}{5} \end{array}$$

49. $$\begin{array}{r} 51\frac{1}{4} = \ 51\frac{1}{4} \\ + \ 29\frac{1}{2} = \ 29\frac{2}{4} \\ \hline 80\frac{3}{4} \end{array}$$

51. $$\begin{array}{r} 32\frac{3}{4} = \ 32\frac{18}{24} \\ 6\frac{1}{3} = \ 6\frac{8}{24} \\ + \ 14\frac{5}{8} = \ 14\frac{15}{24} \\ \hline 52\frac{41}{24} = 52 + 1\frac{17}{24} = 53\frac{17}{24} \end{array}$$

53. $$\begin{array}{r} 89\frac{5}{9} = \ 89\frac{5}{9} \\ 10\frac{1}{3} = \ 10\frac{3}{9} \\ + \ 87\frac{1}{9} = \ 87\frac{1}{9} \\ \hline 186\frac{9}{9} = 186 + 1 = 187 \end{array}$$

55. $\dfrac{7}{8} - \dfrac{3}{8} = \dfrac{7-3}{8} = \dfrac{4}{8} = \dfrac{1}{2}$

57. $\dfrac{2}{3} - \dfrac{1}{6} = \dfrac{4}{6} - \dfrac{1}{6} = \dfrac{4-1}{6} = \dfrac{3}{6} = \dfrac{1}{2}$

59. $\dfrac{5}{12} - \dfrac{1}{16} = \dfrac{20}{48} - \dfrac{3}{48} = \dfrac{17}{48}$

61. $\dfrac{3}{4} - \dfrac{5}{12} = \dfrac{9}{12} - \dfrac{5}{12} = \dfrac{9-5}{12} = \dfrac{4}{12} = \dfrac{1}{3}$

63. $$\begin{array}{r} 16\frac{3}{4} = \ 16\frac{6}{8} \\ - \ 12\frac{3}{8} = \ 12\frac{3}{8} \\ \hline 4\frac{3}{8} \end{array}$$

65. $$\begin{array}{r} 9\frac{7}{8} = \ 9\frac{21}{24} \\ - \ 6\frac{5}{12} = \ 6\frac{10}{24} \\ \hline 3\frac{11}{24} \end{array}$$

67. $$\begin{array}{r} 71\frac{3}{8} = \ 71\frac{9}{24} \\ - \ 62\frac{1}{3} = \ 62\frac{8}{24} \\ \hline 9\frac{1}{24} \end{array}$$

69. $$\begin{array}{r} 19 \ = \ 18\frac{4}{4} \\ -12\frac{3}{4} = \ 12\frac{3}{4} \\ \hline 6\frac{1}{4} \end{array}$$

71. Answers will vary.

73. Answers will vary.

75. $\dfrac{1}{8} + \dfrac{1}{4} + \dfrac{2}{5} = \dfrac{5}{40} + \dfrac{10}{40} + \dfrac{16}{40}$

$$= \dfrac{5+10+16}{40} = \dfrac{31}{40}$$

The total length of the screw is $\frac{31}{40}$ inch.

77. $1\dfrac{7}{8} + \dfrac{1}{2} + 1\dfrac{2}{3} + \dfrac{1}{3} = 1\dfrac{21}{24} + \dfrac{12}{24} + 1\dfrac{16}{24} + \dfrac{8}{24}$

$$= 2\dfrac{57}{24} = 4\dfrac{9}{24} = 4\dfrac{3}{8}$$

The total distance around the wetlands is $4\frac{3}{8}$ miles.

79. $\frac{15}{16} - \left(\frac{3}{8} + \frac{3}{8}\right) = \frac{15}{16} - \frac{6}{8} = \frac{15}{16} - \frac{12}{16} = \frac{3}{16}$

The diameter of the hole is $\frac{3}{16}$ inch.

81. $5\frac{1}{2} + 6\frac{1}{4} + 3\frac{3}{4} + 7 = 5\frac{2}{4} + 6\frac{1}{4} + 3\frac{3}{4} + 7$

$\qquad = 21\frac{6}{4} = 22\frac{2}{4}$

$\qquad = 22\frac{1}{2}$

Hernanda drove $22\frac{1}{2}$ hours.

83.
$$83\frac{5}{8} = 83\frac{15}{24}$$
$$76\frac{3}{4} = 76\frac{18}{24}$$
$$+ \ 182\frac{1}{3} = 182\frac{8}{24}$$
$$341\frac{41}{24} = 342\frac{17}{24}$$

She owns $342\frac{17}{24}$ acres.

85.
$$16\frac{1}{2} = 16\frac{4}{8}$$
$$12\frac{1}{8} = 12\frac{1}{8}$$
$$8\frac{3}{4} = 8\frac{6}{8}$$
$$+ \ 12\frac{5}{8} = 12\frac{5}{8}$$
$$48\frac{16}{8} = 50$$

Comet Auto Supply sold 50 cases.

87. $40 - \left(8\frac{1}{4} + 6\frac{1}{6} + 7\frac{2}{3} + 8\frac{3}{4}\right)$

$\qquad = 40 - \left(8\frac{3}{12} + 6\frac{2}{12} + 7\frac{8}{12} + 8\frac{9}{12}\right)$

$\qquad = 40 - \left(29\frac{22}{12}\right)$

$\qquad = 40 - 30\frac{10}{12}$

$\qquad = 39\frac{12}{12} - 30\frac{10}{12}$

$\qquad = 9\frac{2}{12} = 9\frac{1}{6}$

Pam worked $9\frac{1}{6}$ hours on Friday.

89. $352\frac{1}{8} - \left(71\frac{3}{8} + 18\frac{1}{2} + 143\frac{5}{8}\right)$

$\qquad = 352\frac{1}{8} - \left(71\frac{3}{8} + 18\frac{4}{8} + 143\frac{5}{8}\right)$

$\qquad = 352\frac{1}{8} - \left(232\frac{12}{8}\right)$

$\qquad = 351\frac{9}{8} - 233\frac{4}{8}$

$\qquad = \$118\frac{5}{8}$

The price she paid for the fourth share was $\$118\frac{5}{8}$.

1.3   Multiplication and Division of Fractions

1. $\frac{5}{\overset{}{\underset{4}{8}}} \times \frac{\overset{1}{\cancel{2}}}{3} = \frac{5 \times 1}{4 \times 3} = \frac{5}{12}$

3. $\frac{9}{10} \times \frac{11}{16} = \frac{9 \times 11}{10 \times 16} = \frac{99}{160}$

5. $1\frac{2}{3} \times 2\frac{7}{10} = \frac{\overset{1}{\cancel{5}}}{\underset{1}{\cancel{3}}} \times \frac{\overset{9}{\cancel{27}}}{\underset{2}{\cancel{10}}} = \frac{1 \times 9}{1 \times 2}$

$\qquad = \frac{9}{2} = 4\frac{1}{2}$

7. $4\frac{3}{5} \times 15 = \frac{23}{\underset{1}{\cancel{5}}} \times \frac{\overset{3}{\cancel{15}}}{1}$

$\qquad = \frac{23 \times 3}{1 \times 1} = 69$

9. $\frac{5}{9} \times 2\frac{1}{4} \times 3\frac{2}{3}$

$\qquad = \frac{5}{\underset{1}{\cancel{9}}} \times \frac{\overset{1}{\cancel{9}}}{4} \times \frac{11}{3} = \frac{5 \times 1 \times 11}{1 \times 4 \times 3}$

$\qquad = \frac{55}{12} = 4\frac{7}{12}$

11. $12 \times 2\frac{1}{2} \times 3$

$\qquad = \frac{\overset{6}{\cancel{12}}}{1} \times \frac{5}{\underset{1}{\cancel{2}}} \times \frac{3}{1}$

$\qquad = \frac{6 \times 5 \times 3}{1 \times 1 \times 1} = 90$

13. $\frac{1}{6} \div \frac{1}{3} = \frac{1}{\underset{2}{\cancel{6}}} \times \frac{\overset{1}{\cancel{3}}}{1}$

$\qquad = \frac{1 \times 1}{2 \times 1} = \frac{1}{2}$

15. $\frac{13}{20} \div \frac{26}{30} = \frac{\overset{1}{\cancel{13}}}{\underset{2}{\cancel{20}}} \times \frac{\overset{3}{\cancel{30}}}{\underset{2}{\cancel{26}}}$

$\qquad = \frac{1 \times 3}{2 \times 2} = \frac{3}{4}$

17. $\frac{15}{16} \div \frac{5}{8}$

$\qquad = \frac{\overset{3}{\cancel{15}}}{\underset{2}{\cancel{16}}} \times \frac{\overset{1}{\cancel{8}}}{\underset{1}{\cancel{5}}} = \frac{3 \times 1}{2 \times 1} = \frac{3}{2} = 1\frac{1}{2}$

19. $2\frac{1}{2} \div 3\frac{3}{4}$

$\qquad = \frac{5}{2} \div \frac{15}{4} = \frac{\overset{1}{\cancel{5}}}{\underset{1}{\cancel{2}}} \times \frac{\overset{2}{\cancel{4}}}{\underset{3}{\cancel{15}}} = \frac{1 \times 2}{1 \times 3} = \frac{2}{3}$

21. $3\frac{1}{8} \div \frac{15}{16}$

$$= \frac{25}{8} \div \frac{15}{16} = \frac{\overset{5}{\cancel{25}}}{\cancel{8}} \times \frac{\overset{2}{\cancel{16}}}{\cancel{15}} = \frac{5 \times 2}{1 \times 3} = \frac{10}{3} = 3\frac{1}{3}$$

23. $6 \div 1\frac{1}{4} = 6 \div \frac{5}{4} = \frac{6}{1} \times \frac{4}{5} = \frac{6 \times 4}{1 \times 5} = \frac{24}{5} = 4\frac{4}{5}$

25. Answers will vary.

27. Multiply 80 and $39\frac{5}{8}$.

$$80 \times 39\frac{5}{8} = \frac{\overset{10}{\cancel{80}}}{1} \times \frac{317}{\cancel{8}} = \frac{10 \times 317}{1 \times 1} = 3170$$

The total price for 80 shares of Mattel stock is $3170.

29. Multiply 24 and $74\frac{3}{4}$.

$$24 \times 74\frac{3}{4} = \frac{\overset{6}{\cancel{24}}}{1} \times \frac{299}{\cancel{4}} = \frac{6 \times 299}{1 \times 1} = 1794$$

The total price for 24 shares of Merck stock is $1794.

31. Multiply 32 and $57\frac{1}{8}$.

$$32 \times 57\frac{1}{8} = \frac{\overset{4}{\cancel{32}}}{1} \times \frac{457}{\cancel{8}} = \frac{4 \times 457}{1} = 1828$$

The total price for 32 shares of Ford Motor stock is $1828.

33. $0.8 = \frac{8}{10} = \frac{4}{5}$

35. $0.24 = \frac{24}{100} = \frac{6}{25}$

37. $0.73 = \frac{73}{100}$

39. $0.875 = \frac{875}{1000} = \frac{7}{8}$

41. $0.0375 = \frac{375}{10,000} = \frac{3}{80}$

43. $0.1875 = \frac{1875}{10,000} = \frac{3}{16}$

45. 3.5218 to the nearest tenth is 3.5.
Locate the tenths digit and draw a line.

$$3.5|218$$

Since the digit to the right of the line is 2, leave the tenths digit alone.

3.5218 to the nearest hundredth is 3.52.
Locate the hundredths digit and draw a line.

$$3.52|18$$

Since the digit to the right of the line is 1, leave the hundredths digit alone.

47. 0.0837 to the nearest tenth is 0.1.
Locate the tenths digit and draw a line.

$$0.0|837$$

Since the digit to the right of the line is 8, increase the tenths digit by 1.

0.0837 to the nearest hundredth is 0.08.
Locate the hundredths digit and draw a line.

$$0.08|37$$

Since the digit to the right of the line is 3, leave the hundredths digit alone.

49. 8.643 to the nearest tenth is 8.6.
Locate the tenths digit and draw a line.

$$8.6|43$$

Since the digit to the right of the line is 4, leave the tenths digit alone.

8.643 to the nearest hundredth is 8.64.
Locate the hundredths digit and draw a line.

$$8.64|3$$

Since the digit to the right of the line is 3, leave the hundredths digit alone.

51. 58.956 to the nearest tenth is 59.0.
Locate the tenths digit and draw a line.

$$58.9|56$$

Since the digit to the right of the line is 5, increase the digit by 1.

58.956 to the nearest hundredth is 58.96.
Locate the hundredths digit and draw a line.

$$58.95|6$$

Since the digit to the right of the line is 6, increase the hundredths digit by 1.

53. $\frac{3}{4} = 0.75$

$$\begin{array}{r} 0.75 \\ 4\overline{)3.00} \\ \underline{2\ 8} \\ 20 \\ \underline{20} \\ 0 \end{array}$$

55. $\frac{3}{8} = 0.375$

$$\begin{array}{r} 0.375 \\ 8\overline{)3.000} \\ \underline{2\ 4} \\ 60 \\ \underline{56} \\ 40 \\ \underline{40} \\ 0 \end{array}$$

57. $\frac{1}{6} = 0.167$ or $0.1\overline{6}$

$$\begin{array}{r} 0.1666 \\ 6\overline{)1.0000} \\ \underline{6} \\ 40 \\ \underline{36} \\ 40 \\ \underline{36} \\ 40 \\ \underline{36} \\ 4 \end{array}$$

59. $\frac{13}{16} = 0.813$

$$\begin{array}{r} 0.8125 \\ 16\overline{)13.0000} \\ \underline{12\ 8} \\ 20 \\ \underline{16} \\ 40 \\ \underline{32} \\ 80 \\ \underline{80} \\ 0 \end{array}$$

61. $\frac{8}{25} = 0.32$

$$\begin{array}{r} 0.32 \\ 25\overline{)8.00} \\ \underline{7\ 5} \\ 50 \\ \underline{50} \\ 0 \end{array}$$

63. $\frac{1}{99} = 0.010$

$$\begin{array}{r} 0.0101 \\ 99\overline{)1.0000} \\ \underline{99} \\ 10 \\ \underline{0} \\ 100 \\ \underline{99} \\ 1 \end{array}$$

65. Answers will vary.

67. Answers will vary.

69. Multiply 16 and $2\frac{1}{4}$.

$$16 \times 2\frac{1}{4} = \frac{\overset{4}{\cancel{16}}}{1} \times \frac{9}{\underset{1}{\cancel{4}}} = \frac{4 \times 9}{1 \times 1} = 36$$

Laura needs 36 yards of ribbon.

71. Divide 5025 by $8\frac{3}{8}$.

$$5025 \div 8\frac{3}{8} = \frac{5025}{1} \div \frac{67}{8}$$

$$= \frac{\overset{75}{\cancel{5025}}}{1} \times \frac{8}{\underset{1}{\cancel{67}}}$$

$$= \frac{75 \times 8}{1 \times 1} = 600$$

Bobbi bought 600 shares.

73. Divide 1314 by $109\frac{1}{2}$.

$$1314 \div 109\frac{1}{2} = \frac{1314}{1} \div \frac{219}{2}$$

$$= \frac{\overset{6}{\cancel{1314}}}{1} \times \frac{2}{\underset{1}{\cancel{219}}} = 12$$

12 homes can be fitted with baseboards.

75. Multiply $12\frac{1}{2}$ and $1\frac{3}{4}$.

$$12\frac{1}{2} \times 1\frac{3}{4} = \frac{25}{2} \times \frac{7}{4} = \frac{25 \times 7}{2 \times 4}$$

$$= \frac{175}{8} = 21\frac{7}{8}$$

$21\frac{7}{8}$ ounces of chemical are needed.

77. Multiply $12\frac{3}{4}$ and 28. Then multiply $7\frac{1}{8}$ and 16. Add the totals.

$$12\frac{3}{4} \times 28 = \frac{51}{\cancel{4}_{1}} \times \frac{\cancel{28}^{7}}{1} = \frac{51 \times 7}{1 \times 1} = 357$$

$$7\frac{1}{8} \times 16 = \frac{57}{\cancel{8}_{1}} \times \frac{\cancel{16}^{2}}{1} = \frac{57 \times 2}{1 \times 1} = 114$$

$357 + 114 = 471$

The total number of rolls needed is 471.

79. Divide 40 by $\frac{2}{3}$.

$$40 \div \frac{2}{3} = \frac{\cancel{40}^{20}}{1} \times \frac{3}{\cancel{2}_{1}}$$

$$= \frac{20 \times 3}{1 \times 1} = 60$$

60 trips are needed to deliver the wood.

Chapter 1 Review Exercises

1. $\dfrac{24}{40} = \dfrac{24 \div 8}{40 \div 8} = \dfrac{3}{5}$

2. $\dfrac{32}{64} = \dfrac{32 \div 32}{64 \div 32} = \dfrac{1}{2}$

3. $\dfrac{27}{81} = \dfrac{27 \div 27}{81 \div 27} = \dfrac{1}{3}$

4. $\dfrac{147}{294} = \dfrac{147 \div 147}{294 \div 147} = \dfrac{1}{2}$

5. $\dfrac{63}{70} = \dfrac{63 \div 7}{70 \div 7} = \dfrac{9}{10}$

6. $\dfrac{84}{132} = \dfrac{84 \div 12}{132 \div 12} = \dfrac{7}{11}$

7. $\dfrac{24}{1200} = \dfrac{24 \div 24}{1200 \div 24} = \dfrac{1}{50}$

8. $\dfrac{375}{1000} = \dfrac{375 \div 125}{1000 \div 125} = \dfrac{3}{8}$

9. $\begin{array}{r} 8 \\ 8)\overline{65} \\ \underline{64} \\ 1 \end{array}$ $\dfrac{65}{8} = 8\frac{1}{8}$

10. $\begin{array}{r} 4 \\ 12)\overline{56} \\ \underline{48} \\ 8 \end{array}$ $\dfrac{56}{12} = 4\frac{8}{12} = 4\frac{2}{3}$

11. $\begin{array}{r} 1 \\ 24)\overline{38} \\ \underline{24} \\ 14 \end{array}$ $\dfrac{38}{24} = 1\frac{14}{24} = 1\frac{7}{12}$

12. $\begin{array}{r} 7 \\ 7)\overline{55} \\ \underline{49} \\ 6 \end{array}$ $\dfrac{55}{7} = 7\frac{6}{7}$

13. $\begin{array}{r} 2 \\ 45)\overline{120} \\ \underline{90} \\ 30 \end{array}$ $\dfrac{120}{45} = 2\frac{30}{45} = 2\frac{2}{3}$

14. $\begin{array}{r} 8 \\ 24)\overline{196} \\ \underline{192} \\ 4 \end{array}$ $\dfrac{196}{24} = 8\frac{4}{24} = 8\frac{1}{6}$

15. $\begin{array}{r} 8 \\ 32)\overline{258} \\ \underline{256} \\ 2 \end{array}$ $\dfrac{258}{32} = 8\frac{2}{32} = 8\frac{1}{16}$

16. $\begin{array}{r} 3 \\ 64)\overline{194} \\ \underline{192} \\ 2 \end{array}$ $\dfrac{194}{64} = 3\frac{2}{64} = 3\frac{1}{32}$

17. $\dfrac{5}{8} + \dfrac{7}{12} = \dfrac{15}{24} + \dfrac{14}{24}$

$$= \dfrac{15 + 14}{24} = \dfrac{29}{24} = 1\frac{5}{24}$$

18. $\dfrac{1}{5} + \dfrac{3}{10} + \dfrac{3}{8} = \dfrac{8}{40} + \dfrac{12}{40} + \dfrac{15}{40}$

$$= \dfrac{8 + 12 + 15}{40}$$

$$= \dfrac{35}{40} = \dfrac{7}{8}$$

19. $\dfrac{5}{7} - \dfrac{1}{3} = \dfrac{15}{21} - \dfrac{7}{21} = \dfrac{15 - 7}{21} = \dfrac{8}{21}$

20. $\dfrac{3}{4} - \dfrac{2}{3} = \dfrac{9}{12} - \dfrac{8}{12} = \dfrac{9 - 8}{12} = \dfrac{1}{12}$

21. $\begin{array}{r} 25\frac{1}{6} = 25\frac{1}{6} \\ + 46\frac{2}{3} = 46\frac{4}{6} \\ \hline 71\frac{5}{6} \end{array}$

22.
$$18\tfrac{3}{5} = 18\tfrac{18}{30}$$
$$47\tfrac{7}{10} = 47\tfrac{21}{30}$$
$$+\ 25\tfrac{8}{15} = 25\tfrac{16}{30}$$
$$90\tfrac{55}{30}$$

$$90\tfrac{55}{30} = 90 + 1\tfrac{25}{30} = 91\frac{25}{30} = 91\tfrac{5}{6}$$

23.
$$6\tfrac{7}{12} = 6\tfrac{7}{12}$$
$$-\ 2\tfrac{1}{3} = 2\tfrac{4}{12}$$
$$4\tfrac{3}{12} = 4\tfrac{1}{4}$$

24.
$$92\tfrac{5}{16} = 92\tfrac{5}{16}$$
$$-\ 11\tfrac{1}{4} = 11\tfrac{4}{16}$$
$$81\tfrac{1}{16}$$

25. Add to find the total cost per square.

$$54.52 + 35.75 + 3.65 = 93.92$$

Multiply to find the total cost for 26.3 squares.

$$93.92 \times 26.3 = 2470.096$$

The total cost is $2470.10.

26. Subtract to find the gallons of water saved per flush.

$$3.4 - 1.6 = 1.8$$

Multiply to find the number of flushes in one year.

$$22 \times 365 = 8030$$

Multiply to find the gallons of water saved in one year.

$$1.8 \times 8030 = 14{,}454$$

14,454 gallons of water are saved.

27. Add to find the total hours.

$$5\tfrac{1}{2} + 6\tfrac{1}{4} + 3\tfrac{3}{4} + 7 = 5\tfrac{2}{4} + 6\tfrac{1}{4} + 3\tfrac{3}{4} + 7$$
$$= 21\tfrac{6}{4} = 22\tfrac{2}{4} = 22\tfrac{1}{2}$$

Desiree worked $22\tfrac{1}{2}$ hours.

28. Add to find the total number of gallons used

$$68\tfrac{1}{2} + 37\tfrac{3}{8} + 5\tfrac{3}{4}$$
$$= 68\tfrac{4}{8} + 37\tfrac{3}{8} + 5\tfrac{6}{8} = 110\tfrac{13}{8}$$
$$= 111\tfrac{5}{8}$$

There were $147\tfrac{1}{2}$ gallons of paint. Subtract to find the gallons of paint remaining.

$$147\tfrac{1}{2} = 146\tfrac{12}{8}$$
$$-\ 111\tfrac{5}{8} = 111\tfrac{5}{8}$$
$$35\tfrac{7}{8}$$

There are $35\tfrac{7}{8}$ gallons of paint remaining.

29. Add the measurements of the three sides.

$$202\tfrac{1}{8} = 202\tfrac{1}{8}$$
$$370\tfrac{3}{4} = 370\tfrac{6}{8}$$
$$+\ 274\tfrac{1}{2} = 274\tfrac{4}{8}$$
$$846\tfrac{11}{8} = 846 + 1\tfrac{3}{8} = 847\tfrac{3}{8}$$

Subtract to find the length of the fourth side.

$$1166\tfrac{7}{8}$$
$$-\ 847\tfrac{3}{8}$$
$$319\tfrac{4}{8} = 319\tfrac{1}{2}$$

The length of the fourth side is $319\tfrac{1}{2}$ feet.

30. Add the weights of each type of cheese.

$$12\tfrac{2}{3} = 12\tfrac{16}{24}$$
$$16\tfrac{1}{8} = 16\tfrac{3}{24}$$
$$15\tfrac{1}{2} = 15\tfrac{12}{24}$$
$$+\ 10\tfrac{1}{6} = 10\tfrac{4}{24}$$
$$53\tfrac{35}{24} = 53 + 1\tfrac{11}{24} = 54\tfrac{11}{24}$$

The total weight of the cheese is $54\tfrac{11}{24}$ pounds.

31. $\dfrac{5}{\overset{}{\underset{4}{\cancel{8}}}} \times \dfrac{\overset{1}{\cancel{2}}}{3} = \dfrac{5 \times 1}{4 \times 3} = \dfrac{5}{12}$

32. $\dfrac{1}{\underset{1}{\cancel{3}}} \times \dfrac{7}{8} \times \dfrac{\overset{1}{\cancel{3}}}{5} = \dfrac{1 \times 7 \times 1}{1 \times 8 \times 5} = \dfrac{7}{40}$

33. $\dfrac{1}{6} \div \dfrac{1}{3} = \dfrac{1}{\underset{2}{\cancel{6}}} \times \dfrac{\overset{1}{\cancel{3}}}{1} = \dfrac{1 \times 1}{2 \times 1} = \dfrac{1}{2}$

34. $10 \div \frac{5}{8} = \frac{10}{1} \div \frac{5}{8} = \frac{\overset{2}{\cancel{10}}}{1} \times \frac{8}{\cancel{5}} = \frac{2 \times 8}{1 \times 1} = 16$

35. $2\frac{1}{2} \div 3\frac{3}{4} = \frac{5}{2} \div \frac{15}{4} = \frac{\overset{1}{\cancel{5}}}{\cancel{2}} \times \frac{\overset{2}{\cancel{4}}}{\cancel{15}}$

$\phantom{2\frac{1}{2} \div 3\frac{3}{4}} = \frac{1 \times 2}{1 \times 3} = \frac{2}{3}$

36. $3\frac{3}{4} \div \frac{27}{16} = \frac{15}{4} \div \frac{27}{16}$

$\phantom{3\frac{3}{4} \div \frac{27}{16}} = \frac{\overset{5}{\cancel{15}}}{\underset{1}{\cancel{4}}} \times \frac{\overset{4}{\cancel{16}}}{\underset{9}{\cancel{27}}}$

$\phantom{3\frac{3}{4} \div \frac{27}{16}} = \frac{5 \times 4}{1 \times 9} = \frac{20}{9} = 2\frac{2}{9}$

37. $12\frac{1}{2} \times 1\frac{2}{3}$

$ = \frac{25}{2} \times \frac{5}{3} = \frac{25 \times 5}{2 \times 3} = \frac{125}{6} = 20\frac{5}{6}$

38. $12\frac{1}{3} \div 2 = \frac{37}{3} \div \frac{2}{1} = \frac{37}{3} \times \frac{1}{2}$

$ = \frac{37 \times 1}{3 \times 2} = \frac{37}{6} = 6\frac{1}{6}$

39. Multiply 16.5 and 0.48. Multiply 3 and 1.05. Add to find the total amount spent.

$16.5 \times 0.48 = 7.92$
$3 \times 1.05 = 3.15$
$7.92 + 3.15 = 11.07$

Subtract to find the change.

$(3 \times 5) - 11.07 = 15 - 11.07 = 3.93$

The change was $3.93.

40. Divide 1.4 by 0.39.

$1.4 \div 0.39 \approx 3.5897 \approx 3.6$

There are 3.6 million shares.

41. One-third of the land is sold. So two-thirds of the land is left.

Multiply $\frac{2}{3}$ and $63\frac{3}{4}$.

$\frac{2}{3} \times 63\frac{3}{4} = \frac{\overset{1}{\cancel{2}}}{\underset{1}{\cancel{3}}} \times \frac{\overset{85}{\cancel{255}}}{\underset{2}{\cancel{4}}} = \frac{1 \times 85}{1 \times 2} = \frac{85}{2} = 42\frac{1}{2}$

There are $42\frac{1}{2}$ acres left.

42. Multiply 25 and $8\frac{3}{8}$; multiply 16 and $12\frac{1}{4}$; then add.

$25 \times 8\frac{3}{8} = \frac{25}{1} \times \frac{67}{8} = \frac{25 \times 67}{1 \times 8} = \frac{1675}{8}$

$16 \times 12\frac{1}{4} = \frac{\overset{4}{\cancel{16}}}{1} \times \frac{49}{\underset{1}{\cancel{4}}} = \frac{4 \times 49}{1 \times 1} = 196$

$\frac{1675}{8} + 196 = \frac{1675}{8} + \frac{1568}{8}$

$\phantom{\frac{1675}{8} + 196} = \frac{1675 + 1568}{8} = \frac{3243}{8}$

$\phantom{\frac{1675}{8} + 196} = 405\frac{3}{8} \approx 405.38$

Ellen paid $405.38 for the stock.

43. Divide $157\frac{1}{2}$ by $4\frac{3}{8}$.

$157\frac{1}{2} \div 4\frac{3}{8} = \frac{315}{2} \div \frac{35}{8}$

$\phantom{157\frac{1}{2} \div 4\frac{3}{8}} = \frac{\overset{9}{\cancel{315}}}{\underset{1}{\cancel{2}}} \times \frac{\overset{4}{\cancel{8}}}{\underset{1}{\cancel{35}}}$

$\phantom{157\frac{1}{2} \div 4\frac{3}{8}} = \frac{9 \times 4}{1 \times 1} = 36$

36 pull cords can be made.

44. There are 8 store managers, so each manager will receive $\frac{1}{8}$.

Find $\frac{1}{8}$ of $\frac{2}{3}$.

$\frac{1}{8} \times \frac{2}{3} = \frac{1}{\underset{4}{\cancel{8}}} \times \frac{\overset{1}{\cancel{2}}}{3} = \frac{1 \times 1}{4 \times 3} = \frac{1}{12}$

Each store manager will receive $\frac{1}{12}$ of the profit.

45. $0.25 = \frac{25}{100} = \frac{1}{4}$

46. $0.625 = \frac{625}{1000} = \frac{5}{8}$

47. $0.93 = \frac{93}{100}$

48. $0.005 = \frac{5}{1000} = \frac{1}{200}$

49. 68.433 to the nearest tenth is 68.4.
Locate the tenths digit and draw a line.

68.4|33

Since the digit to the right of the line is 3, leave the tenths digit alone.

68.433 to the nearest hundredth is 68.43.
Locate the hundredths digit and draw a line.

68.43|3

Since the digit to the right of the line is 3, leave the hundredths digit alone.

50. 975.536 to the nearest tenth is 975.5.
Locate the tenths digit and draw a line.

975.5|36

Since the digit to the right of the line is 3, leave the tenths digit alone.

975.536 to the nearest hundredth is 975.54.
Locate the hundredths digit and draw a line.

975.53|6

Since the digit to the right of the line is 6, increase the hundredths digit by one.

51. 0.3549 to the nearest tenth is 0.4.
Locate the tenths digit and draw a line.

0.3|549

Since the digit to the right of the line is 5, increase the tenths digit by one.

0.3549 to the nearest hundredth is 0.35.
Locate the hundredths digit and draw a line.

0.35|49

Since the digit to the right of the line is 4, leave the hundredths digit alone.

52. 8.025 to the nearest tenth is 8.0.
Locate the tenths digit and draw a line.

8.0|25

Since the digit to the right of the line is 2, leave the tenths digit alone.

8.025 to the nearest hundredth is 8.03.
Locate the hundredths digit and draw a line.

8.02|5

Since the digit to the right of the line is 5, increase the hundredths digit by one.

53. 6.965 to the nearest tenth is 7.0.
Locate the tenths digit and draw a line.

6.9|65

Since the digit to the right of the line is 6, increase the tenths digit by one.

6.965 to the nearest hundredth is 6.97.
Locate the hundredths digit and draw a line.

6.96|5

Since the digit to the right of the line is 5, increase the hundredths digit by one.

54. 0.428 to the nearest tenth is 0.4.
Locate the tenths digit and draw a line.

0.4|28

Since the digit to the right of the line is 2, leave the tenths digit alone.

0.428 to the nearest hundredth is 0.43.
Locate the hundredths digit and draw a line.

0.42|8

Since the digit to the right of the line is 8, increase the hundredths digit by one.

55. 0.955 to the nearest tenth is 1.0.
Locate the tenths digit and draw a line.

0.9|55

Since the digit to the right of the line is 5, increase the tenths digit by one.

0.955 to the nearest hundredth is 0.96.
Locate the hundredths digit and draw a line.

0.95|5

Since the digit to the right of the line is 5, increase the hundredths digit by one .

56. 71.249 to the nearest tenth is 71.2.
Locate the tenths digit and draw a line.

71.2|49

Since the digit to the right of the line is 4, leave the tenths digit alone.

71.249 to the nearest hundredth is 71.25.
Locate the hundredths digit and draw a line.

71.24|9

Since the digit to the right of the line is 9, increase the hundredths digit by one.

57. $\dfrac{5}{8} = 0.625$

$$
\begin{array}{r}
0.625 \\
8)\overline{5.000} \\
\underline{4\ 8} \\
20 \\
\underline{16} \\
40 \\
\underline{40} \\
0
\end{array}
$$

58. $\dfrac{3}{4} = 0.75$

$$
\begin{array}{r}
0.75 \\
4)\overline{3.00} \\
\underline{2\ 8} \\
20 \\
\underline{20} \\
0
\end{array}
$$

59. $\dfrac{5}{6} = 0.833$ or $0.8\overline{3}$

$$
\begin{array}{r}
0.8333 \\
6)\overline{5.0000} \\
\underline{4\ 8} \\
20 \\
\underline{18} \\
20 \\
\underline{18} \\
20 \\
\underline{20} \\
0
\end{array}
$$

60. $\dfrac{7}{16} = 0.438$

$$
\begin{array}{r}
0.4375 \\
16)\overline{7.0000} \\
\underline{6\ 4} \\
60 \\
\underline{48} \\
120 \\
\underline{112} \\
80 \\
\underline{80} \\
0
\end{array}
$$

Chapter 1 Summary Exercise

(a) $\dfrac{3}{8} = 0.375;\ \dfrac{3}{16} = 0.1875;\ \dfrac{7}{8} = 0.875;\ \dfrac{13}{16} = 0.8125$

(b) Difference is

$$
\begin{array}{r}
37\frac{7}{8} \ =\ 37\frac{14}{16} \\
-\ 25\frac{3}{16} \ =\ 25\frac{3}{16} \\
\hline
12\frac{11}{16} = 12.6875\ (\$12.69)
\end{array}
$$

(c) Difference is

$$
\begin{array}{r}
38\frac{3}{16} \ =\ 37\frac{19}{16} \\
-\ 22\frac{9}{16} \ =\ 22\frac{9}{16} \\
\hline
15\frac{10}{16} = 15\frac{5}{8} = 15.625\ (\$15.63)
\end{array}
$$

(d) Difference for AT/T is

$$
\begin{array}{r}
68\frac{1}{2} \\
-\ 33 \\
\hline
35\frac{1}{2} = 35.5\ \ \$35.50
\end{array}
$$

(e) Cost

$$80 \times 87\frac{7}{8} = 80 \times 87.875 \qquad \frac{7}{8} = 0.875$$
$$= \$7030$$

(f) Cost

$$100 \times 54\frac{5}{16} = 100 \times 54.3125 \qquad \frac{5}{16} = 0.3125$$
$$= \$5431.25$$

(g) Number of shares

$$2000 \div 32\frac{3}{8} = 2000 \div 32.375$$
$$\approx 61.78\ \ (61\ \text{shares})$$

(h) Coca Cola gain per share

$$
\begin{array}{r}
78\frac{1}{16} \ =\ 78\frac{1}{16} \ =\ 77\frac{17}{16} \\
-\ 63\frac{1}{2} \ =\ 63\frac{8}{16} \ =\ 63\frac{8}{16} \\
\hline
14\frac{9}{16}
\end{array}
$$

Coca Cola total gain

$$50 \times 14\frac{9}{16} = 50 \times 14.5625$$
$$= \$728.13$$

Reynolds Metals loss per share

$$
\begin{array}{r}
76\frac{3}{4} \ =\ 76\frac{12}{16} \ =\ 75\frac{28}{16} \\
-\ 60\frac{13}{16} \ =\ 60\frac{13}{16} \ =\ 60\frac{13}{16} \\
\hline
15\frac{15}{16}
\end{array}
$$

Reynolds Metals total loss

$$30 \times 15\tfrac{15}{16} = 30 \times 15.9375$$
$$= \$478.13$$

Total gain

$$728.13 - 478.13 = \$250.00$$

EQUATIONS AND FORMULAS

2.1 Solving Equations

Check each answer in this section by substituting into the original equation.

1.
$$b + 15 = 74$$
$$b + 15 - 15 = 74 - 15 \quad \textit{Subtract } 15$$
$$b = 59$$

3.
$$r - 45 = 12$$
$$r - 45 + 45 = 12 + 45 \quad \textit{Add } 45$$
$$r = 57$$

5.
$$25 = x + 12$$
$$25 - 12 = x + 12 - 12 \quad \textit{Subtract } 12$$
$$13 = x$$

7.
$$10k = 42$$
$$\frac{10k}{10} = \frac{42}{10} \quad \textit{Divide by } 10$$
$$k = 4.2$$

9.
$$12q = 144$$
$$\frac{12q}{12} = \frac{144}{12} \quad \textit{Divide by } 12$$
$$q = 12$$

11.
$$60 = 30m$$
$$\frac{60}{30} = \frac{30m}{30} \quad \textit{Divide by } 30$$
$$2 = m$$

13.
$$5.9y = 17.7$$
$$\frac{5.9y}{5.9} = \frac{17.7}{5.9} \quad \textit{Divide by } 5.9$$
$$y = 3$$

15.
$$1.54 = 0.7y$$
$$\frac{1.54}{0.7} = \frac{0.7y}{0.7} \quad \textit{Divide by } 0.7$$
$$2.2 = y$$

17.
$$3.92w = 3.136$$
$$\frac{3.92w}{3.92} = \frac{3.136}{3.92} \quad \textit{Divide by } 3.92$$
$$w = 0.8$$

19.
$$0.0002x = 0.08$$
$$\frac{0.0002x}{0.0002} = \frac{0.08}{0.0002} \quad \textit{Divide by } 0.0002$$
$$x = 400$$

21.
$$\frac{s}{7} = 42$$
$$\frac{s}{7} \cdot 7 = 42 \cdot 7 \quad \textit{Multiply by } 7$$
$$\frac{s}{7} \cdot 7 = 294$$
$$s = 294$$

23.
$$\frac{r}{7} = 1$$
$$\frac{r}{7} \cdot 7 = 1 \cdot 7 \quad \textit{Multiply by } 7$$
$$r = 7$$

25.
$$\frac{2}{3}b = 8$$
$$\frac{3}{2} \cdot \frac{2}{3}b = \frac{3}{2} \cdot 8 \quad \textit{Multiply by } \tfrac{3}{2}$$
$$\frac{\cancel{3}^{1}}{\cancel{2}_{1}} \cdot \frac{\cancel{2}^{1}}{\cancel{3}_{1}}b = \frac{\cancel{8}^{4}}{1} \cdot \frac{3}{\cancel{2}_{1}}$$
$$b = 12$$

27.
$$35 = \frac{7}{5}t$$
$$\frac{5}{7} \cdot 35 = \frac{5}{7} \cdot \frac{7}{5}t \quad \textit{Multiply by } \tfrac{5}{7}$$
$$\frac{5}{\cancel{7}_{1}} \cdot \frac{\cancel{35}^{5}}{1} = \frac{\cancel{5}^{1}}{7} \cdot \frac{\cancel{7}^{1}}{\cancel{5}_{1}}t$$
$$25 = t$$

29. $2x = \dfrac{5}{3}$

   $\dfrac{1}{2} \cdot 2x = \dfrac{1}{2} \cdot \dfrac{5}{3}$ *Multiply by $\frac{1}{2}$*

   $x = \dfrac{5}{6}$

31. $3p = \dfrac{5}{12}$

   $\dfrac{1}{3} \cdot 3p = \dfrac{1}{3} \cdot \dfrac{5}{12}$ *Multiply by $\frac{1}{3}$*

   $p = \dfrac{5}{36}$

33. $7b + 9 = 37$

   $7b + 9 - 9 = 37 - 9$ *Subtract 9*

   $7b = 28$

   $\dfrac{7b}{7} = \dfrac{28}{7}$ *Divide by 7*

   $b = 4$

35. $7y - 23 = 58$

   $7y - 23 + 23 = 58 + 23$ *Add 23*

   $7y = 81$

   $\dfrac{7y}{7} = \dfrac{81}{7}$ *Divide by 7*

   $y = \dfrac{81}{7} = 11\dfrac{4}{7}$

37. $6p + 41.5 = 69.4$

   $6p + 41.5 - 41.5 = 69.4 - 41.5$ *Subtract 41.5*

   $6p = 27.9$

   $\dfrac{6p}{6} = \dfrac{27.9}{6}$ *Divide by 6*

   $p = 4.65$

39. $6c + \dfrac{3}{4} = 8$

   $6c + \dfrac{3}{4} - \dfrac{3}{4} = 8 - \dfrac{3}{4}$ *Subtract $\frac{3}{4}$*

   $6c = \dfrac{29}{4}$

   $\dfrac{1}{6} \cdot 6c = \dfrac{1}{6} \cdot \dfrac{29}{4}$ *Multiply by $\frac{1}{6}$*

   $c = \dfrac{29}{24} = 1\dfrac{5}{24}$

41. $7q - \dfrac{2}{3} = 4$

   $7q - \dfrac{2}{3} + \dfrac{2}{3} = 4 + \dfrac{2}{3}$ *Add $\frac{2}{3}$*

   $7q = \dfrac{14}{3}$

   $\dfrac{1}{7} \cdot 7q = \dfrac{1}{\overset{}{\underset{1}{7}}} \cdot \dfrac{\overset{2}{14}}{3}$ *Multiply by $\frac{1}{7}$*

   $q = \dfrac{2}{3}$

43. $5.2z - 4 = 1.2$

   $5.2z - 4 + 4 = 1.2 + 4$ *Add 4*

   $5.2z = 5.2$

   $\dfrac{5.2z}{5.2} = \dfrac{5.2}{5.2}$ *Divide by 5.2*

   $z = 1$

45. $27.85 = 3 + 7.1p$

   $27.85 - 3 = 3 - 3 + 7.1p$ *Subtract 3*

   $24.85 = 7.1p$

   $\dfrac{24.85}{7.1} = \dfrac{7.1p}{7.1}$ *Divide by 7.1*

   $3.5 = p$

47. $7m + 4m - 5m = 78$

   $6m = 78$ *Combine like terms*

   $\dfrac{6m}{6} = \dfrac{78}{6}$ *Divide by 6*

   $m = 13$

49. $2s + s + 3s = 12$

   $6s = 12$ *Combine like terms*

   $\dfrac{6s}{6} = \dfrac{12}{6}$ *Divide by 6*

   $s = 2$

51. $5y + 2 = 3(y + 4)$

   $5y + 2 = 3y + 12$ Use the distributive
property on the right side.

   $5y + 2 - 2 = 3y + 12 - 2$ *Subtract 2*

   $5y = 3y + 10$

   $5y - 3y = 3y - 3y + 10$ *Subtract 3y*

   $2y = 10$

   $\dfrac{2y}{2} = \dfrac{10}{2}$ *Divide by 2*

   $y = 5$

53. $3(m-4) = m+2$

$3m - 12 = m + 2$ *Use the distributive property on the left side.*

$3m - 12 + 12 = m + 2 + 12$ *Add 12*

$3m = m + 14$

$3m - m = m - m + 14$ *Subtract m*

$2m = 14$

$\dfrac{2m}{2} = \dfrac{14}{2}$ *Divide by 2*

$m = 7$

55. $4(y+8) = 3(y+14)$ *Use the distributive property.*

$4y + 32 = 3y + 42$

$4y + 32 - 32 = 3y + 42 - 32$ *Subtract 32*

$4y = 3y + 10$

$4y - 3y = 3y - 3y + 10$ *Subtract 3y*

$y = 10$

57. $\dfrac{3}{4}s + \dfrac{1}{5}s = \dfrac{4}{5}$

$\dfrac{15}{20}s + \dfrac{4}{20}s = \dfrac{4}{5}$ *Combine like terms*

$\dfrac{19}{20}s = \dfrac{4}{5}$

$\dfrac{\overset{1}{\cancel{20}}}{\underset{1}{\cancel{19}}} \cdot \dfrac{\overset{1}{\cancel{19}}}{\underset{1}{\cancel{20}}}s = \dfrac{4}{\underset{1}{\cancel{5}}} \cdot \dfrac{\overset{4}{\cancel{20}}}{19}$ *Multiply by $\frac{20}{19}$*

$s = \dfrac{16}{19}$

59. $\dfrac{3}{8}y + \dfrac{1}{4} = \dfrac{9}{8}y - \dfrac{1}{4}$

$\dfrac{3}{8}y + \dfrac{1}{4} + \dfrac{1}{4} = \dfrac{9}{8}y - \dfrac{1}{4} + \dfrac{1}{4}$ *Add $\frac{1}{4}$*

$\dfrac{3}{8}y + \dfrac{2}{4} = \dfrac{9}{8}y$

$\dfrac{3}{8}y - \dfrac{3}{8}y + \dfrac{1}{2} = \dfrac{9}{8}y - \dfrac{3}{8}y$ *Subtract $\frac{3}{8}y$*

$\dfrac{1}{2} = \dfrac{6}{8}y$

$\dfrac{1}{2} = \dfrac{3}{4}y$

$\dfrac{\overset{2}{\cancel{4}}}{3} \cdot \dfrac{1}{\underset{1}{\cancel{2}}} = \dfrac{\overset{1}{\cancel{4}}}{\underset{1}{\cancel{3}}} \cdot \dfrac{\overset{1}{\cancel{3}}}{\underset{1}{\cancel{4}}}y$ *Multiply by $\frac{4}{3}$*

$\dfrac{2}{3} = y$

61. $2(y+1) = 4(4 - 2.5y)$

$2y + 2 = 16 - 10y$ *Use the distributive property*

$2y + 10y + 2 = 16 - 10y + 10y$ *Add 10y*

$12y + 2 = 16$

$12y + 2 - 2 = 16 - 2$ *Subtract 2*

$12y = 14$

$\dfrac{12y}{12} = \dfrac{14}{12}$ *Divide by 12*

$y = \dfrac{14}{12} = 1\frac{2}{12} = 1\frac{1}{6}$

63. $0.7452(3k - 1) = 3.94956$

$2.2356k - 0.7452 = 3.94956$

Use the distributive property on the left side.

$2.2356k - 0.7452 + 0.7452 = 3.94956 + 0.7452$

Add 0.7452.

$2.2356k = 4.69476$

$\dfrac{2.2356k}{2.2356} = \dfrac{4.69476}{2.2356}$

Divide by 2.2356.

$k = 2.1$

65. $1.2(2 + 3r) = 0.8(2r + 5)$

$2.4 + 3.6r = 1.6r + 4$ *Use the distributive property*

$2.4 + 3.6r - 1.6r = 1.6r - 1.6r + 4$ *Subtract 1.6r*

$2.4 + 2r = 4$

$2.4 - 2.4 + 2r = 4 - 2.4$ *Subtract 2.4*

$2r = 1.6$

$\dfrac{2r}{2} = \dfrac{1.6}{2}$ *Divide by 2*

$r = 0.8$

67. Answers will vary.

2.2   Applications of Equations

1. 27 plus a number

$27 + x$

3. a number added to 22

$22 + x$

5. 4 less than a number

$x - 4$

7. subtract $3\frac{1}{2}$ from a number

$x - 3\frac{1}{2}$

9. triple a number

$$3x$$

11. three fifths of a number

$$\frac{3}{5}x$$

13. the quotient of 9 and a number

$$\frac{9}{x}$$

15. 16 divided by a number

$$\frac{16}{x}$$

17. the product of 2.1 and the sum of 4 plus a number

$$2.1(4 + x)$$

19. 7 times the difference of a number and 3

$$7(x - 3)$$

21. The cost of 12 CDs at y dollars each is $12y$.

23. The amount that should be ordered is $472 - x$.

25. $73 - x$ are nonunion.

27. One crate of berries costs $\frac{172}{x}$.

29. Robin has a $21 - x$ books left.

31. Let $n =$ the number.

3 times 7	plus a number	is	36
3×7	$+ n$	$=$	36

Solve the equation.

$$\begin{aligned}
(3 \times 7) + n &= 36 \\
21 + n &= 36 \\
21 - 21 + n &= 36 - 21 \quad \textit{Subtract 21} \\
n &= 15
\end{aligned}$$

33. Let $n =$ the number.

6	times	4 minus a number	is	15
6	$\times$	$4 - n$	$=$	15

Solve the equation.

$$\begin{aligned}
6(4 - n) &= 15 \\
24 - 6n &= 15 \quad \textit{Distributive property} \\
24 - 24 - 6n &= 15 - 24 \quad \textit{Subtract 24} \\
-6n &= -9 \\
\frac{-6n}{-6} &= \frac{-9}{-6} \quad \textit{Divide by } -6 \\
n &= \frac{3}{2} = 1.5
\end{aligned}$$

35. Let $n =$ the number.

6	added to	a number	result is	7	times	the number
6	$+$	n	$=$	7	$\times$	n

Solve the equation.

$$\begin{aligned}
6 + n &= 7n \\
6 + n - n &= 7n - n \quad \textit{Subtract } n \\
6 &= 6n \\
\frac{6}{6} &= \frac{6n}{6} \quad \textit{Divide by 6} \\
1 &= n
\end{aligned}$$

37. Let $n =$ the number.

5	times	a number	added to	twice the number	result is	10
5	$\times$	n	$+$	$2n$	$=$	10

Solve the equation.

$$\begin{aligned}
5n + 2n &= 10 \\
7n &= 10 \quad \textit{Combine like terms} \\
\frac{7n}{7} &= \frac{10}{7} \\
n &= \frac{10}{7} = 1\frac{3}{7}
\end{aligned}$$

39. Let $x =$ the number of computers sold by Bill Thompson, and

$x - 18 =$ the number of computers sold by the other salesperson.

Thompson	$+$	other salesperson	$=$	total
x	$+$	$(x - 18)$	$=$	72

Solve the equation.

$$\begin{aligned}
x + x - 18 &= 72 \\
2x - 18 &= 72 \quad \textit{Combine terms} \\
2x &= 90 \quad \textit{Add 18} \\
x &= 45 \quad \textit{Divide by 2}
\end{aligned}$$

Bill Thompson sold 45 computers.

41. Let $x =$ the number of workers in the packaging department, and
$x + 15 =$ the number of workers in the production department.

$$\text{packaging} \; + \; \text{production} \; = \; \text{total}$$
$$x \qquad + \quad (x+15) \quad = \quad 277$$

Solve the equation.

$$x + x + 15 = 277$$
$$2x + 15 = 277 \quad \textit{Combine terms}$$
$$2x = 262 \quad \textit{Subtract 15}$$
$$x = 131 \quad \textit{Divide by 2}$$
$$x + 15 = 146$$

There were 146 people in the production department.

43. Let $p =$ the original price of the car.

$$\frac{9}{10} \quad \text{of} \quad \genfrac{}{}{0pt}{}{\text{original}}{\text{price}} \quad = \quad \text{sale price}$$
$$\frac{9}{10} \quad \cdot \quad p \qquad = \qquad 18{,}450$$

Solve the equation.

$$\frac{9}{10}p = 18{,}450$$
$$\frac{10}{9} \cdot \frac{9}{10}p = 18{,}450 \cdot \frac{10}{9} \quad \textit{Multiply by } \tfrac{10}{9}$$
$$p = 20{,}500$$

The original price of the automobile was \$20,500.

45. Let $x =$ the number of deluxe model homes, and
$\frac{3}{2}x =$ the number of economy model homes.

$$\text{economy} \; + \; \text{deluxe} \; = \; 105$$
$$\frac{3}{2}x \quad + \quad x \quad = \quad 105$$

Solve the equation.

$$\frac{3}{2}x + x = 105$$
$$\frac{3}{2}x + \frac{2}{2}x = 105$$
$$\frac{5}{2}x = 105 \qquad \textit{Combine terms}$$
$$\frac{2}{5} \cdot \frac{5}{2}x = 105 \cdot \frac{2}{5} \qquad \textit{Multiply by } \tfrac{2}{5}$$
$$x = 42$$
$$\frac{3}{2}x = \frac{3}{2} \cdot 42 = 63$$

There were 42 deluxe model homes and 63 economy model homes.

47. Let $a =$ the amount spent on all other employees, and
$\frac{3}{5}a =$ the amount spent on announcers.

$$\text{announcers} \; + \; \genfrac{}{}{0pt}{}{\text{all other}}{\text{employees}} \; = \; 10{,}500$$
$$\frac{3}{5}a \qquad + \qquad a \qquad = \quad 10{,}500$$

Solve the equation.

$$\frac{3}{5}a + a = 10{,}500$$
$$\frac{3}{5}a + \frac{5}{5}a = 10{,}500$$
$$\frac{8}{5}a = 10{,}500 \qquad \textit{Combine terms}$$
$$\frac{5}{8} \cdot \frac{8}{5}a = 10{,}500 \cdot \frac{5}{8} \qquad \textit{Multiply by } \tfrac{5}{8}$$
$$a = 6562.5$$
$$\frac{3}{5}s = \frac{3}{5} \cdot 6562.5$$
$$= 3937.50$$

The amount spent on announcers is \$3937.50 and the amount spent on all other employees is \$6562.50.

49. Let $x =$ the rent from offices, and
$3\frac{1}{2}x =$ the rent from retail stores.

$$\text{offices} \; + \; \text{retail stores} \; = \; \$67{,}500$$
$$x \quad + \quad 3\tfrac{1}{2}x \qquad = \quad 67{,}500$$

Solve the equation.

$$x + 3\tfrac{1}{2}x = 67{,}500$$
$$x + \frac{7}{2}x = 67{,}500$$
$$\frac{2}{2}x + \frac{7}{2}x = 67{,}500$$
$$\frac{9}{2}x = 67{,}500 \qquad \textit{Combine terms}$$
$$\frac{2}{9} \cdot \frac{9}{2}x = 67{,}500 \cdot \frac{2}{9} \qquad \textit{Multiply by } \tfrac{2}{9}$$
$$x = 15{,}000$$
$$3\tfrac{1}{2}x = \frac{7}{2} \cdot 15{,}000$$
$$= 52{,}500$$

\$15,000 comes from office rent and \$52,500 comes from retail stores.

51. Let $x =$ the number of experienced workers, and $63 - x =$ the new workers.

Set up a table.

	Number of Workers	Pay per Hour	Total Cost
Experienced Workers	x	$9	$9x$
New Workers	$63 - x$	$6	$6(63 - x)$
Totals	63		483

$9 per hour workers		$6 per hour workers		total cost
$+$				$=$
$9x$	$+$	$6(63 - x)$	$=$	483

Solve the equation.

$$9x + 6(63 - x) = 483$$

Use the distributive property.

$$9x + 378 - 6x = 483$$
$$3x + 378 = 483 \qquad \textit{Combine terms}$$
$$3x = 105 \qquad \textit{Subtract 378}$$
$$x = 35 \qquad \textit{Divide by 3}$$
$$63 - x = 63 - 35 = 28$$

There are 28 new workers and 35 experienced workers.

53. Let $s =$ the number of Altimas sold, and $120 - s =$ the number of Sentras sold.

Set up a table.

	Number sold	Profit for each car	Total Profit
Altimas	s	$1200	$1200s$
Sentras	$120 - s$	$850	$850(120 - s)$
Totals	120		130,350

profit on Altima sales		profit on Sentra sales		total profit
$1200s$	$+$	$850(120 - s)$	$=$	$130,350$

Solve the equation.

$$1200s + 850(120 - s) = 130{,}350$$

Use the distributive property

$$1200s + 102{,}000 - 850s = 130{,}350$$
$$350s + 102{,}000 = 130{,}350 \qquad \textit{Combine terms}$$
$$350s = 28{,}350 \qquad \textit{Subtract 102,000}$$
$$s = 81 \qquad \textit{Divide by 350}$$
$$120 - s = 120 - 81 = 39$$

81 Altimas and 39 Sentras were sold.

55. Answers will vary.

2.3   Formulas

1. $I = PRT; P = 4600, R = 0.085, T = 1\frac{1}{2} = 1.5$
$$I = 4600 \cdot 0.085 \cdot 1.5$$
$$I = 586.50$$

3. $P = \dfrac{nRT}{V}; n = 6, R = 0.0821, T = 315, V = 10$
$$P = \frac{6 \cdot 0.0821 \cdot 315}{10}$$
$$P = \frac{155.169}{10}$$
$$P = 15.5169$$

5. $R = \dfrac{D}{1 - DT}; \quad D = 0.05, \quad T = 4$
$$R = \frac{0.05}{1 - 0.05 \cdot 4}$$
$$R = \frac{0.05}{1 - 0.2}$$
$$R = \frac{0.05}{0.8} = 0.0625$$

7. $P = 2L + 2W; P = 40, W = 6$
$$40 = 2L + 2 \cdot 6$$
$$40 = 2L + 12$$
$$40 - 12 = 2L + 12 - 12 \qquad \textit{Subtract 12}$$
$$28 = 2L$$
$$\frac{28}{2} = \frac{2L}{2} \qquad \textit{Divide by 2}$$
$$14 = L$$

9. $P = \dfrac{I}{RT}; T = 3, I = 540, P = 2250$
$$2250 = \frac{540}{R \cdot 3}$$
$$3R \cdot 2250 = 3R \cdot \frac{540}{3R} \qquad \textit{Multiply by 3R}$$
$$6750R = 540$$
$$\frac{6750R}{6750} = \frac{540}{6750} \qquad \textit{Divide by 6750}$$
$$R = 0.08$$

11. $y = mx^2 + c; m = 3, x = 7, c = 4.2$
$$y = 3 \cdot 7^2 + 4.2$$
$$y = 3 \cdot 49 + 4.2$$
$$y = 147 + 4.2$$
$$y = 151.2$$

13. $M = P(1+i)^n; P = 640, i = 0.02, n = 8$
$M = 640(1 + 0.02)^8$
$M = 640(1.02)^8$
$M = 640(1.171659381)$
$M \approx 749.86$

15. $E = mc^2; m = 7.5, c = 1$
$E = 7.5 \cdot 1^2$
$E = 7.5$

17. $A = \frac{1}{2}(b+B)h; A = 105, b = 19, B = 11$
$105 = \frac{1}{2}(19 + 11)h$
$105 = \frac{1}{\cancel{2}}(\overset{15}{\cancel{30}})h$
$105 = 15h$
$\frac{105}{15} = \frac{15h}{15}$ *Divide by 15*
$7 = h$

19. $P = \frac{S}{1+RT}; S = 24{,}600, R = 0.06, T = \frac{5}{12}$
$P = \frac{24{,}600}{1 + (0.06)(\frac{5}{12})}$
$P = \frac{24{,}600}{1 + 0.025}$
$P = \frac{24{,}600}{1.025}$
$P = 24{,}000$

21. $A = LW$; for L
$\frac{A}{W} = \frac{LW}{W}$ *Divide by W*
$\frac{A}{W} = L$

23. $PV = nRT$; for V
$\frac{PV}{P} = \frac{nRT}{P}$ *Divide by P*
$V = \frac{nRT}{P}$

25. $M = P(1+i)^n$; for P
$\frac{M}{(1+i)^n} = \frac{P(1+i)^n}{(1+i)^n}$ *Divide by* $(1+i)^n$
$\frac{M}{(1+i)^n} = P$

27. $P = \frac{A}{1+i}$; for i
$P(1+i) = \frac{A}{1+i}(1+i)$ *Multiply by* $(1+i)$
$P(1+i) = A$
$P + Pi = A$ *Distributive property*
$P - P + Pi = A - P$ *Subtract P*
$Pi = A - P$
$\frac{Pi}{P} = \frac{A-P}{P}$ *Divide by P*
$i = \frac{A-P}{P}$

29. $P = M(1 - DT)$; for D
$P = M - MDT$ *Distributive property*
$P - M = M - M - MDT$ *Subtract M*
$P - M = -MDT$
$\frac{P-M}{-MT} = \frac{-MDT}{-MT}$ *Divide by* $-MT$
$\frac{P-M}{-MT} = D$
or $\frac{M-P}{MT} = D$

31. $A = \frac{1}{2}(b+B)h$; for h
$2A = 2 \cdot \frac{1}{2}(b+B)h$ *Multiply by 2*
$2A = (b+B)h$
$\frac{2A}{(b+B)} = \frac{(b+B)h}{(b+B)}$ *Divide by* $(b+B)$
$\frac{2A}{b+B} = h$

33. $M = Pe^{ni}$; for P
$\frac{M}{e^{ni}} = \frac{Pe^{ni}}{e^{ni}}$ *Divide by* e^{ni}
$\frac{M}{e^{ni}} = P$

35. Let x = average cost per pair
$82x = 3883.52$
$\frac{82x}{82} = \frac{3883.52}{82}$ *Divide by 82*
$x = 47.36$

The average cost per pair is \$47.36.

37. Let x = cost for a set of bongo drums.

$$6x + 7 \cdot 269 = 2445.80$$
$$6x + 1883 = 2445.80$$
$$6x = 562.80 \qquad \text{\textit{Subtract} 1883}$$
$$x = 93.80 \qquad \text{\textit{Divide by} 6}$$

The cost for a set of bongo drums is \$93.80.

39. Use the formula $S = 160 + 0.03x$.

(a) $S = 160 + 0.03x$

Let $x = \$1152$.

$$S = 160 + 0.03(1152)$$
$$= 160 + 34.56$$
$$= \$194.56$$

(b) Let $x = \$1796$.

$$S = 160 + 0.03(1796)$$
$$= 160 + 53.88$$
$$= \$213.88$$

(c) Let $x = \$2314$.

$$S = 160 + 0.03(2314)$$
$$= 160 + 69.42$$
$$= \$229.42$$

41. Let x = gross sales, and

$\dfrac{1}{40}x$ = returns.

$$\begin{array}{ccccc} \text{Net Sales} & = & \text{Gross sales} & - & \text{Returns} \\ 230 & = & x & - & \dfrac{1}{40}x \end{array}$$

$$230 = \frac{40}{40}x - \frac{1}{40}x$$
$$230 = \frac{39}{40}x \qquad \text{\textit{Combine terms}}$$
$$\frac{40}{39} \cdot 230 = \frac{40}{39} \cdot \frac{39}{40}x \qquad \text{\textit{Multiply by} } \frac{40}{39}$$
$$236 \approx x$$

Gross sales are approximately \$236 million.

43. Let x = the cost of the item, and

$\dfrac{3}{4}x$ = the markup.

$$\begin{array}{ccccc} \text{selling price} & = & \text{cost} & + & \text{markup} \\ 84 & = & x & + & \dfrac{3}{4}x \end{array}$$

$$84 = \frac{4}{4}x + \frac{3}{4}x$$
$$84 = \frac{7}{4}x \qquad \text{\textit{Combine terms}}$$
$$\frac{4}{7} \cdot 84 = \frac{4}{7} \cdot \frac{7}{4}x \qquad \text{\textit{Multiply by} } \frac{4}{7}$$
$$48 = x$$

The cost of the item is \$48.

45. Let x = the revenue, and

$\dfrac{5}{6}x$ = the expenses.

$$\begin{array}{ccccc} \text{profit} & = & \text{revenue} & - & \text{expenses} \\ 15{,}000 & = & x & - & \dfrac{5}{6}x \end{array}$$

$$15{,}000 = \frac{6}{6}x - \frac{5}{6}x$$
$$15{,}000 = \frac{1}{6}x \qquad \text{\textit{Combine terms}}$$
$$6 \cdot 15{,}000 = 6 \cdot \frac{1}{6}x \qquad \text{\textit{Multiply by} 6}$$
$$90{,}000 = x$$

The revenue was \$90,000.

47. $I = PRT$; $\quad P = 5200, \quad R = 0.075, \quad T = 1$
$$I = 5200 \cdot 0.075 \cdot 1$$
$$I = 390$$

The interest would be \$390.

49. $I = PRT$; $\quad P = 22{,}000, \quad T = 2, \quad I = 5720$

$$I = PRT$$
$$5720 = 22{,}000 \cdot R \cdot 2$$
$$5720 = 44{,}000R$$
$$\frac{5720}{44{,}000} = \frac{44{,}000R}{44{,}000} \qquad \text{\textit{Divide by} 44{,}000}$$
$$0.13 = R$$

The rate of interest was 13%.

51. $I = PRT$; $\quad P = 18{,}200, \quad R = 0.11, \quad I = 8008$

$$I = PRT$$
$$8008 = 18{,}200 \cdot 0.11 \cdot T$$
$$8008 = 2002T$$
$$\frac{8008}{2002} = \frac{2002T}{2002} \qquad \text{\textit{Divide by} 2002}$$
$$4 = T$$

The time for the loan is 4 years.

53. $M = P(1 + RT)$; $M = 4560, R = 0.07, T = 2$

$$M = P(1 + RT)$$
$$4560 = P(1 + 0.07 \cdot 2)$$
$$4560 = P(1 + 0.14)$$
$$4560 = P(1.14)$$
$$\frac{4560}{1.14} = \frac{1.14P}{1.14} \quad \text{Divide by 1.14}$$
$$4000 = P$$

Initially John deposited $4000.

55. $M = P(1 + i)^n$; M; $M = 5989.50, i = 0.10, n = 3$

$$M = P(1 + i)^n$$
$$5989.50 = P(1 + 0.10)^3$$
$$5989.50 = P(1.1)^3$$
$$5989.50 = P(1.331)$$
$$\frac{5989.50}{1.331} = \frac{P(1.331)}{1.331} \quad \text{Divide by 1.331}$$
$$4500 = P$$

Bill borrowed $4500.

57. Answers will vary.

2.4 Ratios and Proportions

1. 18 kilometers to 64 kilometers
$$\frac{18}{64} = \frac{9}{32}$$

3. 216 students to 8 faculty
$$\frac{216}{8} = \frac{27}{1}$$

5. 8 men to 6 women
$$\frac{8}{6} = \frac{4}{3}$$

7. 30 kilometers to 8 meters

Convert 30 kilometers to meters.

30 kilometers = 30,000 meters
$$\frac{30,000}{8} = \frac{3750}{1}$$

9. 90 dollars to 40 cents

Convert 90 dollars to cents.

90 dollars = 9000 cents
$$\frac{9000}{40} = \frac{225}{1}$$

11. 4 dollars to 10 quarters

Convert 4 dollars to quarters.
4 dollars = 16 quarters.
$$\frac{16}{10} = \frac{8}{5}$$

13. 20 hours to 5 days

Convert 5 days to hours.
5 days = 120 hours.
$$\frac{20}{120} = \frac{1}{6}$$

15. $0.80 to $3
$$\frac{0.8}{3} = \frac{8}{30} = \frac{4}{15}$$

17. $3.24 to $0.72
$$\frac{3.24}{0.72} = \frac{324}{72} = \frac{9}{2}$$

19. $\dfrac{3}{5} = \dfrac{21}{35}$

The cross products are $3 \cdot 35 = 105$
and $5 \cdot 21 = 105$
The cross products are equal, so the proportion is true.

21. $\dfrac{9}{7} = \dfrac{720}{480}$

The cross products are $9 \cdot 480 = 4320$
and $7 \cdot 720 = 5040$
The cross products are not equal, so the proportion is false.

23. $\dfrac{69}{320} = \dfrac{7}{102}$

The cross products are $69 \cdot 102 = 7038$
and $320 \cdot 7 = 2240$
The cross products are not equal, so the proportion is false

25. $\dfrac{19}{32} = \dfrac{33}{77}$

The cross products are $19 \cdot 77 = 1463$
and $32 \cdot 33 = 1056$
The cross products are not equal, so the proportion is false.

27. $\dfrac{110}{18} = \dfrac{160}{27}$

The cross products are $110 \cdot 27 = 2970$
and $18 \cdot 160 = 2880$.
The cross products are not equal, so the proportion is false.

29. $\dfrac{32}{75} = \dfrac{61}{108}$

The cross products are $32 \cdot 108 = 3456$
and $\quad 75 \cdot 61 = 4575$
The cross products are not equal, so the proportion is false.

31. $\dfrac{7.6}{10} = \dfrac{76}{100}$

The cross products are $7.6 \cdot 100 = 760$
and $\quad 10 \cdot 76 = 760.$
The cross products are equal, so the proportion is true.

33. $\dfrac{2\frac{1}{4}}{5} = \dfrac{9}{20}$

The cross products are $2\frac{1}{4} \cdot 20 = 45$
and $\quad 5 \cdot 9 = 45.$
The cross products are equal, so the proportion is true.

35. $\dfrac{4\frac{1}{5}}{6\frac{1}{8}} = \dfrac{27}{41}$

The cross products are $4\frac{1}{5} \cdot 41 = 172.2$
and $6\frac{1}{8} \cdot 27 = 165.375.$
The cross products are not equal, so the proportion is false.

37. $\dfrac{8.15}{2.03} = \dfrac{61.125}{15.225}$

The cross products are

$8.15 \cdot 15.225 = 124.08375 \quad$ and
$2.03 \cdot 61.125 = 124.08375.$

The cross products are equal so the proportion is true.

39. $\dfrac{x}{15} = \dfrac{49}{105}$

$x \cdot 105 = 15 \cdot 49 \quad$ *Cross multiply*
$105x = 735$

$\dfrac{105x}{105} = \dfrac{735}{105} \quad$ *Divide by 105*

$x = 7$

41. $\dfrac{6}{9} = \dfrac{r}{108}$

$6 \cdot 108 = 9 \cdot r \quad$ *Cross multiply*
$648 = 9r$

$\dfrac{648}{9} = \dfrac{9r}{9} \quad$ *Divide by 9*

$72 = r$

43. $\dfrac{63}{s} = \dfrac{3}{5}$

$63 \cdot 5 = 3 \cdot s \quad$ *Cross multiply*
$315 = 3s$

$\dfrac{315}{3} = \dfrac{3s}{3} \quad$ *Divide by 3*

$105 = s$

45. $\dfrac{1}{2} = \dfrac{r}{7}$

$1 \cdot 7 = 2 \cdot r \quad$ *Cross multiply*
$7 = 2r$

$\dfrac{7}{2} = \dfrac{2r}{2} \quad$ *Divide by 2*

$\dfrac{7}{2} = r \quad \left(\text{or } 3\frac{1}{2} = r\right)$

47. $\dfrac{\frac{3}{4}}{6} = \dfrac{3}{x}$

$\dfrac{3}{4} \cdot x = 6 \cdot 3 \quad$ *Cross multiply*

$\dfrac{3}{4}x = 18$

$\dfrac{4}{3} \cdot \dfrac{3}{4}x = \dfrac{4}{3} \cdot 18 \quad$ *Multiply by $\frac{4}{3}$*

$x = 24$

49. $\dfrac{12}{p} = \dfrac{23.571}{15.714}$

$12 \cdot 15.714 = 23.571p \quad$ *Cross multiply*
$188.568 = 23.571p$

$\dfrac{188.568}{23.571} = \dfrac{23.571p}{23.571} \quad$ *Divide by 23.571*

$8 = p$

51. Answers will vary.

53. Let x be the number of tickets it can expect to sell in 9 days.
Set up a proportion.

$\dfrac{2}{9} = \dfrac{350}{x}$

$2 \cdot x = 9 \cdot 350 \quad$ *Cross multiply*
$2x = 3150$
$x = 1575 \quad$ *Divide by 2*

It can expect to sell 1575 tickets.

55. Let x be the cost for a 12-unit apartment house.
Set up a proportion.

$$\frac{5}{12} = \frac{172,000}{x}$$

$5 \cdot x = 12 \cdot 172,000$ *Cross multiply*
$5x = 2,064,000$
$x = 412,800$ *Divide by 5*

The cost for a 12-unit apartment house is \$412,800.

57. Let x be the cost of 12 dresses.
Set up a proportion.

$$\frac{22}{12} = \frac{176}{x}$$

$22 \cdot x = 12 \cdot 176$ *Cross multiply*
$22x = 2112$
$x = 96$ *Divide by 22*

The cost of 12 dresses is \$96.

59. Let x be the amount of material needed for 12
dresses.
Set up a proportion.

$$\frac{5}{12} = \frac{15}{x}$$

$5 \cdot x = 12 \cdot 15$ *Cross multiply*
$5x = 180$
$x = 36$ *Divide by 5*

36 yards of material are needed.

61. Let x be the distance between the two other cities.
Set up a proportion.

$$\frac{2}{17} = \frac{120}{x}$$

$2 \cdot x = 17 \cdot 120$
$2x = 2040$
$x = 1020$

The cities are 1020 miles apart.

63. Let x be the sales for the 52-week year.
Set up a proportion.

$$\frac{20}{52} = \frac{274,312}{x}$$

$20 \cdot x = 52 \cdot 274,312$
$20x = 14,264,224$
$x = 713,211.2$

The sales for the entire year are \$713,211.20.

65. Let x be the profits of the second partner.
Set up a proportion.

$$\frac{5}{8} = \frac{15,000}{x}$$

$5 \cdot x = 8 \cdot 15,000$ *Cross multiply*
$5x = 120,000$
$x = 24,000$ *Divide by 5*

The second partner received \$24,000.

67. Let x be the distance ducks migrate in the amount
of time it takes the songbirds.
Set up a proportion.

$$\frac{20}{59} = \frac{500}{x}$$

$20 \cdot x = 59 \cdot 500$ *Cross multiply*
$20x = 29,500$
$x = 1475$ *Divide by 20*

Ducks migrate 1475 miles.

69. Let x be the amount of the iceberg that is under
the water.
Set up a proportion.

$$\frac{\frac{1}{8}}{\frac{7}{8}} = \frac{500,000}{x}$$

$$\frac{1}{7} = \frac{500,000}{x}$$ *Simplify*

$1 \cdot x = 7 \cdot 500,000$ *Cross multiply*
$x = 3,500,000$

The amount of the iceberg that is under the water
is 3,500,000 cubic meters.

71. Let x be the number of U.S. dollars he will receive.
Set up a proportion.

$$\frac{105}{20,355} = \frac{1}{x}$$

$105 \cdot x = 20,355 \cdot 1$ *Cross multiply*
$105x = 20,355$
$x \approx 193.86$ *Divide by 105*

He will receive \$193.86.

Chapter 2 Review Exercises

1. $x + 45 = 96$
$x + 45 - 45 = 96 - 45$ *Subtract* 45
$x = 51$

2. $r - 36 = 14.7$
$r - 36 + 36 = 14.7 + 36$ *Add* 36
$r = 50.7$

3. $8t + 45 = 175.4$
$8t + 45 - 45 = 175.4 - 45$ *Subtract* 45
$8t = 130.4$
$$\frac{8t}{8} = \frac{130.4}{8} \qquad \textit{Divide by } 8$$
$t = 16.3$

4. $4t - 6 = 15$
$4t - 6 + 6 = 15 + 6$ *Add* 6
$4t = 21$
$$\frac{4t}{4} = \frac{21}{4} \qquad \textit{Divide by } 4$$
$$t = \frac{21}{4} = 5\tfrac{1}{4}$$

5. $\dfrac{s}{6} = 42$
$$\frac{s}{6} \cdot 6 = 42 \cdot 6 \quad \textit{Multiply by } 6$$
$s = 252$

6. $\dfrac{5z}{8} = 85$
$$\frac{8}{5} \cdot \frac{5z}{8} = \frac{8}{5} \cdot 85 \quad \textit{Multiply by } \tfrac{8}{5}$$
$z = 136$

7. $\dfrac{m}{4} - 5 = 9$
$$\frac{m}{4} - 5 + 5 = 9 + 5 \quad \textit{Add } 5$$
$$\frac{m}{4} = 14$$
$$4 \cdot \frac{m}{4} = 4 \cdot 14 \quad \textit{Multiply by } 4$$
$m = 56$

8. $5(x - 3) = 3(x + 4)$

Use the distributive property

$5x - 15 = 3x + 12$
$5x - 3x - 15 = 3x - 3x + 12$ *Subtract* $3x$
$2x - 15 = 12$
$2x - 15 + 15 = 12 + 15$ *Add* 15
$2x = 27$
$$\frac{2x}{2} = \frac{27}{2} \qquad\qquad \textit{Divide by } 2$$
$$x = \frac{27}{2} = 13\tfrac{1}{2}$$

9. $6y = 2y + 28$
$6y - 2y = 2y - 2y + 28$ *Subtract* $2y$
$4y = 28$
$$\frac{4y}{4} = \frac{28}{4} \qquad\qquad \textit{Divide by } 4$$
$y = 7$

10. $3r - 7 = 2(4 - 3r)$

Use the distributive property on the right.

$3r - 7 = 8 - 6r$
$3r + 6r - 7 = 8 - 6r + 6r$ *Add* $6r$
$9r - 7 = 8$
$9r - 7 + 7 = 8 + 7$ *Add* 7
$9r = 15$
$$\frac{9r}{9} = \frac{15}{9} \qquad\qquad \textit{Divide by } 9$$
$$r = \frac{15}{9} = 1\tfrac{6}{9} = 1\tfrac{2}{3}$$

11. $0.15(2x - 3) = 5.85$

Use the distributive property on the left.

$0.3x - 0.45 = 5.85$
$0.3x - 0.45 + 0.45 = 5.85 + 0.45$ *Add* 0.45
$0.3x = 6.3$
$$\frac{0.3x}{0.3} = \frac{6.3}{0.3} \qquad \textit{Divide by } 0.3$$
$x = 21$

12. $0.6(y - 3) = 0.1y$

Use distributive property on the left.

$0.6y - 1.8 = 0.1y$
$0.6y - 0.6y - 1.8 = 0.1y - 0.6y$ *Subtract* 0.6y
$-1.8 = -0.5y$
$$\frac{-1.8}{-0.5} = \frac{-0.5y}{-0.5} \qquad \textit{Divide by } -0.5$$
$3.6 = y$

13. 94 times a number

$$94x$$

14. $\frac{1}{2}$ times a number

$$\frac{1}{2}x$$

15. 6 times a number is added to the number

$$x + 6x$$

16. 5 times a number is decreased by 11

$$5x - 11$$

17. The sum of 3 times a number and 7

$$3x + 7$$

18. Multiply 3 and 14.95 and add the product to 95.

$$(3 \times 14.95) + 95 = 44.85 + 95$$
$$= 139.85$$

Subtract 47.50 from 139.85

$$139.85 - 47.50 = 92.35$$

Molly needs to save $92.35 more.

19.
$$P = 2.775A + 4.5$$
$$60 = 2.775A + 4.5$$
$$60 - 4.5 = 2.775A + 4.5 - 4.5 \quad \textit{Subtract 4.5}$$
$$55.5 = 2.775A$$

$$\frac{55.5}{2.775} = \frac{2.775A}{2.775} \qquad \textit{Divide by 2.775}$$

$$20 = A$$

They must spend $20,000 (20 thousand) on advertising.

20. Let $x =$ the water bill, and
$4x =$ the phone bill.

$$x + 4x = 540$$
$$5x = 540 \qquad \textit{Combine terms}$$
$$\frac{5x}{5} = \frac{540}{5} \qquad \textit{Divide by 5}$$
$$x = 108$$
$$4x = 4 \cdot 108$$
$$= 432$$

The water bill was $108, and the phone bill was $432.

21. Let $E =$ the number of employees, and
$\frac{1}{4}E + 5 =$ the employees with 25 or more
years of service.

$$\frac{1}{4}E + 5 = 24$$
$$\frac{1}{4}E + 5 - 5 = 24 - 5 \quad \textit{Subtract 5}$$
$$\frac{1}{4}E = 19$$
$$4 \cdot \frac{1}{4}E = 4 \cdot 19 \quad \textit{Multiply by 4}$$
$$E = 76$$

The company has 76 employees.

22. Let $C =$ the number of tickets for
children, and
$100 - C =$ the number of adult tickets.

$$3C + 6(100 - C) = 390$$
$$3C + 600 - 6C = 390$$
$$-3C + 600 = 390 \quad \textit{Combine like terms}$$
$$-3C + 600 - 600 = 390 - 600 \quad \textit{Subtract 600}$$
$$-3C = -210$$
$$\frac{-3C}{-3} = \frac{-210}{-3} \qquad \textit{Divide by } -3$$
$$C = 70$$
$$100 - C = 100 - 70 = 30$$

The theater sold 70 tickets for children, and 30 adult tickets.

23. $I = PRT; \quad I = \$960, \quad R = 0.12, \quad T = 2$

$$960 = P \cdot 0.12 \cdot 2$$
$$960 = 0.24P$$
$$\frac{960}{0.24} = \frac{0.24P}{0.24} \quad \textit{Divide by 0.24}$$
$$4000 = P$$

24. $M = P(1 + RT); \quad M = \$3770, \quad R = 0.04,$
$T = 4$

$$3770 = P(1 + 0.04 \cdot 4)$$
$$3770 = P(1 + 0.16)$$
$$3770 = P(1.16)$$
$$\frac{3770}{1.16} = \frac{1.16P}{1.16} \quad \textit{Divide by 1.16}$$
$$3250 = P$$

25. $M = P(1+i)^n$; $M = \$14,526.80$,
$i = 0.1$, $n = 6$

$14{,}526.80 = P(1 + 0.1)^6$
$14{,}526.80 = P(1.1)^6$
$14{,}526.80 = P(1.771561)$

$\dfrac{14{,}526.80}{1.771561} = \dfrac{1.771561P}{1.771561}$ *Divide by 1.771561*

$8200 \approx P$

26. $I = PRT$; for R

$\dfrac{I}{PT} = \dfrac{PRT}{PT}$ *Divide by PT*

$\dfrac{I}{PT} = R$

27. $M = P(1 + RT)$; for T
$M = P + PRT$ *Distributive property*
$M - P = P - P + PRT$ *Subtract P*
$M - P = PRT$

$\dfrac{M - P}{PR} = \dfrac{PRT}{PR}$ *Divide by PR*

$\dfrac{M - P}{PR} = T$

28. $R = \dfrac{D}{1 - DT}$; for T

Multiply both sides by $1 - DT$.

$R(1 - DT) = \dfrac{D}{1 - DT}(1 - DT)$
$R(1 - DT) = D$
$R - RDT = D$ *Distributive property*
$R - R - RDT = D - R$ *Subtract R*
$-RDT = D - R$

$\dfrac{-RDT}{-RD} = \dfrac{D - R}{-RD}$ *Divide by $-RD$*

$T = \dfrac{D - R}{-RD}$ or $T = \dfrac{R - D}{RD}$

29. $\$17$ to 50 cents
Convert $\$17$ to cents

$\$17 = 1700$ cents

$\dfrac{1700}{50} = \dfrac{34}{1}$

30. 9 days to 12 hours
Convert 9 days to hours.

9 days $= 216$ hours

$\dfrac{216}{12} = \dfrac{18}{1}$

31. $\$5000$ to $\$250$

$\dfrac{5000}{250} = \dfrac{20}{1}$

32. 3 years to 15 months
Convert 3 years to months.

3 years $= 36$ months

$\dfrac{36}{15} = \dfrac{12}{5}$

33. $\$2$ dollars to 75 cents
Convert $\$2$ to cents

$\$2 = 200$ cents

$\dfrac{200}{75} = \dfrac{8}{3}$

34. $\dfrac{v}{14} = \dfrac{27}{126}$

$v \cdot 126 = 14 \cdot 27$ *Cross multiply*
$126v = 378$

$\dfrac{126v}{126} = \dfrac{378}{126}$ *Divide by 126*

$v = 3$

35. $\dfrac{5}{y} = \dfrac{20}{27}$

$5 \cdot 27 = y \cdot 20$ *Cross multiply*
$135 = 20y$

$\dfrac{135}{20} = \dfrac{20y}{20}$ *Divide by 20*

$\dfrac{27}{4} = y$ (or $6\frac{3}{4} = y$)

36. $\dfrac{3}{8} = \dfrac{z}{12}$

$3 \cdot 12 = 8 \cdot z$ *Cross multiply*
$36 = 8z$

$\dfrac{36}{8} = \dfrac{8z}{8}$ *Divide by 8*

$\dfrac{9}{2} = z$ (or $4\frac{1}{2} = z$)

37. $\dfrac{6}{11} = \dfrac{90}{s}$

$6 \cdot s = 11 \cdot 90$ *Cross multiply*
$6s = 990$

$\dfrac{6s}{6} = \dfrac{990}{6}$ *Divide by 6*

$s = 165$

38.
$$\frac{20}{r} = \frac{60}{72}$$

$20 \cdot 72 = r \cdot 60$ *Cross multiply*

$1440 = 60r$

$$\frac{1440}{60} = \frac{60r}{60} \qquad \textit{Divide by } 60$$

$24 = r$

39. Let x be the number of bass in the lake with parasites.
Set up a proportion.

$$\frac{60}{18,400} = \frac{14}{x}$$

$60 \cdot x = 18,400 \cdot 14$ *Cross multiply*

$60x = 257,600$

$x \approx 4293$ *Divide by* 60

There are 4293 bass with parasites.

40. Let x be the pressure at 9850 feet.
Set up a proportion.

$$\frac{6700}{9850} = \frac{3220}{x}$$

$6700 \cdot x = 9850 \cdot 3220$ *Cross multiply*

$6700x = 31,717,000$

$x \approx 4734$ *Divide by* 6700

The pressure at 9850 feet is approximately 4734 pounds per square inch.

41. Let x be the cost for 12.5 gallons.
Set up a proportion.

$$\frac{1.40}{1} = \frac{x}{12.5}$$

$1.40 \cdot 12.5 = 1 \cdot x$

$17.5 = x$

The gas costs $17.50.

42. Let x be the number of pages John proofreads in 3 hours.

3 hours = 180 minutes

Set up a proportion.

$$\frac{12}{180} = \frac{7}{x}$$

$12 \cdot x = 180 \cdot 7$ *Cross multiply*

$12x = 1260$

$$\frac{12x}{12} = \frac{1260}{12} \qquad \textit{Divide by } 12$$

$x = 105$

In 3 hours, John proofreads 105 pages.

43. Let x be the amount spent on training.
Set up a proportion.

$$\frac{3}{1} = \frac{x}{19,000}$$

$3 \cdot 19,000 = 1 \cdot x$ *Cross multiply*

$57,000 = x$

$57,000 is spent on training.

44. Let c be the cost for 5 shirts.
Set up a proportion.

$$\frac{8}{5} = \frac{223.20}{c}$$

$8 \cdot c = 5 \cdot 223.20$ *Cross multiply*

$8c = 1116$

$$\frac{8c}{8} = \frac{1116}{8} \qquad \textit{Divide by } 8$$

$c = 139.5$

The cost for 5 shirts is $139.50.

45. Let x be the number of video cassettes that Kim bought.
She paid $468.75 ($500 − $31.25) for the cassettes.

Set up a proportion.

$$\frac{75}{468.75} = \frac{4}{x}$$

$75 \cdot x = 468.75 \cdot 4$ *Cross multiply*

$75x = 1875$

$$\frac{75x}{75} = \frac{1875}{75} \qquad \textit{Divide by } 75$$

$x = 25$

Kim bought 25 video cassettes.

Chapter 2 Summary Exercise

(a) Add all the expenses.

$2800 + $2200 + $250 + $200 + $650 = $6100

(b) The price of the book is $10.60.
70% covers the expenses, so 30% is profit.

30% of $10.60 is $3.18.
Profit = 3.18N$ − $6100

(c) Expenses are \$6100.

Profit for each book is \$3.18.

Divide \$6100 by \$3.18.

$$
\begin{array}{r}
1918\,.2 \\
3.18_\wedge\,)\overline{6,100.00_\wedge 0} \\
\underline{3\ 18} \\
2\ 920 \\
\underline{2\ 862} \\
58\ 0 \\
\underline{31\ 8} \\
26\ 20 \\
\underline{25\ 44} \\
76\ 0 \\
\underline{63\ 6} \\
12\ 4
\end{array}
$$

They must sell 1919 books to break even.

(d) The owner would probably receive less salary.

(e)
$$
\begin{aligned}
P &= 3.18N - 6100 \\
2500 &= 3.18N - 6100 \\
2500 + 6100 &= 3.18N - 6100 + 6100 \\
8600 &= 3.18N \\
\frac{8600}{3.18} &= \frac{3.18N}{3.18} \\
2704.4 &= N
\end{aligned}
$$

To reach a profit of \$2500, they must sell 2705 books. (2704 won't quite give them the desired profit.)

Chapter 3

PERCENT

3.1   Writing Decimals and Fractions as Percents

For Exercises 1-15, move the decimal point two places to the right and attach a percent sign.

1. $0.2 = 20\%$

3. $0.72 = 72\%$

5. $1.4 = 140\%$

7. $0.375 = 37.5\%$

9. $4.625 = 462.5\%$

11. $0.0025 = 0.25\%$

13. $0.0015 = 0.15\%$

15. $3.45 = 345\%$

17. $\frac{1}{4} = \frac{25}{100} = 0.25 = 25\%$

19. $\frac{1}{10} = \frac{10}{100} = 0.10 = 10\%$

21. $\frac{1}{50} = \frac{2}{100} = 0.02 = 2\%$

23. $\frac{3}{8} = \frac{3 \times 125}{8 \times 125} = \frac{375}{1000} = 0.375 = 37.5\%$

25. $\frac{1}{8} = \frac{1 \times 125}{8 \times 125} = \frac{125}{1000} = 0.125 = 12.5\%$

27. $\frac{1}{200} = \frac{1 \times 5}{200 \times 5} = \frac{5}{1000} = 0.005 = 0.5\%$

29. $\frac{7}{8} = \frac{7 \times 125}{8 \times 125} = \frac{875}{1000} = 0.875 = 87.5\%$

31. $\frac{3}{50} = \frac{6}{100} = 0.06 = 6\%$

For Exercises 33-47, move the decimal two places to the left and drop the percent sign.

33. $65\% = 0.65$

35. $75\% = 0.75$

37. $0.6\% = 0.006$

39. $0.25\% = 0.0025$

41. $315\% = 3.15$

43. $200.6\% = 2.006$

45. $540.6\% = 5.406$

47. $0.07\% = 0.0007$

49. Answers will vary.

51. Answers will vary.

	Fraction	Decimal	Percent
53.	$\frac{1}{2}$	0.5	50%
55.	$\frac{3}{20}$	0.15	15%
57.	$\frac{1}{4}$	0.25	25%
59.	$6\frac{1}{8}$	6.125	612.5%
61.	$7\frac{1}{4}$	7.25	725%
63.	$\frac{1}{400}$	0.0025	0.25%
65.	$\frac{1}{3}$	$0.33\overline{3}$	$33\frac{1}{3}\%$
67.	$\frac{3}{400}$	0.0075	0.75%
69.	$\frac{1}{8}$	0.125	12.5%
71.	$2\frac{1}{2}$	2.5	250%
73.	$10\frac{767}{2000}$	10.3835	1038.35%
75.	$4\frac{3}{8}$	4.375	437.5%
77.	$\frac{27}{400}$	0.0675	6.75%

3.2   Finding the Part

1. 20% of 80 guests

$P = B \times R$
$P = 80 \times 0.20$
$P = 16$ guests

3. 22.5% of $1086

$P = B \times R$
$P = \$1086 \times 0.225$
$P = \$244.35$

5. 4% of 120 feet

$P = B \times R$
$P = 120 \times 0.04$
$P = 4.8$ feet

7. 175% of 5820 miles

$P = B \times R$
$P = 5820 \times 1.75$
$P = 10{,}185$ miles

9. 17.5% of 1040 homes

$P = B \times R$
$P = 1040 \times 0.175$
$P = 182$ homes

11. 118% of 125.8 yards

$P = B \times R$
$P = 125.8 \times 1.18$
$P = 148.444$ yards

13. $90\frac{1}{2}\%$ of $5930

$P = B \times R$
$P = \$5930 \times 0.905$
$P = \$5366.65$

15. 0.5% of $1300

$P = R \times B$
$P = \$1300 \times 0.005$
$P = \$6.50$

17. Answers will vary.

19. $P = BR$
$P = \$240 \times 22\%$
$P = \$240 \times 0.22$
$P = \$52.80$

The amount withheld is $52.80.

21. $P = BR$
$= \$3100 \times 20\%$
$= \$3100 \times 0.2$
$= \$620$
$P + 25 = \$645$

The total charge is $645.

23. $P = BR$
$P = 3680 \times 55\%$
$P = 3680 \times 0.55$
$P = 2024$

You would expect 2024 shoppers to have a written shopping list.

25. $P = BR$
$P = \$148 \times 62.2\%$
$P = \$148 \times 0.622$
$P \approx \$92.06$

The total annual sales of the Hormel products was $92.06 million.

27. **(a)** 39% $(100\% - 61\%)$ are female.

(b) $P = BR$
$P = 3.8$ million $\times\ 61\%$
$P = 3.8$ million $\times\ 0.61$
$P = 2.318$ million

In Cuba, 2.318 million or 2,318,000 workers in the labor force are male.

29. $P = BR$
$P = 377 \times 62\%$
$P = 377 \times 0.62$
$P \approx 234$

The number of executives that were upbeat about global business growth was 234.

31. 14% $(100\% - 86\%)$ of the products did reach the objectives.

$P = BR$
$P = 15{,}401 \times 14\%$
$P = 15{,}401\ \times\ 0.14$
$P = 2156.14$

The number of products that did reach the objectives is 2156.

33. $P = BR$
$P = \$100 \text{ million} \times 35\%$
$P = \$100 \text{ million} \times 0.35\%$
$P = \$35 \text{ million}$

The increase in sales is \$35 million.

\$100 million + \$35 million = \$135 million

Estimated orange juice sales next year are \$135 million.

35. $P = BR$
$P = \$48,680 \times 6\frac{1}{2}\%$

$P = \$48,680 \times 0.065$
$P = \$3164.20$

The tax is \$3164.20.

\$48,680 + \$3164.20 = \$51,844.20

The combined amount is \$51,844.20.

37. $P = BR$
$P = \$385,200 \times 32\frac{1}{2}\%$

$P = \$385,200 \times 0.325$
$P = \$125,190$

The increase in sales over last year is \$125,190.

\$385,200 + \$125,190 = \$510,390

The volume of parts sold this year is \$510,390.

39. $P = BR$
$P = \$174,900 \times 6\%$
$P = \$174,900 \times 0.06$
$P = \$10,494$

The commission of the sale was \$10,494.

$P = BR$
$P = \$10,494 \times 60\%$
$P = \$10,494 \times 0.6$
$P = \$6296.40$

Dugally receives a commission of \$6296.40.

41. Pamela

$P = BR$
$P = \$3150 \times 10\%$
$P = \$3150 \times 0.10$
$P = \$315$

$\$315 \times 12 = \$3780$

Peter

$P = BR$
$P = \$28,400 \times 10\%$
$P = \$28,400 \times 0.10$
$P = \$2840$

\$3780 + \$2840 = \$6620

They will save \$6620 in one year.

3.3 Finding the Base

1. P is 265; R is 25%.

$$R \times B = P$$
$$0.25 \times B = 265$$
$$\frac{0.25B}{0.25} = \frac{265}{0.25}$$
$$B = 1060$$

265 bowlers is 25% of <u>1060</u> bowlers.

3. P is 75; R is 40%.

$$R \times B = P$$
$$0.40 \times B = 75$$
$$\frac{0.40B}{0.40} = \frac{75}{0.40}$$
$$B = 187.5$$

75 miles is 40% of <u>187.5</u> miles.

5. P is 55; R is 5.5%.

$$R \times B = P$$
$$0.055 \times B = 55$$
$$\frac{0.055B}{0.055} = \frac{55}{0.055}$$
$$B = 1000$$

55 packages is 5.5% of <u>1000</u> packages.

7. P is 36; R is 0.75%.

$$R \times B = P$$
$$0.0075 \times B = 36$$
$$\frac{0.0075B}{0.0075} = \frac{36}{0.0075}$$
$$B = 4800$$

36 employees is 0.75% of <u>4800</u> employees.

9. P is 33; R is 0.15%.

$$R \times B = P$$
$$0.0015 \times B = 33$$
$$\frac{0.0015B}{0.0015} = \frac{33}{0.0015}$$
$$B = 22,000$$

33 rolls is 0.15% of 22,000 rolls.

11. P is 50; R is 0.25%.

$$R \times B = P$$
$$0.0025 \times B = 50$$
$$\frac{0.0025B}{0.0025} = \frac{50}{0.0025}$$
$$B = 20,000$$

50 doors is 0.25% of 20,000 doors.

13. P is \$33,870; R is $37\frac{1}{2}$%.

$$R \times B = P$$
$$0.375 \times B = \$33,870$$
$$\frac{0.375B}{0.375} = \frac{\$33,870}{0.375}$$
$$B = \$90,320$$

\$33,870 is $37\frac{1}{2}$% of \$90,320.

15. P is 350; R is 20%.

$$R \times B = P$$
$$0.20 \times B = 350$$
$$\frac{0.20B}{0.20} = \frac{350}{0.20}$$
$$B = 1750$$

20% of 1750 sacks is 350 sacks.

17. P is 375; R is 0.12%.

$$R \times B = P$$
$$0.0012 \times B = 375$$
$$\frac{0.0012B}{0.0012} = \frac{375}{0.0012}$$
$$B = 312,500$$

375 crates is 0.12% of 312,500 crates.

19. P is 327; R is 0.5%.

$$R \times B = P$$
$$0.005 \times B = 327$$
$$\frac{0.005B}{0.005} = \frac{327}{0.005}$$
$$B = 65,400$$

0.5% of 65,400 homes is 327 homes.

21. P is 12; R is 0.03%.

$$R \times B = P$$
$$0.0003 \times B = 12$$
$$\frac{0.0003B}{0.0003} = \frac{12}{0.0003}$$
$$B = 40,000$$

12 audits is 0.03% of 40,000 audits.

23. Answers will vary.

25. P is 3000; R is 10%.

$$R \times B = P$$
$$0.10 \times B = 3000$$
$$\frac{0.10B}{0.10} = \frac{3000}{0.10}$$
$$B = 30,000$$

The total size of its global workforce is 30,000 workers.

27. P is 1785; R is 23%.

$$R \times B = P$$
$$0.23 \times B = 1785$$
$$\frac{0.23B}{0.23} = \frac{1785}{0.23}$$
$$B = 7760.87$$

The total enrollment is 7761 students.

29. P is \$840; R is 30%.

$$R \times B = P$$
$$0.30 \times B = \$840$$
$$\frac{0.30B}{0.30} = \frac{\$840}{0.30}$$
$$B = \$2800$$

The buyer's minimum monthly income is \$2800.

31. P is 2.8 million; R is 14%.

$$R \times B = P$$
$$0.14 \times B = 2.8$$
$$\frac{0.14B}{0.14} = \frac{2.8}{0.14}$$
$$B = 20$$

The total number of adolescents is 20 million or 20,000,000.

33. P is 5.2 million; R is 4%.

$$R \times B = P$$
$$0.04 \times B = 5.2$$
$$\frac{0.04B}{0.04} = \frac{5.2}{0.04}$$
$$B = 130$$

The total number of returns filed is 130 million.

35. 2.6% (100% − 97.4%) is the percent retained by the casinos.

$$P = \$4823; \ R = 2.6\%$$

$$R \times B = P$$
$$0.026 \times B = \$4823$$
$$\frac{0.026B}{0.026} = \frac{\$4823}{0.026}$$
$$B = \$185,500$$

The total amount played on the slot machines was $185,500.

Supplementary Exercises: Base and Part

1. P is $202 million; R is 1%.

$$R \times B = P$$
$$0.01 \times B = \$202$$
$$\frac{0.01B}{0.01} = \frac{\$202}{0.01}$$
$$B = \$20,200 \text{ million}$$

The total foreign invested in China last year was $20,200 million (or 20,200,000,000)

3. B is $423,750; R is 68%.

$$P = B \times R$$
$$P = \$423,750 \times 0.68$$
$$P = \$288,150$$

The amount of insurance coverage is $288,150.

5. P is 220,917; R is 46.2%.

$$R \times B = P$$
$$0.462 \times B = 220,917$$
$$\frac{0.462B}{0.462} = \frac{220,917}{0.462}$$
$$B = 478,175.3$$

In 1967, 478,175 Mustangs were sold.

7. B is $594 million; R is 15.8%.

$$P = B \times R$$
$$P = \$594 \times 0.158$$
$$P = \$93.85 \text{ million}$$

The amount of sales that were private label is $93.9 million.

9. P is $308.75; R is 9.5%.

$$R \times B = P$$
$$0.095 \times B = \$308.75$$
$$\frac{0.095B}{0.095} = \frac{\$308.75}{0.095}$$
$$B = \$3250$$

$$\$3250 \times 12 = \$39,000$$
Nancy Barre's annual earnings are $39,000.

11. B is 2200; R is 38%.

$$P = B \times R$$
$$P = 2200 \times 0.38$$
$$P = 836$$

836 drivers were wearing seat belts.

13. P is $1.9 million; R is 3%.

$$R \times B = P$$
$$0.03 \times B = \$1.9$$
$$\frac{0.03B}{0.03} = \frac{\$1.9}{0.03}$$
$$B = \$63.3 \text{ million}$$

Sales of Whiskas cat food last year were $63.3 million.

$$\$63.3 - \$1.9 = \$61.4$$

Sales of Whiskas cat food after the decrease were $61.4 million.

3.4 Finding the Rate

1. ___% of 2760 is 276.

$$R \times 2760 = 276$$

$$\frac{2760R}{2760} = \frac{276}{2760}$$

$$R = 0.1 = 10\%$$

3. 310 is ___% of 248.

___% of 248 is 310.

$$R \times 248 = 310$$

$$\frac{248R}{248} = \frac{310}{248}$$

$$R = 1.25 = 125\%$$

5. ___% of 78.57 is 22.2.

$$R \times 78.57 = 22.2$$

$$\frac{78.57R}{78.57} = \frac{22.2}{78.57}$$

$$R = 0.2825 = 28.3\%$$

7. 73.1 is ___% of 786.8

___% of 786.8 is 73.1.

$$R \times 786.8 = 73.1$$

$$\frac{786.8R}{786.8} = \frac{73.1}{786.8}$$

$$R = 0.0929 = 9.3\%$$

9. ___% of $53.75 is $2.20.

$$R \times \$53.75 = \$2.20$$

$$\frac{53.75R}{53.75} = \frac{2.20}{53.75}$$

$$R = 0.0409 = 4.1\%$$

11. 46 is ___% of 780

___% of 780 is 46.

$$R \times 780 = 46$$

$$\frac{780R}{780} = \frac{46}{780}$$

$$R = 0.0589 = 5.9\%$$

13. ___% of 2 is 2.05.

$$R \times 2 = 2.05$$

$$\frac{2R}{2} = \frac{2.05}{2}$$

$$R = 1.025 = 102.5\%$$

15. 13,830 is ___% of 78,400.

___% of 78,400 is 13,830.

$$R \times 78,400 = 13,830$$

$$\frac{78,400R}{78,400} = \frac{13,830}{78,400}$$

$$R = 0.1764 = 17.6\%$$

17. ___% of $330 is $91.74.

$$R \times \$330 = \$91.74$$

$$\frac{330R}{330} = \frac{91.74}{330}$$

$$R = 0.278 = 27.8\%$$

19. Answers will vary.

21. ___% of $132,900 is $7442.40.

$$R \times \$132,900 = \$7442.40$$

$$\frac{132,900R}{132,900} = \frac{7442.40}{132,900}$$

$$R = 0.056 = 5.6\%$$

23. ___% of 48,000 is 960.

$$R \times 48,000 = 960$$

$$\frac{48,000R}{48,000} = \frac{960}{48,000}$$

$$R = 0.02 = 2\%$$

25. Total advertising expenditures are

$2250 + $954 + $1950 + $1425 + $1605 + $2775 = $10,959.

___% of $10,959 is $954.

$$R \times \$10,959 = \$954$$

$$\frac{10,959R}{10,959} = \frac{954}{10,959}$$

$$R = 0.0870 = 8.7\%$$

27. The amount of increase is

$$145,000 - 131,000 = 14,000.$$

___% of 131,000 is 14,000.

$$R \times 131,000 = 14,000$$

$$\frac{131,000R}{131,000} = \frac{14,000}{131,000}$$

$$R = 0.1068 = 10.7\%$$

29. The amount of decrease is

$$\$0.135 - \$0.105 = \$0.030.$$

____% of $0.135 is $0.030.

$$R \times 0.135 = \$0.030$$

$$\frac{\$0.135R}{\$0.135} = \frac{\$0.030}{\$0.135}$$

$$R = 0.2222 = 22.2\%$$

Supplementary Exercises: Rate, Base, and Part

1. B is 20 million; R is 3%.

$$P = B \times R$$
$$P = 20 \times 0.03$$
$$P = 0.6 \text{ million}$$

Medical doctors referred 0.6 million (or 600,000) American patients to chiropractors.

3. ____% of 3200 is 450.

$$R \times 3200 = 450$$

$$\frac{3200R}{3200} = \frac{450}{3200}$$

$$R = 0.1406 = 14.1\%$$

They have planned 14.1% more additional stores..

5. B is 44,500; R is 38%.

$$P = B \times R$$
$$P = 44,500 \times 0.38$$
$$P = 16,910$$

There are 16,910 economy hotels and motels.

7. P is 17,126; R is 41%.

$$R \times B = P$$
$$0.41 \times B = 17,126$$
$$\frac{0.41B}{0.41} = \frac{17,126}{0.41}$$
$$B = 41,771$$

There were 41,771 traffic deaths last year.

9. B is $398; R is 7%.

$$P = B \times R$$
$$P = \$398 \times 0.07$$
$$P = \$27.86$$

The new price is

$$\$398 - \$27.86 = \$370.14$$

$$P = B \times R$$
$$P = \$370.14 \times 0.07$$
$$P = \$25.91$$

The total cost is

$$\$370.14 + \$25.91 = \$396.05$$

11. (a) ____% of 50 is 15.

$$R \times 50 = 15$$

$$\frac{50R}{50} = \frac{15}{50}$$

$$R = 0.3 = 30\%$$

30% of states have a blood alcohol limit of 0.08%.

(b) $100\% - 30\% = 70\%$

70% have a limit of 0.10%.

13. P is 51,156; R is 9.8%.

$$R \times B = P$$
$$0.098 \times B = 51,156$$
$$\frac{0.098B}{0.098} = \frac{51,156}{0.098}$$
$$B = 522,000$$

The number of workers after the layoffs is

$$522,000 - 51,156 = 470,844$$

15. ____% of $1220 is $298.

$$R \times \$1220 = \$298$$

$$\frac{1220R}{1220} = \frac{298}{1220}$$

$$R = 0.2442 = 24.4\%$$

24.4% of his payment is interest.

17. The amount of increase is

$$36,000 - 32,000 = 4000.$$

____% of 32,000 is 4000.

$$R \times 32,000 = 4000$$

$$\frac{32,000R}{32,000} = \frac{4000}{32,000}$$

$$R = 0.125 = 12.5\%$$

The percent of increase is 12.5%.

19. The amount of decrease is

$$82,000 - 77,500 = 4500.$$

____% of 82,000 is 4500.

$$R \times 82,000 = 4500$$

$$\frac{82,000R}{82,000} = \frac{4500}{82,000}$$

$$R = 0.0548 = 5.5\%$$

The percent of decrease is 5.5%.

21. R is 25%; P is 240.

$$R \times B = P$$
$$0.25 \times B = 240$$

$$\frac{0.25B}{0.25} = \frac{240}{0.25}$$

$$B = 960$$

There were 960 candy bars in the machine.

23. They spend 83% (25% + 30% + 8% + 20%) of their income. They save 17% (100% − 83%). Their annual earnings are

$$\$28,500 + (\$1950 \times 12) = \$51,900$$

$$P = B \times R$$
$$P = \$51,900 \times 0.17$$
$$P = \$8823$$

The couple will save \$8823.

25. ____% of 11 is 2.6.

$$R \times 11 = 2.6$$

$$\frac{11R}{11} = \frac{2.6}{11}$$

$$R = 0.2363 = 23.6\%$$

There are 23.6% of the buildings insured for flooding.

27. B is 9000; R is 63.8%.

$$P = B \times R$$
$$P = 9000 \times 0.638$$
$$P = 5742$$

5742 deaths would have been prevented.

29. ____% of 50 is 18.

$$R \times 50 = 18$$

$$\frac{50R}{50} = \frac{18}{50}$$

$$R = 0.36 = 36\%$$

36% of the top 50 companies were Japanese companies.

3.5   Increase and Decrease Problems

1. $100\% \times B + 20\% \times B = \$450$
$$120\% \times B = \$450$$

$$\frac{1.2B}{1.2} = \frac{\$450}{1.2}$$

$$B = \$375$$

3. $100\% \times B + 10\% \times B = \$30.70$
$$110\% \times B = \$30.70$$

$$\frac{1.1B}{1.1} = \frac{\$30.70}{1.1}$$

$$B = \$27.91$$

5. $100\% \times B - 20\% \times B = \$20$
$$80\% \times B = \$20$$

$$\frac{0.8B}{0.8} = \frac{\$20}{0.8}$$

$$B = \$25$$

7. $100\% \times B - 30\% \times B = \$598.15$
$$70\% \times B = \$598.15$$

$$\frac{0.7B}{0.7} = \frac{\$598.15}{0.7}$$

$$B = \$854.50$$

9. Answers will vary.

11. $100\% \times B + 8\% \times B = \$178,740$
$$108\% \times B = \$178,740$$

$$\frac{1.08B}{1.08} = \frac{\$178,740}{1.08}$$

$$B = \$165,500$$

Last year's selling price was \$165,500.

13. $100\% \times B + 9\% \times B = \$94$ million
$$109\% \times B = \$94$$

$$\frac{1.09B}{1.09} = \frac{\$94}{1.09}$$

$$B = \$86.2 \text{ million}$$

Last year's sales were \$86.2 million or \$86,200,000.

15. $100\% \times B - 28.75\% \times B = 114$

$71.25\% \times B = 114$

$$\frac{0.7125B}{0.7125} = \frac{114}{0.7125}$$

$B = 160 \text{ feet}$

The distance needed to stop without the antilock braking system is 160 feet.

17. $100\% \times B + 26\% \times B = \$19,104$

$126\% \times B = \$19,104$

$$\frac{1.26B}{1.26} = \frac{\$19,104}{1.26}$$

$B = \$15,161.90$

The average cost of a wedding five years ago was $15,161.90

19. $100\% \times B + 20\% \times B = \$170,035.20$

$120\% \times B = \$170,035.20$

$$\frac{1.2B}{1.2} = \frac{\$170,035.20}{1.2}$$

$B = \$141,696$

The sales last year were $141,696.

$100\% \times B + 20\% \times B = \$141,696$

$120\% \times B = \$141,696$

$$\frac{1.2B}{1.2} = \frac{\$141,696}{1.2}$$

$B = \$118,080$

The sales two years ago were $118,080.

21. $100\% \times B + 12\% \times B = \$66 \text{ billion}$

$112\% \times B = \$66$

$$\frac{1.12B}{1.12} = \frac{\$66}{1.12}$$

$B = \$58.92 \text{ billion}$

This year the cost of nursing home care is $58.9 billion.

23. The total sales were $5750 + $4186 = $9936.

$100\% \times B - 28\% \times B = \$9936$

$72\% \times B = \$9936$

$$\frac{0.72B}{0.72} = \frac{\$9936}{0.72}$$

$B = \$13,800$

$13,800 - $9936 = $3864

The value of the equipment left to sell is $3864.

25. $100\% \times B - 2.4\% \times B = \$1.052 \text{ billion}$

$97.6\% \times B = \$1.052$

$$\frac{0.976B}{0.976} = \frac{\$1.052}{0.976}$$

$B = \$1.078 \text{ billion}$

Last year's sales were $1.078 billion or $1,078,000,000.

27. $100\% \times B - 2\% \times B = 50.2 \text{ million}$

$98\% \times B = 50.2$

$$\frac{0.98B}{0.98} = \frac{50.2}{0.98}$$

$B = 51.22 \text{ million}$

Last year 51.2 million acres were planted.

29. Find the enrollment in 2000.

$100\% \times B + 8\% \times B = 23,328$

$108\% \times B = 23,328$

$$\frac{1.08B}{1.08} = \frac{23,328}{1.08}$$

$B = 21,600 \text{ students}$

The enrollment in 2000 was 21,600 students.

$100\% \times B + 8\% \times B = 21,600$

$108\% \times B = 21,600$

$$\frac{1.08B}{1.08} = \frac{21,600}{1.08}$$

$B = 20,000 \text{ students}$

The student enrollment in 1999 was 20,000 students.

31. $100\% \times B + 15\% = 79.81 \text{ million}$

$115\% \times B = 79.81$

$$\frac{1.15B}{1.15} = \frac{79.81}{1.15}$$

$B = 69.4 \text{ million}$

69.4 personal computers were shipped last year.

33. $100\% \times B - 14\% \times B = 5645$

$86\% \times B = 5645$

$$\frac{0.86B}{0.86} = \frac{5645}{0.86}$$

$B = 6563.95$

Last year 6564 new homes were sold.

Chapter 3 Review Exercises

1. P is 18; R is 12%.

$$R \times B = P$$
$$0.12 \times B = 18$$
$$\frac{0.12B}{0.12} = \frac{18}{0.12}$$
$$B = 150$$

18 members is 12% of 150 members.

2. B is 480; R is 5%

$$P = B \times R$$
$$P = 480 \times 0.05$$
$$P = 24 \text{ vans}$$

3. P is 33; R is 3%.

$$R \times B = P$$
$$0.03 \times B = 33$$
$$\frac{0.03B}{0.03} = \frac{33}{0.03}$$
$$B = 1100$$

33 shippers is 3% of 1100 shippers.

4. P is 36; B is 1440.

$$R \times B = P$$
$$R \times 1440 = 36$$
$$\frac{1440R}{1440} = \frac{36}{1440}$$
$$R = 0.025 = 2.5\%\hat{}$$

36 accounts is 2.5% of 1440 accounts.

5. $\frac{1}{4}\%$ of $1500

$$P = B \times R$$
$$P = \$1500 \times 0.0025$$
$$P = \$3.75$$

6. $24\% = 0.24 = \frac{24}{100} = \frac{6}{25}$

7. P is 24; R is $2\frac{1}{2}\%$.

$$R \times B = P$$
$$0.025 \times B = 24$$
$$\frac{0.025B}{0.025} = \frac{24}{0.025}$$
$$B = 960$$

24 loads of $2\frac{1}{2}\%$ is 960 loads.

8. $87.5\% = 0.875 = \frac{875}{1000} = \frac{875 \div 125}{1000 \div 125} = \frac{7}{8}$

9. P is $70.55; B is $830.

$$R \times B = P$$
$$R \times \$830 = \$70.55$$
$$\frac{\$830R}{\$830} = \frac{\$70.55}{\$830}$$
$$R = 0.085 = 8.5\%$$

$70.55 is 8.5% of $830.

10. $\frac{1}{2}\% = 0.005 = \frac{5}{1000} = \frac{5 \div 5}{1000 \div 5} = \frac{1}{200}$

11. 2.1% of $79.25

$$P = B \times R$$
$$P = \$79.25 \times 0.021$$
$$P = \$1.66$$

The dividend per share is $1.66.

12. R is 0.5%; P is 1120.

$$R \times B = P$$
$$0.005 \times B = 1120$$
$$\frac{0.005B}{0.005} = \frac{1120}{0.005}$$
$$B = 224{,}000$$

The total monthly production is 224,000 units.

13. B is 3.83; R is 29.2%.

$$P = B \times R$$
$$P = 3.83 \times 0.292$$
$$P = 1.12 \text{ million}$$

$$3.83 - 1.12 = 2.71$$

2.71 million vehicles were sold this year.

14. P is 3 million; R is 14%.

$$R \times B = P$$
$$0.14 \times B = 3$$
$$\frac{0.14B}{0.14} = \frac{3}{0.14}$$
$$B = 21.43 \text{ million}$$

There are 21.43 million people in this age bracket.

15. (a) $22\% + 38\% + 14\% + 15\% = 89\%$

They plan to spend $100\% - 89\% = 11\%$ of the total budget on bumper stickers.

(b) B is \$3400; R is 11%

$P = B \times R$
$P = \$3400 \times 0.11$
$P = \$374$
$\$374 \times 12 = \$4488$

They plan to spend \$4488 ($12 \times \$374$) on bumper stickers for the entire year.

16. ___% of \$25,000 is \$17,000.

$R \times \$25,000 = \$17,000$

$$\frac{25,000R}{25,000} = \frac{\$17,000}{25,000}$$

$$R = 0.68$$

The employee will actually receive 68% of the bonus.

17. $100\% \times B - 25\% \times B = \$637.50$
$75\% \times B = \$637.50$

$$\frac{0.75B}{0.75} = \frac{\$637.50}{0.75}$$

$$B = \$850$$

The original price of the digital camera is \$850.

18. Find the number of back packs sold last year.

$100\% \times B + 10\% \times B = 1452$
$110\% \times B = 1452$

$$\frac{1.1B}{1.1} = \frac{1452}{1.1}$$

$$B = 1320$$

1320 backpacks were sold last year.

$100\% \times B + 10\% \times B = 1320$
$110\% \times B = 1320$

$$\frac{1.1B}{1.1} = \frac{1320}{1.1}$$

$$B = 1200$$

Two years ago 1200 backpacks were sold.

19. The increase in stock shares was
$449.5 - 12.3 = 437.2$ pence.

___% of 437.2 is 12.3.

$$R \times 437.2 = 12.3$$

$$\frac{437.2R}{437.2} = \frac{12.3}{437.2}$$

$$R = 0.028 = 2.8\%$$

The percent of increase is 2.8%.

20. $100\% \times B + 12.5\% \times B = 8.4$
$112.5\% \times B = 8.4$

$$\frac{1.125B}{1.125} = \frac{8.4}{1.125}$$

$$B = 7.5 \text{ million}$$

7.5 million (or 7,500,000) people paid income tax on their Social Security income last year.

21. ___% of \$1258.5 million is \$553.7 million.

$$R \times \$1258.5 = \$553.7$$

$$\frac{\$1258.5R}{\$1258.5} = \frac{\$553.7}{\$1258.5}$$

$$R = 0.44 = 44\%$$

44% of sales were Star-Kist.

22. B is 16; R is 78%.

$P = B \times R$
$P = 16 \times 0.78$
$P = 12.48$

In a 16-ounce bottle of shampoo, there are 12.5 ounces of water.

23. $100\% \times B + 13.7\% \times B = 111{,}150$
$113.7\% \times B = 111{,}150$

$$\frac{1.137B}{1.137} = \frac{111{,}150}{1.137}$$

$$B = 97{,}757.2$$

Last year's sales were 97,757 copies.

24. $100\% \times B - 11.8\% \times B = \$35{,}138.88$
$88.2\% \times B = \$35{,}138.88$

$$\frac{0.882B}{0.882} = \frac{\$35{,}138.88}{0.882}$$

$$B = \$39{,}840$$

The total amount of her sales was \$39,840.

25. ___% of 230,000 is 112,091.

$$R \times 230{,}000 = 112{,}091$$

$$\frac{230{,}000R}{230{,}000} = \frac{112{,}091}{230{,}000}$$

$$R = 0.4873 = 48.7\%$$

48.7% of the patent applications resulted in patents.

26. ____% of 33,000,000 is 240,000.

$$R \times 33{,}000{,}000 = 240{,}000$$

$$\frac{33{,}000{,}000R}{33{,}000{,}000} = \frac{240{,}000}{33{,}000{,}000}$$

$$R = 0.0072$$

0.7% of the Argentines will have plastic surgery this year.

27. 32% of 5.75 billion

$P = B \times R$
$P = 5.75 \times 0.32$
$P = 1.84$ billion people

28. 39% of $3.5 billion

$P = B \times R$
$P = \$3.5 \times 0.39$
$P = \$1.365$ billion

29. The amount of increase is
155 − 15 = 140 birds.

____% of 15 is 140.

$$R \times 15 = 140$$

$$\frac{15R}{15} = \frac{140}{15}$$

$$R = 9.333\overline{3}$$

The percent of increase in the number of whooping cranes is $933\frac{1}{3}\%$.

30. The amount of increase is
55,600 − 36,000 = 19,600.

____% × 36,000 = 19,600

$$R \times 36{,}000 = 19{,}600$$

$$36{,}000R = 19{,}600$$

$$\frac{36{,}000R}{36{,}000} = \frac{19{,}600}{36{,}000}$$

$$R = 0.544\overline{4}$$

The percent of increase in the number of guaranteed loans is 54.4%.

31. The amount of decrease is
64,743 − 64,031 = 712.

____% × 64,743 = 712

$$R \times 64{,}743 = 712$$

$$\frac{64{,}743R}{64{,}743} = \frac{712}{64{,}743}$$

$$R = 0.0109$$

The percent of decrease was 1.1%.

32. B is 2.4 million; R is 25%

$P = B \times R$
$P = 2.4 \times 0.25$
$P = 0.6$

$2.4 - 0.6 = 1.8$

The number of Canadian tourists visiting **Florida** this year is 1.8 million or 1,800,000.

33.
$$100\% \times B + 40\% \times B = 350{,}000$$
$$140\% \times B = 350{,}000$$
$$\frac{1.4B}{1.4} = \frac{350{,}000}{1.4}$$
$$B = 250{,}000$$

The production last year was 250,000.

34.
$$100\% \times B + 2\% \times B = 151{,}477$$
$$102\% \times B = 151{,}477$$
$$\frac{1.02B}{1.02} = \frac{151{,}477}{1.02}$$
$$B = 148{,}506.8$$

Last year 148,507 units were sold.

35.
$$100\% \times B + 4.2\% \times B = \$12.40$$
$$104.2\% \times B = \$12.40$$
$$\frac{1.042B}{1.042} = \frac{\$12.40}{1.042}$$
$$B = \$11.90$$

The average hourly wages last year were $11.90.

36.
$$100\% \times B - 26\% \times B = 72$$
$$74\% \times B = 72$$
$$\frac{0.74B}{0.74} = \frac{72}{0.74}$$
$$B = 97.2$$

Before the cutback, daily production was 97 units.

Chapter 3 Summary Exercise

Chipper Jones

There is a 0% change from last year, so the card price this year is $40.

Cal Ripken, Jr.

The amount of decrease is

$75 - $70 = $5.

___% of $75 is $5.

$R \times \$75 = \$5$

$$\frac{\$75R}{\$75} = \frac{\$5}{\$75}$$

$$R = 0.06\overline{6} = 7\%$$

The percent change from last year is −7%.

Nomar Garciapara

$$100\% \times B + 650\% \times B = \$30$$
$$750\% \times B = \$30$$

$$\frac{7.5B}{7.5} = \frac{\$30}{7.5}$$

$$B = \$4$$

The card price last year was $4.

Nolan Ryan

Find the decrease.

$P = B \times R$
$P = \$1000 \times 10\%$
$P = \$1000 \times 0.1$
$P = \$100$

The card price this year is $900.

Ken Griffey, Jr.

$$100\% \times B + 33\% \times B = \$100$$
$$133\% \times B = \$100$$

$$\frac{1.33B}{1.33} = \frac{\$100}{1.33}$$

$$B = \$75$$

The card price last year was $75.

Frank Thomas

The amount of decrease is $90 − $80 = $10.

___% of $90 is $10

$R \times \$90 = \$10$

$$\frac{90R}{90} = \frac{\$10}{\$90}$$

$$R = 0.111 = 11\%$$

The percent of change from last year is 11%.

Will Clark

Find the decrease.

$P = B \times R$
$P = \$6 \times 17\%$
$P = \$6 \times 0.17$
$P = \$1$

$6 − $1 = $5

The card price this year is $5.

Mark McGuire

$$100\% \times B + 650\% \times B = \$150$$
$$750\% \times B = \$150$$

$$\frac{7.5B}{7.5} = \frac{\$150}{\$7.5}$$

$$B = \$20$$

The card price last year was $20.

Sammy Sosa

The amount of increase is
$150 − $3 = $147.

___% of $3 is $147.

$R \times \$3 = \$147$

$$\frac{\$3B}{\$3} = \frac{\$147}{\$3}$$

$$B = 49 = 4900\%$$

The percent change from last year is 4900%.

Alex Rodriquez

The amount of decrease is
$40 − $30 = $10.

_____% of $40 is $10.

$R \times \$40 = \$10$

$$\frac{\$40R}{\$40} = \frac{\$10}{\$40}$$

$$R = 0.25 = 25\%$$

The percent change from last year is −25%.

Cumulative Review Exercises (Chapters 1-3)

1. $450 +$325 + $320 + $182 +$150 = $1427
   $1620 − $1427 = $193

   Bryan Gripka's savings are $193.

2. $(6 \times \$1256) + (15 \times \$895) = \$20{,}961$

   The total cost is $20,961.

3. $29,742.18+($14,096.18 + $6,529.42)−$18,709.51
   = $31,658.27

   The firm's checking account balance at the end of April is $31,658.27.

4. $1436.13 ÷ $53.19 = 27

   It will take Christine Grexa 27 months to pay off $1436.13.

5. $\dfrac{48}{54} = \dfrac{48 \div 6}{54 \div 6} = \dfrac{8}{9}$

6. $8\frac{1}{8} = \dfrac{(8 \times 8) + 1}{8} = \dfrac{65}{8}$

7. $15\overline{)107} \quad \dfrac{105}{2} \qquad \dfrac{107}{15} = 7\frac{2}{15}$

8. $1\frac{2}{3} + 2\frac{3}{4} = 1\frac{8}{12} + 2\frac{9}{12} = 3\frac{17}{12}$
   $= 3 + 1\frac{5}{12} = 4\frac{5}{12}$

9. $5\frac{7}{8} + 7\frac{2}{3} = 5\frac{21}{24} + 7\frac{16}{24} = 12\frac{37}{24}$
   $= 12 + 1\frac{13}{24} = 13\frac{13}{24}$

10. $6\frac{1}{3} - 4\frac{7}{12} = 6\frac{4}{12} - 4\frac{7}{12}$
    $= 5\frac{16}{12} - 4\frac{7}{12} = 1\frac{9}{12} = 1\frac{3}{4}$

11. $8\frac{1}{2} \times \dfrac{9}{17} \times \dfrac{2}{3} = \dfrac{\overset{1}{\cancel{17}}}{\underset{1}{\cancel{2}}} \times \dfrac{\overset{3}{\cancel{9}}}{\underset{1}{\cancel{17}}} \times \dfrac{\overset{1}{\cancel{2}}}{\underset{1}{\cancel{3}}}$
    $= \dfrac{1 \times 3 \times 1}{1 \times 1 \times 1} = \dfrac{3}{1} = 3$

12. $3\frac{3}{4} \div \dfrac{27}{16} = \dfrac{15}{4} \div \dfrac{27}{16} = \dfrac{\overset{5}{\cancel{15}}}{\underset{1}{\cancel{4}}} \times \dfrac{\overset{4}{\cancel{16}}}{\underset{9}{\cancel{27}}}$
    $= \dfrac{5 \times 4}{1 \times 9} = \dfrac{20}{9} = 2\frac{2}{9}$

13. If $\frac{1}{3}$ of the land is sold, then the amount left is
    $1 - \dfrac{1}{3} = \dfrac{3}{3} - \dfrac{1}{3} = \dfrac{2}{3}.$

    Multiply $63\frac{3}{4}$ by $\dfrac{2}{3}$.

    $63\frac{3}{4} \times \dfrac{2}{3} = \dfrac{\overset{85}{\cancel{255}}}{\underset{2}{\cancel{4}}} \times \dfrac{\overset{1}{\cancel{2}}}{\underset{1}{\cancel{3}}}$
    $= \dfrac{85 \times 1}{2 \times 1} = \dfrac{85}{2} = 42\frac{1}{2}$

    The area of the land that is left is $42\frac{1}{2}$ acres.

14. $5\frac{1}{2} + 6\frac{1}{4} + 3\frac{3}{4} + 7$
    $= 5\frac{2}{4} + 6\frac{1}{4} + 3\frac{3}{4} + 7$
    $= 21\frac{6}{4} = 22\frac{2}{4} = 22\frac{1}{2}$

    Bonnie studied $22\frac{1}{2}$ hours altogether.

15. $527\frac{1}{24} - \left(107\frac{2}{3} + 150\frac{3}{4} + 138\frac{5}{8}\right)$
    $= 527\frac{1}{24} - \left(107\frac{16}{24} + 150\frac{18}{24} + 138\frac{15}{24}\right)$
    $= 527\frac{1}{24} - \left(395\frac{49}{24}\right) = 527\frac{1}{24} - 397\frac{1}{24}$
    $= 130$

    The length of the fourth side is 130 feet.

16. There are 8 store managers, so each manager will receive $\frac{1}{8}$.

    Find $\frac{1}{8}$ of $\frac{2}{3}$.

    $\dfrac{1}{8} \times \dfrac{2}{3} = \dfrac{1}{\underset{4}{\cancel{8}}} \times \dfrac{\overset{1}{\cancel{2}}}{3} = \dfrac{1 \times 1}{4 \times 3} = \dfrac{1}{12}$

    Each store manager will receive $\frac{1}{12}$ of the profit sharing funds.

17. $0.35 = \dfrac{35}{100} = \dfrac{35 \div 5}{100 \div 5} = \dfrac{7}{20}$

18.
$$\begin{array}{r} 0.6666 \\ 3)\overline{2.0000} \\ \underline{1\,8} \\ 20 \\ \underline{18} \\ 20 \\ \underline{18} \\ 20 \\ \underline{18} \\ 20 \end{array} \qquad \frac{2}{3} = 0.667$$

19. 78.572 is the nearest hundredth is 78.57.
Locate the hundredths digit and draw a line.

$$78.57|2$$

Since the digit to the right of the line is 2, leave the hundredths digit alone.

20. 4732.489 to the nearest hundredth is 4732.49.
Locate the hundredths digit and draw a line.

$$4732.48|9$$

Since the digit to the right of the line is 9, increase the hundredths digit by 1.

21. 62.65 to the nearest tenth is 62.7.
Locate the tenths digit and draw a line.

$$62.6|5$$

Since the digit to the right of the line is 5, increase the tenths digit by 1.

22. 215.6749 to the nearest thousandth is 215.675.
Locate the thousandth digit and draw a line.

$$215.674|9$$

Since the digit to the right of the line is 9, increase the thousandth digit by 1.

23.
$$\begin{aligned} x + 17 &= 43 \\ x + 17 - 17 &= 43 - 17 \quad \textit{Subtract } 17 \\ x &= 26 \end{aligned}$$

24.
$$\begin{aligned} y - 33 &= 52.4 \\ y - 33 + 33 &= 52.4 + 33 \quad \textit{Add } 33 \\ y &= 85.4 \end{aligned}$$

25.
$$\begin{aligned} \frac{z}{4} - 10 &= 18 \\ \frac{z}{4} - 10 + 10 &= 18 + 10 \quad \textit{Add } 10 \\ \frac{z}{4} &= 28 \\ 4 \cdot \frac{z}{4} &= 4 \cdot 28 \quad \textit{Multiply by } 4 \\ z &= 112 \end{aligned}$$

26.
$$\begin{aligned} 4(r - 2) &= 2(r + 8) \\ 4r - 8 &= 2r + 16 \quad \textit{Distributive property} \\ 4r - 8 + 8 &= 2r + 16 + 8 \quad \textit{Add } 8 \\ 4r &= 2r + 24 \\ 4r - 2r &= 2r - 2r + 24 \quad \textit{Subtract } 2r \\ 2r &= 24 \\ \frac{2r}{2} &= \frac{24}{2} \qquad \textit{Divide by } 2 \\ r &= 12 \end{aligned}$$

27. $\frac{3}{4}$ times a number

$$\frac{3}{4}x$$

28. 5 times a number is added to the number

$$5x + x$$

29. 8 times a number is decreased by 8

$$8x - 8$$

30. The sum of 6 times a number and 5

$$6x + 5$$

31. $I = PRT;\ I = \$2880, R = 0.08, P = \$12,000$

$$\begin{aligned} 2880 &= 12,000 \cdot 0.08 \cdot T \\ 2880 &= 960T \\ \frac{2880}{960} &= \frac{960T}{960} \\ 3 &= T \end{aligned}$$

32. $M = P(1 + RT);\ M = \$2035, R = 0.05, T = 2$

$$\begin{aligned} 2035 &= P(1 + 0.05 \cdot 2) \\ 2035 &= P(1 + 0.1) \\ 2035 &= P(1.1) \\ \frac{2035}{1.1} &= \frac{1.1P}{1.1} \\ \$1850 &= P \end{aligned}$$

33. \$2000 to \$400

$$\frac{2000}{400} = \frac{5}{1}$$

34. 21 feet to 5 yards
Convert 5 yards to feet.
5 yards = 15 feet.

$$\frac{21}{15} = \frac{7}{5}$$

35. $\dfrac{3}{x} = \dfrac{14}{42}$

$3 \cdot 42 = x \cdot 14$ *Cross multiply*

$126 = 14x$

$\dfrac{126}{14} = \dfrac{14x}{14}$ *Divide by 14*

$9 = x$

36. $\dfrac{5}{8} = \dfrac{22}{y}$

$5 \cdot y = 8 \cdot 22$ *Cross multiply*

$5y = 176$

$\dfrac{5y}{5} = \dfrac{176}{5}$ *Divide by 5*

$y = \dfrac{176}{5} = 35.2$

37. Let x be the number of first-time home buyers. Set up a proportion.

$\dfrac{80}{2,480} = \dfrac{24}{x}$

$80 \cdot x = 2480 \cdot 24$ *Cross multiply*

$80x = 59,520$

$x = 744$ *Divide by 80*

There were 744 first-time buyers.

38. Let x be the amount spent on product development. Set up a proportion.

$\dfrac{1}{4} = \dfrac{38,500}{x}$

$1 \cdot x = 4 \cdot 38,500$ *Cross multiply*

$x = 154,000$

$154,000 is spent on product development.

39. $\dfrac{5}{8} = \dfrac{5 \times 125}{8 \times 125} = \dfrac{625}{1000} = 0.625 = 62.5\%$

40. $0.25\% = 0.0025$

41. 18% of 2500 prospects

$P = B \times R$

$P = 2500 \times 0.18$

$P = 450$ prospects

42. 134% of $80

$P = B \times R$

$P = 80 \times 1.34$

$P = \$107.20$

43. 275 is _____% of 1100.
_____% of 1100 is 275.

$R \times 1100 = 275$

$\dfrac{1100R}{1100} = \dfrac{275}{1100}$

$R = 0.25 = 25\%$

44. 375 is _____% of 250.
_____% of 250 is 375.

$R \times 250 = 375$

$\dfrac{250R}{250} = \dfrac{375}{250}$

$R = 1.5 = 150\%$

45. _____% of 32,340 is 20,860.

$R \times 32,340 = 20,860$

$\dfrac{32,340R}{32,340} = \dfrac{20,860}{32,340}$

$R = 0.6450 = 64.5\%$

64.5% of applicants were approved.

46. $B = 76,800; \ R = 28.5\%$

$P = B \times R$

$P = 76,800 \times 0.285$

$P = 21,888$

There are 21,888 people under the age of 18.

47. $100\% \times B + 9.5\% \times B = 64,040$

$109.5\% \times B = 64,040$

$\dfrac{1.095B}{1.095} = \dfrac{64,040}{1.095}$

$B = 58,484$

Last year's enrollment was 58,484 students.

48. $100\% \times B - 25\% \times B = \$12,570$

$75\% \times B = \$12,570$

$\dfrac{0.75B}{0.75} = \dfrac{\$12,570}{0.75}$

$B = \$16,760$

Total sales were $16,760.

Chapter 4

BANKING SERVICES

4.1 Checking Accounts and Check Registers

1. $5.00 + 92($0.10) = $14.20

3. $12.00 + 40($0.20) = $20.00

5. $7.50 + 48($0.20) = $17.10

7. $7.50 + 72($0.20) = $21.90

9. Date: Mar. 8, 20__
Amount: $380.71

Bal Bro't For'd:	$3971.28
Am't Deposited:	$ 79.26
Total:	$4050.54
Am't this Check:	$ 380.71
Balance For'd:	$3669.83

11. Date: Dec. 4, 20__
Amount: $37.52

Bal Bro't For'd:	$1126.73
Am't Deposited:	$0
Total:	$1126.73
Am't this Check:	$ 37.52
Balance For'd:	$1089.21

13. Answers will vary.

15. Answers will vary.

17. Date: Oct. 10, 20__
Amount: $39.12

Bal Bro't For'd:	$5972.89
Am't Deposited:	$ 775.50
Total:	$6748.39
Am't this Check:	$ 39.12
Balance For'd:	$6709.27

19. Bal. Bro't For'd $1629.86
$1629.86 − $250.45 = $1379.41
$1379.41 − $149.00 = $1230.41
$1230.41 + $117.73 = $1348.14
$1348.14 − $ 69.80 = $1278.34
$1278.34 + $329.86 = $1608.20
$1608.20 + $418.30 = $2026.50
$2026.50 − $109.76 = $1916.74
$1916.74 − $614.12 = $1302.62
$1302.62 − $ 32.18 = $1270.44
$1270.44 + $520.95 = $1791.39

21. Bal. Bro't For'd $3852.48
$3852.48 − $ 143.16 = $3709.32
$3709.32 − $ 118.40 = $3590.92
$3590.92 + $ 286.32 = $3877.24
$3877.24 − $ 80.00 = $3797.24
$3797.24 − $ 986.22 = $2811.02
$2811.02 − $ 375.50 = $2435.52
$2435.52 + $ 1201.82 = $3637.34
$3637.34 − $ 735.68 = $2901.66
$2901.66 − $ 223.94 = $2677.72
$2677.72 + $ 498.01 = $3175.73
$3175.73 − $ 78.24 = $3097.49

4.2 Checking Services and Depositing Credit Card Transactions

1. **(a)** Charges: $2419.76 (total sales)

(b) Credits: $203.86 (total credits)

(c) Gross:

$$\$2419.76 - \$203.86 = \$2215.90$$

(d) Discount charge:

$$\$2215.90 \times 0.03(3\%) = \$66.48$$

(e) Credit:

$$\$2215.90 - \$66.48 = \$2149.42$$

3. (a) Charges: \$1591.44 (total sales)

(b) Credits: \$189.39 (total credits)

(c) Gross amount:

$$1591.44 - \$189.39 = \$1402.05$$

(d) Discount charge:

$$\$1402.05 \times 0.04(4\%) = \$56.08$$

(e) Credit:

$$\$1402.05 - \$56.08 = \$1345.97$$

5. (a) Charges: \$1064.72 (total sales)

(b) Credits: \$72.83 (total credits)

(c) Gross amount:

$$\$1064.72 - \$72.83 = \$991.89$$

(d) Discount charge:

$$\$991.89 \times 0.03(3\%) = \$29.76$$

(e) Credit:

$$\$991.89 - \$29.76 = \$962.13$$

7. Answers will vary.

4.3   Reconciliation

1. Current balance
= balance + deposits − checks outstanding
= \$4572.15 + \$1387.42 − \$864.10
= \$5095.47

3. Current balance
= balance + deposits − checks outstanding
= \$7911.42 + \$531.52 − \$752.32
= \$7690.62

5. Current balance
= balance + deposits − checks outstanding
= \$19,523.20 + \$6803.74 − \$8012.22
= \$18,314.72

7. (1) \$6875.09
(2) \$701.56; \$421.78; \$689.35
(3) \$6875.09 + \$701.56 + \$421.78 + \$689.35
     = \$8687.78
(4) \$371.52 + \$429.07 + \$883.69 + \$35.62
     = \$1719.90
(5) \$6967.88
(6) \$6965.92
(7) \$8.75
(8) \$6957.17
(9) \$10.71
(10) \$6967.88

9. Bank statement balance   \$6237.44

Add: Deposits not recorded

\$1442.44	
+ 479.50	+ 1921.94
	\$8159.38

Less: Outstanding checks

\$146.36	
91.52	
43.78	
+ 379.52	− 661.18

Adjusted balance   \$7498.20

Checkbook balance   \$7779.00

Less: Bank charges

\$246.70	
15.60	
+ 18.50	− 280.80

Adjusted balance   \$7498.20

11. Bank statement balance   \$4074.65

Add: Deposits not recorded

\$ 907.82	
+ 1784.15	+ 2691.97
	6766.62

Less: Outstanding checks

\$642.55	
1082.98	
73.25	
+ 471.83	− 2270.61

Adjusted balance   \$4496.01

Checkbook balance   \$4661.31

Add: Interest credit   + 10.18

              \$4671.49

Less: Bank charges

$168.40
+   7.08
                                                    − 175.48
Adjusted balance $4496.01

13. Answers will vary.

15. Answers will vary.

17. Bank statement balance $6380.86

Add: Deposits not recorded

                                    +   830.75
                                        $7211.61

Less: Outstanding checks

$100.50
315.62
+ 67.29   − 483.41
Adjusted balance $6728.20

Checkbook balance $6800.57
Add: Interest credit   +   22.48
                                        $6823.05

Less: Bank charges

$82.15
+ 12.70   −   94.85
Adjusted balance $6728.20

Chapter 4 Review Exercises

1. $7.50 + 42(\$0.20) = \$15.90$

2. $12.00 + 35(\$0.20) = \$19.00$

3. $5.00 + 52(\$0.10) = \$10.20$

4. Date: Aug 6, 20__
Amount: $6892.12
Bal Bro't For'd: $16,409.82
Am't Deposited: $0
Total: $16,409.82
Am't this Check: $ 6892.12
Balance For'd: $ 9517.70

5. Date: Aug 8, 20__
Amount: $1258.36
Bal Bro't For'd: $ 9517.70
Am't Deposited: $ 1572.00
Total: $11,089.70
Am't this Check: $ 1258.36
Balance For'd: $ 9831.34

6. Date: Aug 14, 20__
Amount: $416.14
Bal Bro't For'd: $ 9831.34
Am't Deposited: $10,000.00
Total: $19,831.34
Am't this Check: $ 416.14
Balance For'd: $19,415.20

7. To find the total charges, add all the sales.
Total charges: $1064.72

8. Add the credits.
Total credits: $72.83

9. Gross amount: Total charges − total credits

$1064.72 − \$72.83 = \$991.89$

10. Discount charge:

$991.89 \times 0.04(4\%) = \$39.68$

11. Credit:

$991.89 − \$39.68 = \$952.21$

12. (1) $4964.52
(2) $1912.72; $1436.48
(3) $4964.52 + \$1912.72 + \$1436.48$
= 8313.72
(4) $1520 + \$146.64 + \$31.16 + \$572.76$
= 2270.56
(5) $6043.16
(6) $6204.64
(7) $140.68 + \$30.84 = \$171.52$
(8) $6033.12
(9) $10.04
(10) $6043.16

13. Bank statement balance $8149.30

Add: Deposits not recorded

$1815.64
+ 3568.30   5383.94
                                        $13,533.24

Less: Outstanding checks

$1285.10
2165.96
146.50
+ 943.66   − 4541.22
Adjusted balance $8992.02

Checkbook balance $9322.62
Add: Interest credit + 20.36
$9342.98

Less: Bank charges

$336.80
+ 14.16 − 350.96
Adjusted balance $8992.02

14. Bank statement balance $1302.43

Add: Deposits not recorded + 418.35
$1720.78

Less: Outstanding checks

$ 68.17
215.84
+ 169.56 − 453.57
Adjusted balance $1267.21

Checkbook balance $1347.42
Add: Interest credit + 6.52
$1353.94

Less: Bank charges

$78.93
+ 7.80 − 86.73
Adjusted balance $1267.21

Chapter 4 Summary Exercise

(a) Gross deposit:

$6438.50 − $336.81 = $6101.69

(b) Credit:

$6101.69 × 0.035(3.5\%) = $213.56
$6101.69 − $213.56 = $5888.13

(c) Total checks outstanding:

$758.14 + $38.37 + $1671.88 + $120.13
+ $2264.75 + $78.11 + $3662.73 + $816.25
+ $400 = $9810.36

(d) Deposits not recorded:

$458.23 + $771.18 + $235.71 + $1278.55
+ $663.52 + $1475.39 = $4882.58

(e) Current balance:

$4228.34 + $5888.13 − $9810.36
+ $4882.58 = $5188.69

PAYROLL

5.1 Gross Earnings (Wages and Salaries)

1. $7 + 4 + 7 + 10 + 8 + 4 = 40$ hours
40 regular hours
0 overtime hours

Overtime rate

$1\frac{1}{2} \times \$8.10 = \$12.15$

3. $3 + 6 + 8.25 + 8 + 8.5 + 5 = 38.75$ hours
38.75 regular hours
0 overtime hours

Overtime rate

$1\frac{1}{2} \times \$7.80 = \$11.70$

5. $9.5 + 7 + 9 + 9.25 + 10.5 = 45.25$ hours
40 regular hours
5.25 overtime hours

Overtime rate

$1\frac{1}{2} \times \$11.48 = \$17.22$

7. 40 hours $\times \$8.10 =$ $324.00
   0 overtime hours $=$ $\underline{+\qquad 0}$
   gross earnings $324.00

9. 38.75 hours $\times \$7.80 =$ $302.25
   0 overtime hours $=$ $\underline{+\qquad 0}$
   gross earnings $302.25

11.   40 hours $\times \$11.48 =$ $459.20
   5.25 hours $\times \$17.22 =$ $\underline{+\quad 90.41}$
   gross earnings $549.61

13. Overtime rate

$1\frac{1}{2} \times \$8.80 = \$13.20$

Regular earnings

39.5 hours $\times \$8.80 = \$347.60$

Overtime earnings$= \$0$

Total gross earnings

$347.60 + \$0 = \$347.60$

15. Overtime rate

$1\frac{1}{2} \times \$7.20 = \$10.80$

Regular earnings

40 hours $\times \$7.20 = \$288$

Overtime earnings

4.5 hours $\times \$10.80 = \$48.60$

Total gross earnings

$288 + \$48.60 = \$336.60$

17. Overtime rate

$1\frac{1}{2} \times \$9.18 = \$13.77$

Regular earnings

40 hours $\times \$9.18 = \$367.20$

Overtime earnings

4.25 hours $\times \$13.77 = \$58.52$

Total gross earnings

$367.20 + \$58.52 = \$425.72$

For Exercises 19-23, use

Total hours $\times$ Regular rate
   $=$ straight-time earnings
 $+$ Over-time hours $\times \frac{1}{2}$ regular rate
   $=$ overtime premium
   $=$ total gross earnings

19. $10 + 9 + 8 + 5 + 12 + 7 = 51$ hours

Overtime hours

$51 - 40 = 11$ hours

Over-time premium rate

$\frac{1}{2} \times \$7.40 = \$3.70$

Regular earnings

51 hours $\times \$7.40 = \$377.40$

Overtime earnings

11 hours $\times \$3.70 = \$40.70$

Total gross earnings

$377.40 + \$40.70 = \$418.10$

21. $12 + 11 + 8 + 8.25 + 11 = 50.25$ hours

Overtime hours

$50.25 - 40 = 10.25$ hours

Overtime premium rate

$\frac{1}{2} \times \$8.60 = \$4.30$

Regular earnings

50.25 hours $\times \$8.60 = \$432.15$

Overtime earnings

10.25 hours $\times \$4.30 = \$44.08$

Total gross earnings

$\$432.15 + \$44.08 = \$476.23$

23. $10 + 9.25 + 9.5 + 11.5 + 10 = 50.25$ hours

Overtime hours

$50.25 - 40 = 10.25$ hours

Overtime premium rate

$\frac{1}{2} \times \$10.20 = \$5.10$

Regular earnings

50.25 hours $\times \$10.20 = \$512.55$

Overtime earnings

10.25 hours $\times \$5.10 = \$52.28$

Total gross earnings

$\$512.55 + \$52.28 = \$564.83$

25. Regular hours

$8 + 8 + 8 + 6 + 5 = 35$ hours

Overtime hours

$2 + 1 + 3 = 6$ hours

Overtime rate

$1\frac{1}{2} \times \$6.70 = \$10.05$

Regular earnings

35 hours $\times \$6.70 = \$234.50$

Overtime earnings

6 hours $\times \$10.05 = \$60.30$

Total gross earnings

$\$234.50 + \$60.30 = \$294.80$

27. Regular hours

$7.5 + 8 + 8 + 8 + 8 = 39.5$ hours

Overtime hours

$1 + 2.75 = 3.75$ hours

Overtime rate

$1\frac{1}{2} \times \$6.70 = \$10.05$

Regular earnings

39.5 hours $\times \$6.70 = \$264.65$

Overtime earnings

3.75 hours $\times \$10.05 = \$37.69$

Total gross earnings

$\$264.65 + \$37.69 = \$302.34$

29. Regular hours

$8 + 8 + 7.75 + 8 + 8 = 39.75$ hours

Overtime hours

$1.5 + 0.5 + 1.5 = 3.5$ hours

Overtime rate

$1\frac{1}{2} \times \$10.20 = \$15.30$

Regular earnings

39.75 hours $\times \$10.20 = \$405.45$

Overtime earnings

3.5 hours $\times \$15.30 = \$53.55$

Total gross earnings

$\$405.45 + \$53.55 = \$459$

31. Answers will vary.

33. $\$248 \times 52 = \$12,896$ annual
$\$12,896 \div 12 = \$1074.67$ monthly
$\$12,896 \div 24 = \$537.33$ semimonthly
$\$12,896 \div 26 = \$496$ biweekly

35. $\$852 \times 26 = \$22,152$ annual
$\$22,152 \div 12 = \$1846$ monthly
$\$22,152 \div 24 = \$923$ semimonthly
$\$22,152 \div 52 = \$426$ weekly

37. $\$1087.50 \times 24 = \$26,100$ annual
$\$26,100 \div 12 = \$2175$ monthly
$\$26,100 \div 26 = \$1003.85$ biweekly
$\$26,100 \div 52 = \$501.92$ weekly

39. $\$2680 \times 12 = \$32{,}160$ annual
$\$32{,}160 \div 24 = \$1340$ semimonthly
$\$32{,}160 \div 26 = \$1236.92$ biweekly
$\$32{,}160 \div 52 = \$618.46$ weekly

41. $\$21{,}580 \div 12 = \$1798.33$ monthly
$\$21{,}580 \div 24 = \$899.17$ semimonthly
$\$21{,}580 \div 26 = \$830$ biweekly
$\$21{,}580 \div 52 = \$415$ weekly

43. $\$520 \div 40 = \$13$ per hour

$1\frac{1}{2} \times \$13 = \$19.50$ overtime rate

$16 \times \$19.50 = \$312$ overtime
$\$520 + \$312 = \$832$

45. $\$418 \div 45 = \$9.29$ per hour

$1\frac{1}{2} \times \$9.29 = \$13.94$ overtime rate

$5 \times \$13.94 = \$69.70$ overtime
$\$418 + \$69.70 = \$487.70$

47. $\$450 \div 32 = \$14.06$ per hour

$1\frac{1}{2} \times \$14.06 = \$21.09$ overtime rate

$12 \times \$21.09 = \$253.08$ overtime
$\$450 + \$253.08 = \$703.08$

49. Overtime hours

$48 - 40 = 8$ overtime hours

Overtime rate

$1\frac{1}{2} \times \$7.40 = \$11.10$

Regular earnings

40 hours $\times \$7.40 = \$296$

Overtime earnings

8 hours $\times \$11.10 = \$88.80$

Total gross earnings

$\$296 + \$88.80 = \$384.80$

51. Regular hours

$8 + 7 + 8 + 4.5 + 8 = 35.5$ hours

Overtime hours

$1.5 + 2.75 + 0.75 = 5$ hours

Overtime rate

$1\frac{1}{2} \times \$7.80 = \$11.70$

Regular earnings

35.5 hours $\times \$7.80 = \$276.90$

Overtime earnings

5 hours $\times \$11.70 = \$58.50$

Total gross earnings

$\$276.90 + \$58.50 = \$335.40$

53. Hourly rate

$\$648 \div 40 = \$16.20$

Overtime hours

$46 - 40 = 6$ hours

Overtime rate

$1\frac{1}{2} \times \$16.20 = \$24.30$

Overtime earnings

6 hours $\times \$24.30 = \$145.80$

Total gross earnings

$\$648 + \$145.80 = \$793.80$

55. Hourly rate

$\$638 \div 40 = \$15.95$

Overtime hours

$52 - 40 = 12$ hours

Overtime rate

$1\frac{1}{2} \times \$15.95 = \$23.93$

Overtime earnings

12 hours $\times \$23.93 = \$287.16$

Total gross earnings

$\$638 + \$287.16 = \$925.16$

57. $\$630 \times 52 = \$32{,}760$

(a) $\$32{,}760 \div 26 = \$1260$ biweekly

(b) $\$32{,}760 \div 24 = \$1365$ semimonthly

(c) $\$32{,}760 \div 12 = \$2730$ monthly

(d) $\$630 \times 52 = \$32{,}760$ annual

59. Answers will vary.

61. (a)   High School Graduate

$27,038 \div 52 = \$519.96$ weekly

$27,038 \div 26 = \$1039.92$ biweekly

$27,038 \div 24 = \$1126.58$ semimonthly

$27,038 \div 12 = \$2253.17$ monthly

(b)   College Graduate

$44,523 \div 52 = \$856.21$ weekly

$44,523 \div 26 = \$1712.42$ biweekly

$44,523 \div 24 = \$1855.13$ semimonthly

$44,523 \div 12 = \$3710.25$ monthly

5.2   Gross Earnings (Commission)

1. Gross earnings

$= (\$2810 - \$208) \times 8\%$

$= \$2602 \times 0.08$

$= \$208.16$

3. Gross earnings

$= (\$2875 - \$64) \times 15\%$

$= \$2811 \times 0.15$

$= \$421.65$

5. Gross earnings

$= (\$25,658 - \$4083) \times 9\%$

$= \$21,575 \times 0.09$

$= \$1941.75$

7. Gross earnings

$= (\$45,618 - \$2281) \times 1\%$

$= \$43,337 \times 0.01$

$= \$433.37$

9.
$$
\begin{array}{ll}
\$18,550 & \text{(total sales)} \\
\underline{-\ \ \ 7500} & \$7500 \times 0.06 = \$450 \\
\$11,050 & \\
\underline{-\ \ \ 7500} & \$7500 \times 0.08 = \$600 \\
\$\ \ 3550 & \$3550 \times 0.10 = \underline{\$355} \\
& \$1405
\end{array}
$$

The gross earnings are $1405.

11.
$$
\begin{array}{ll}
\$10,480 & \text{(total sales)} \\
\underline{-\ \ \ 7500} & \$7500 \times 0.06 = \$450.00 \\
\$\ \ 2980 & \$2980 \times 0.08 = \underline{\$238.40} \\
& \$688.40
\end{array}
$$

The gross earnings are $688.40.

13.
$$
\begin{array}{ll}
\$11,225 & \text{(total sales)} \\
\underline{-\ \ 7500} & \$7500 \times 0.06 = \$450 \\
\$\ \ 3725 & \$3725 \times 0.08 = \underline{\$298} \\
& \$748
\end{array}
$$

The gross earnings are $748.

15.
$$
\begin{array}{ll}
\$25,860 & \text{(total sales)} \\
\underline{-\ \ \ 7500} & \$7500 \times 0.06 = \$450 \\
\$18,360 & \\
\underline{-\ \ \ 7500} & \$7500 \times 0.08 = \$600 \\
\$10,860 & \$10,860 \times 0.10 = \underline{\$1086} \\
& \$2136
\end{array}
$$

The gross earnings are $2136.

17. Answers will vary.

19. Commission sales

$\$5250 - \$220 = \$5030$

Gross commission

$0.04 \times \$5030 = \$201.20$

Gross earnings

$\$290 + \$201.20 = \$491.20$

21. Commission sales

$\$6380 - \$295 - \$2000 = \$4085$

Gross commission

$0.06 \times \$4085 = \$245.10$

Gross earnings $= \$245.10$

23. Commission sales

$\$12,420 - \$390 - \$2500 = \$9530$

Gross commission

$0.03 \times \$9530 = \$285.90$

Gross earnings $= \$285.90$

25. Commission sales

$\$4215 - \$318 - \$1000 = \$2897$

Gross commission

$0.05 \times \$2897 = \$144.85$

Gross earnings

$\$210 + \$144.85 = \$354.85$

27. Gross earnings
$$= \text{Commissions} - \text{Draw}$$
$$= (0.08 \times \$9850) - \$350$$
$$= \$788 - \$350$$
$$= \$438$$

29. Gross earnings
$$= \text{fixed earning} + \text{commission}$$
$$= \$1750 + (0.02 \times \$194{,}800)$$
$$= \$1750 + \$3896$$
$$= \$5646$$

31. (a)

$$
\begin{array}{ll}
\$27{,}700 & \text{(total sales)} \\
\underline{-\ \ 6000} & \$6000 \times 0.06 = \ \ \ \$360 \\
\$21{,}700 & \\
\underline{-16{,}000} & \$16{,}000 \times 0.08 = \$1280 \\
\$5700 & \$5700 \times 0.15 = \underline{\ \ \$855} \\
& \$2495
\end{array}
$$

Total commission = $2495

(b) Gross earnings
$$= \text{commissions} - \text{draw}$$
$$= \$2495 - \$800$$
$$= \$1695$$

33. Personal commission sales

$$\$2825 - (\$84 + \$1000) = \$1741$$

Personal commission

$$0.05 \times \$1741 = \$87.05$$

Override commission sales

$$\$8656 - \$317 = \$8339$$

Override commission

$$0.015 \times \$8339 = \$125.09$$

Gross earnings

$$\$200 + \$87.05 + \$125.09 = \$412.14$$

5.3 Gross Earnings (Piecework)

1. Total pieces

$$150 + 124 + 172 + 110 + 96 = 652$$

Gross earnings

$$652 \times \$0.39 = \$254.28$$

3. Total pieces

$$98 + 86 + 79 + 108 + 80 = 451$$

Gross earnings

$$451 \times \$0.75 = \$338.25$$

5. Total pieces

$$118 + 124 + 143 + 132 + 148 = 665$$

Gross earnings

$$665 \times \$0.68 = \$452.20$$

7. Total pieces

$$125 + 118 + 115 + 132 + 98 = 588$$

Gross earnings

$$558 \times \$0.46 = \$270.48$$

9. Total pieces

$$149 + 135 + 118 + 125 + 143 = 670$$

Gross earnings

$$670 \times \$0.78 = \$522.60$$

11. Gross earnings

$$
\begin{array}{ll}
829 & \text{(total units)} \\
\underline{-\ 500} & 500 \times \$0.10 = \$50.00 \\
329 & \\
\underline{-\ 200} & 200 \times \$0.12 = \$24.00 \\
129 & 129 \times \$0.14 = \underline{\$18.06} \\
& \$92.06
\end{array}
$$

13. Gross earnings

$$
\begin{array}{ll}
1182 & \text{(total units)} \\
\underline{-\ 500} & 500 \times \$0.10 = \$\ 50.00 \\
682 & \\
\underline{-\ 200} & 200 \times \$0.12 = \$\ 24.00 \\
482 & \\
\underline{-\ 300} & 300 \times \$0.14 = \$\ 42.00 \\
182 & 182 \times \$0.16 = \underline{\$\ 29.12} \\
& \$145.12
\end{array}
$$

15. Gross earnings

$$
\begin{array}{ll}
1250 & \text{(total units)} \\
\underline{-\ 500} & 500 \times \$0.10 = \$\ 50.00 \\
750 & \\
\underline{-\ 200} & 200 \times \$0.12 = \$\ 24.00 \\
550 & \\
\underline{-\ 300} & 300 \times \$0.14 = \$\ 42.00 \\
250 & 250 \times \$0.16 = \underline{\$\ 40.00} \\
& \$156.00
\end{array}
$$

17. $8 \times \$6.18 = \$49.44$ per day

Monday	$66 \times \$0.75 =$	$\$49.50$
Tuesday	$75 \times \$0.75 =$	$\$56.25$
Wednesday	hourly $=$	$\$49.44$
Thursday	$72 \times \$0.75 =$	$\$54.00$
Friday	$68 \times \$0.75 =$	$\underline{\$51.00}$
gross earnings		$\$260.19$

19. $8 \times \$6.80 = \$54.40$ per day

Monday	$80 \times \$0.75 =$	$\$60.00$
Tuesday	hourly $=$	$\$54.40$
Wednesday	$75 \times \$0.75 =$	$\$56.25$
Thursday	$78 \times \$0.75 =$	$\$58.50$
Friday	$74 \times \$0.75 =$	$\underline{\$55.50}$
gross earnings		$\$284.65$

21. $8 \times \$6.75 = \$54.00$ per day

Monday	$75 \times \$0.75 =$	$\$56.25$
Tuesday	$84 \times \$0.75 =$	$\$63.00$
Wednesday	$72 \times \$0.75 =$	$\$54.00$
Thursday	$93 \times \$0.75 =$	$\$69.75$
Friday	hourly $=$	$\underline{\$54.00}$
gross earnings		$\$297.00$

23. $8 \times \$6.30 = \$50.40$ per day

Monday	$73 \times \$0.75 =$	$\$54.75$
Tuesday	hourly $=$	$\$50.40$
Wednesday	$78 \times \$0.75 =$	$\$58.50$
Thursday	hourly $=$	$\$50.40$
Friday	$81 \times \$0.75 =$	$\underline{\$60.75}$
gross earnings		$\$274.80$

For Exercises 25-31, use

Gross earnings = piecework earnings
   − (spoiled items × chargeback rate).

25. Piecework earnings

$$
\begin{aligned}
510 \times \$0.72 &= \$367.20 \\
1\tfrac{1}{2} \times 74 \times \$0.72 &= \underline{+\ \ \ 79.92} \\
&\quad\ \ \$447.12
\end{aligned}
$$

Gross earnings
$= \$447.12 - (20 \times \$0.38)$
$= \$447.12 - \$7.60$
$= \$439.52$

27. Piecework earnings

$$
\begin{aligned}
493 \times \$0.86 &= \$423.98 \\
1\tfrac{1}{2} \times 74 \times \$0.86 &= \underline{+\ \ \ 95.46} \\
&\quad\ \ \$519.44
\end{aligned}
$$

Gross earnings
$= \$519.44 - (34 \times \$0.46)$
$= \$519.44 - \$15.64$
$= \$503.80$

29. Piecework earnings

$$
\begin{aligned}
286 \times \$0.95 &= \$271.70 \\
1\tfrac{1}{2} \times 38 \times \$0.95 &= \underline{+\ \ \ 54.15} \\
&\quad\ \ \$325.85
\end{aligned}
$$

Gross earnings
$= \$325.85 - (4 \times \$0.82)$
$= \$325.85 - \$3.28$
$= \$322.57$

31. Piecework earnings

$$
\begin{aligned}
403 \times \$0.68 &= \$274.04 \\
1\tfrac{1}{2} \times 72 \times \$0.68 &= \underline{+\ \ \ 73.44} \\
&\quad\ \ \$347.48
\end{aligned}
$$

Gross earnings
$= \$347.48 - (15 \times \$0.45)$
$= \$347.48 - \$6.75$
$= \$340.73$

33. Answers will vary.

35. Piecework earnings

$$
\begin{aligned}
142 \times \$4.75 &= \$674.50 \\
1\tfrac{1}{2} \times 26 \times \$4.75 &= \underline{+\ \ 185.25} \\
&\quad\ \ \$859.75
\end{aligned}
$$

Gross earnings
$= \$859.75 - (7 \times \$2.25)$
$= \$859.75 - \$15.75$
$= \$844$

37. Piecework earnings

$$
\begin{aligned}
310 \times \$1.35 &= \$418.50 \\
1\tfrac{1}{2} \times 110 \times \$1.35 &= \underline{+\ \ 222.75} \\
&\quad\ \ \$641.25
\end{aligned}
$$

Gross earnings

$$= \$641.25 - (20 \times \$0.85)$$
$$= \$641.25 - \$17$$
$$= \$624.25$$

39. Total production

$$136 + 112 + 108 + 96 + 122 = 574$$

Total chargebacks

$$6 + 3 + 5 + 8 + 4 = 26$$

Gross earnings

$$= (574 \times \$0.85) - (26 \times \$0.35)$$
$$= \$487.90 - \$9.10$$
$$= \$478.80$$

5.4   Social Security, Medicare, and Other Taxes

1. Social security tax

$$\$324.72 \times 0.062 = \$20.13$$

Medicare tax

$$\$324.72 \times 0.0145 = \$4.71$$

3. Social security tax

$$\$463.24 \times 0.062 = \$28.72$$

Medicare tax

$$\$463.24 \times 0.0145 = \$6.72$$

5. Social security tax

$$\$854.71 \times 0.062 = \$52.99$$

Medicare tax

$$\$854.71 \times 0.0145 = \$12.39$$

7. Social security tax

$$\$1086.25 \times 0.062 = \$67.35$$

Medicare tax

$$\$1086.25 \times 0.0145 = \$15.75$$

9.
$$\begin{array}{r} \$80,000.00 \\ - \ \$77,871.24 \\ \hline \$2128.76 \end{array}$$

$$\$2128.76 \times 0.062 = \$131.98$$

11.
$$\begin{array}{r} \$80,000.00 \\ - \ \$75,721.59 \\ \hline \$4278.41 \end{array}$$

$$\$4278.41 \times 0.062 = \$265.26$$

13.
$$\begin{array}{r} \$80,000.00 \\ - \ \$79,819.75 \\ \hline \$180.25 \end{array}$$

$$\$180.25 \times 0.062 = \$11.18$$

15.

$40 \times \$9.22 = \$368.80$	regular
$1\frac{1}{2} \times \$9.22 \times 5.5 = \$76.07$	overtime
$\$368.80 + \$76.07 = \$444.87$	gross
$\$444.87 \times 6.2\% = \$27.58$	FICA
$\$444.87 \times 1.45\% = \$6.45$	medicare
$\$444.87 \times 1\% = \$4.45$	SDI

17.

$40 \times \$10.30 = \$412$	regular
$1\frac{1}{2} \times \$10.30 \times 4 = \$61.80$	overtime
$\$412 + \$61.80 = \$473.80$	gross
$\$473.80 \times 6.2\% = \$29.38$	FICA
$\$473.80 \times 1.45\% = \$6.87$	medicare
$\$473.80 \times 1\% = \$4.74$	SDI

19.

$40 \times \$8.18 = \$327.20$	regular
$1\frac{1}{2} \times \$8.18 \times 5 = \$61.35$	overtime
$\$327.20 + \$61.35 = \$388.55$	gross
$\$388.55 \times 6.2\% = \$24.09$	FICA
$\$388.55 \times 1.45\% = \$5.63$	medicare
$\$388.55 \times 1\% = \$3.89$	SDI

21.

$40 \times \$6.24 = \$249.60$	regular
$1\frac{1}{2} \times \$6.24 \times 6\frac{3}{4} = \$63.18$	overtime
$\$249.60 + \$63.18 = \$312.78$	gross
$\$312.78 \times 6.2\% = \$19.39$	FICA
$\$312.78 \times 1.45\% = \$4.54$	medicare
$\$312.78 \times 1\% = \$3.13$	SDI

23.

$$40 \times \$8.58 = \$343.20$$
$$1\frac{1}{2} \times \$8.58 \times 3.5 = \$45.05$$
$$\$343.20 + \$45.05 = \$388.25$$

(a) $\$388.25 \times 6.2\% = \$24.07$ FICA
(b) $\$388.25 \times 1.45\% = \$5.63$ medicare

25. $\$19,482 - \$193 = \$19,289$
$\quad \$19,289 \times 8\% = \$1543.12$ commission

(a) $\$1543.12 \times 6.2\% = \$95.67$ FICA
(b) $\$1543.12 \times 1.45\% = \$22.38$ medicare
(c) $\$1543.12 \times \ \ 1\% = \$15.43$ SDI

27. Social security tax

$36,852.80 \times 0.124 = \$4569.75$

Medicare tax

$36,852.80 \times 0.029 = \$1068.73$

29. Social security tax

$34,817.16 \times 0.124 = \$4317.33$

Medicare tax

$34,817.16 \times 0.029 = \$1009.70$

31. Social security tax

$26,843.60 \times 0.124 = \$3328.61$

Medicare tax

$26,843.60 \times 0.029 = \$778.46$

33. Answers will vary.

5.5 Income Tax Withholding

1. $241   **3.** $54   **5.** $172

7. $74   **9.** $69

11. $0.044 \times \$188.60 = \$8.30$

13. $0.028 \times \$317.43 = \$8.89$

15. $0.06 \times \$1476.32 = \$88.58$

17. withholding allowance

$52.88 \times 4 = \$211.52$
$417.58 - \$211.52 = \$206.06$
$206.06 - \$124 = \$82.06$
$82.06 \times 15\% = \$12.31$

federal tax	$12.31
FICA tax ($6.2\% \times \$417.58$)	$25.89
medicare ($1.45\% \times \$417.58$)	$ 6.05
total deductions	$44.25

Net pay

$417.58 - \$44.25 = \$373.33$

19. withholding allowance

$229.17 \times 1 = \$229.17$
$1532.18 - \$229.17 = \$1303.01$
$1303.01 - \$221 = \$1082.01$
$1082.01 \times 15\% = \$162.30$

federal tax	$162.30
FICA tax ($6.2\% \times \$1532.18$)	$ 95.00
medicare ($1.45\% \times \$1532.18$)	$ 22.22
total deductions	$279.52

Net pay

$1532.18 - \$279.52 = \$1252.66$

21. withholding allowance

$114.58 \times 3 = \$343.74$
$1938.76 - \$343.74 = \$1595.02$
$1595.02 - \$269 = \$1326.02$
$1326.02 \times 15\% = \$198.90$

federal tax	$198.90
FICA tax ($6.2\% \times \$1938.76$)	$120.20
medicare ($1.45\% \times \$1938.76$)	$ 28.11
total deductions	$347.21

Net pay

$1938.76 - \$347.21 = \$1591.55$

23. withholding allowance

$114.58 \times 6 = \$687.48$
$1971.06 - \$687.48 = \$1283.58$
$1283.58 - \$269 = \$1014.58$
$1014.58 \times 15\% = \$152.19$

federal tax	$152.19
FICA tax ($6.2\% \times \$1971.06$)	$122.21
medicare ($1.45\% \times \$1971.06$)	$ 28.58
total deductions	$302.98

Net pay

$1971.06 - \$302.98 = \$1668.08$

25. withholding allowance

$105.77 \times 3 = \$317.31$
$710.56 - \$317.31 = \$393.25$
$393.25 - \$102 = \$291.25$
$291.25 \times 15\% = \$43.69$

federal tax	$43.69
FICA tax ($6.2\% \times \$710.56$)	$44.05
medicare ($1.45\% \times \$710.56$)	$10.30
total deductions	$98.04

Net pay

$710.56 - \$98.04 = \$612.52$

27. withholding allowance

$52.88 \times 1 = \$52.88$

$915.34 - \$52.88 = \$862.46$

$862.46 - \$525 = \$337.46$

$71.10 + (\$337.46 \times 28\%) = \$165.59$

federal tax	$165.59
FICA tax ($6.2\% \times \$915.34$)	$ 56.75
medicare ($1.45\% \times \$915.34$)	$ 13.27
total deductions	$235.61

Net pay

$915.34 - \$235.61 = \$679.73$

29. withholding allowance

$229.17 \times 3 = \$687.51$

$5312.59 - \$687.51 = \$4625.08$

$4625.08 - \$3958 = \$667.08$

$513 + (\$667.08 \times 28\%) = \$699.78$

federal tax	$ 699.78
FICA tax ($6.2\% \times \$5312.59$)	$ 329.38
medicare ($1.45\% \times \$5312.59$)	$ 77.03
total deductions	$1106.19

Net pay

$5312.59 - \$1106.19 = \$4206.40$

31. withholding allowance

$52.88 \times 2 = \$105.76$

$431.25 - \$105.76 = \$325.49$

$325.49 - \$51 = \$274.49$

$274.49 \times 15\% = \$41.17$

federal tax	$41.17
FICA tax ($6.2\% \times \$431.25$)	$26.74
medicare ($1.45\% \times \$431.25$)	$ 6.25
total deductions	$74.16

Net pay

$431.25 - \$74.16 = \$357.09$

33. Answers will vary.

35. Gilbert, 4, m $783 weekly

federal withholding

$783 - (4 \times \$52.88) = \$571.48$

$(\$571.48 - \$124) \times 15\% = \$67.12$

federal tax	=	$ 67.12
state tax ($3.4\% \times \$783$)	=	$ 26.62
FICA tax ($6.2\% \times \$783$)	=	$ 48.55
medicare ($1.45\% \times \$783$)	=	$ 11.35
SDI ($1\% \times \$783$)	=	$ 7.83
union dues	=	$ 15.50
credit union savings	=	$100.00
total deductions	=	$276.97

net pay

$783 - \$276.97 = \$506.03$

37. Salesperson, 2, s

gross earnings

$(\$11,284 - \$5000 - \$424.50) \times 7\% + \$410$

$= (\$5859.50 \times 0.07) + \$410$

$= \$820.17$

federal withholding

$820.17 - (2 \times \$52.88) = \$714.41$

$714.41 - \$525 = \$189.41$

$71.10 + (\$189.41 \times 28\%) = \$124.13$

federal tax	=	$124.13
state tax ($3.4\% \times \$820.17$)	=	$ 27.89
FICA tax ($6.2\% \times \$820.17$)	=	$ 50.85
medicare ($1.45\% \times \$820.17$)	=	$ 11.89
SDI ($1\% \times \$820.17$)	=	$ 8.20
credit union savings	=	$ 50.00
Salvation Army	=	$ 10.00
professional dues		$ 15.00
total deductions	=	$297.96

net pay

$820.17 - \$297.96 = \$522.21$

39. Minkner, 3, m

gross earnings

$4200 + (1.5\% \times \$42,618) = \$4,839.27$

federal withholding

$4839.27 - (3 \times \$229.17) = \$4151.76$

$4151.76 - \$3958 = \$193.76$

$513 + (\$193.76 \times 28\%) = \$567.25$

federal tax   = $ 567.25
FICA tax (6.2% × $4839.27)   = $ 300.03
medicare (1.45% × $4839.27)   = $ 70.17
SDI (1% × $4839.27)   = $ 48.39
credit union savings   = $ 150.00
charitable contributions   = $ 25.00
savings bond   = $ 50.00
total deductions   = $1210.84

net pay

$4839.27 − $1210.84 = $3628.43

5.6   Payroll Records and Quarterly Returns

1. Social security

$15,634.18 × 6.2% = $969.32

Medicare

$15,634.18 × 1.45 = $226.70
$969.32 + $226.70 = $1196.02

Total

$1196.02 + $1196.02 = $2392.04

3. Social security

$17,462.10 × 6.2% = $1082.65

Medicare

$17,462.10 × 1.45% = $253.20
$1082.65 + $253.20 = $1335.85

Total

$1335.85 + $1335.85 = $2671.70

5. Social security

$14,131.59 × 6.2% = $876.16

Medicare

$14,131.59 × 1.45% = $204.91
$876.16 + $204.91 = $1081.07

Total

$1081.07 + $1081.07 = $2162.14

7. Social security

$62,475.80 × 6.2% = $3873.50

Medicare

$62,475.80 × 1.45% = $905.90
$3873.50 + $905.90 = $4779.40

Total

$4779.40 + $4779.40 = $9558.80

9. $6150.82 × 6.2% = $381.35
$6150.82 × 1.45% = $89.19

$381.35	employee FICA
381.35	employer FICA
89.19	employee medicare
89.19	employer medicare
+ 629.18	withholding tax
$1570.26	total due

11. $32,121.85 × 6.2% = $1991.55
$32,121.85 × 1.45% = $465.77

$1991.55	employee FICA
1991.55	employer FICA
465.77	employee medicare
465.77	employer medicare
+ 8215.08	withholding tax
$13,129.72	total due

13. $37,271.39 × 6.2% = $2310.83
$37,271.39 × 1.45% = $540.44

$2310.83	employee FICA
$2310.83	employer FICA
540.44	employee medicare
540.44	employer medicare
+ 7128.64	withholding tax
$12,831.18	total due

15. $34,547.86 × 6.2% = $2141.97
$34,547.86 × 1.45% = $500.94

$2141.97	employee FICA
2141.97	employer FICA
500.94	employee medicare
500.94	employer medicare
+ 12,628.19	withholding tax
$17,914.01	total due

17. Answers will vary.

19. $21,928.10 × 6.2% = $1359.54
$21,928.10 × 1.45% = $317.96

$1359.54   employee FICA
1359.54   employer FICA
317.96   employee medicare
+ 317.96   employer medicare
$3355.00   total due

21. $7622.84 × 6.2% = $472.62
$7622.84 × 1.45% = $110.53

$472.62   employee FICA
472.62   employer FICA
110.53   employee medicare
110.53   employer medicare
+ 1625.68   withholding tax
$2791.98   total due

23. **(a)** $3280 + $2600 = $5880
$7000 − $5880 = $1120
subject to tax

(b) $1120 × 6.2% = $69.44   *tax due*

25. **(a)** $1800 × 3 = $5400 earned in
first quarter
$7000 − $5400 = $1600

(b) $1600 × 6.2% = $99.20   *tax due*

Chapter 5 Review Exercises

1. Overtime rate

$1\frac{1}{2} × $9.14 = 13.71

Regular earnings

40 hours × $9.14 = $365.60

Overtime earnings

8.5 hours × $13.71 = $116.54

Gross earnings

$365.60 + $116.54 = $482.14

2. Overtime rate

$1\frac{1}{2} × $8.50 = 12.75

Regular earnings

40 hours × $8.50 = $340

Overtime earnings

8 hours × $12.75 = $102

Gross earnings

$340 + $102 = $442

3. Regular earnings

38.25 hours × $7.40 = $283.05

Overtime earnings: $0
Gross earnings: $283.05

4. Overtime rate

$1\frac{1}{2} × $6.80 = 10.20

Regular earnings

40 hours × $6.80 = $272

Overtime earnings

17.25 hours × $10.20 = $175.95
Gross earnings

$272 + $175.95 = $447.95

5. $410.80 × 52 = $21,361.60   annual
$21,361.60 ÷ 12 = $1780.13   monthly
$21,361.60 ÷ 24 = $890.07   semimonthly
$21,361.60 ÷ 26 = $821.60   biweekly

6. $1060 × 26 = $27,560   annual
$27,560 ÷ 12 = $2296.67   monthly
$27,560 ÷ 24 = $1148.33   semimonthly
$27,560 ÷ 52 = $530   weekly

7. $18,000 ÷ 12 = $1500   monthly
$18,000 ÷ 24 = $750   semimonthly
$18,000 ÷ 26 = $692.31   biweekly
$18,000 ÷ 52 = $346.15   weekly

8. $875 × 24 = $21,000   annual
$21,000 ÷ 12 = $1750   monthly
$21,000 ÷ 26 = $807.69   biweekly
$21,000 ÷ 52 = $403.85   weekly

9. $640 ÷ 40 = $16 per hour

$1\frac{1}{2} × $16 = 24 overtime rate

5 hours × $24 = $120 overtime

$640 + $120 = $760

10. $342 ÷ 36 = $9.50 per hour

$1\frac{1}{2} × $9.50 = 14.25 overtime rate

6 hours × $14.25 = $85.50 overtime

Weekly gross earnings

$342 + $85.50 = $427.50

11. Commission sales

$48,620 - $3106 = $45,514

Gross earnings

$0.08 \times $45,514 = $3641.12

12. Commission sales

$38,740 - $1245 = $37,495

Gross earnings

$0.09 \times $37,495 = $3374.55

13.

Oil Change	=	$63 \times $1.25	=	$78.75
Lubrication	=	$46 \times $1.50	=	$69.00
Oil/Lube	=	$38 \times $2.50	=	$95.00
Gross earnings				$242.75

14.

$$
\begin{array}{rll}
28 \text{ units:} & \text{(total units)} & \\
-\ 20 \text{ units:} & 20 \times \$4.50 = & \$90 \\
\hline
8 \text{ units:} & 8 \times \$5.50 = & +\ 44 \\
\hline
\text{Gross earnings} & & \$134
\end{array}
$$

15. Total commission

6% of $135,000 = $8100

Commission for broker

$\frac{1}{2} \times $8100 = $4050

Commission for Fisher

$\frac{1}{2} \times $4050 = $2025

16.

$$
\begin{array}{ll}
\$5850 & \text{(total sales)} \\
-\ \$2000 & \$2000 \times 0.06 = \$120 \\
\hline
\$3850 & \\
-\ \$2000 & \$2000 \times 0.08 = \$160 \\
\hline
\$1850 & \$1850 \times 0.10 = \underline{\$185} \\
& \$465
\end{array}
$$

The gross earnings are $465.

17.

$$
\begin{array}{ll}
\$7200 & \text{(total sales)} \\
-\ \$2000 & \$2000 \times 0.06 = \$120 \\
\hline
\$5200 & \\
-\ \$2000 & \$2000 \times 0.08 = \$160 \\
\hline
\$3200 & \$3200 \times 0.10 = \underline{\$320} \\
& \$600
\end{array}
$$

The gross earnings are $600.

18. Piecework earnings

$$
\begin{array}{l}
1850 \times \$0.12 = \$222.00 \\
1\frac{1}{2} \times 285 \times \$0.12 = \underline{\$\ 51.30} \\
\hphantom{1\frac{1}{2} \times 285 \times \$0.12 = } \$273.30
\end{array}
$$

Gross earnings

$= $273.30 - (92 \times 0.09)$
$= $273.30 - 8.28
$= 265.02

19. At the end of October, the employee has earned

$7855 \times 10 = $78,550

(a) FICA

$7855 \times 6.2\% = $487.01

(b) medicare

$7855 \times 1.45\% = $113.90

20. At the end of November, the employee has earned

$7855 \times 11 = $86,405
$86,405 - $80,000 = $6405

(a) FICA

$7855 - $6405 = $1450
$1450 \times 6.2\% = $89.90

(b) medicare

$7855 \times 1.45\% = $113.90

21. $18

22. $72

23. $151

24. $93

25. $104

26. $27

27. Precilo, 1, s $1852.75 monthly

federal withholding

$1852.75 - (1 \times $229.17) = $1623.58
($1623.58 - $221) \times 15\% = $210.39

federal tax	=	$210.39
FICA (6.2% × $1852.75)	=	$114.87
medicare (1.45% × $1852.75)	=	$ 26.86
SDI (1% × $1852.75)	=	$ 18.53
other deductions	=	$ 37.80
total deductions		$408.45

Net pay

$1852.75 - $408.45 = $1444.30

28. Colley, 4, m $522.11 weekly

federal withholding

$$\$522.11 - (4 \times \$52.88) = \$310.59$$
$$(\$310.59 - \$124) \times 15\% = \$27.99$$

federal withholding	= $ 27.99
state withholding	= $ 15.34
FICA (6.2% × $522.11)	= $ 32.37
medicare (1.45% × $522.11)	= $ 7.57
SDI (1% × $522.11)	= $ 5.22
credit union savings	= $ 20.00
educational television	= $ 7.50
total deductions	$115.99

Net pay

$$\$522.11 - \$115.99 = \$406.12$$

29. Harper, 6, m $677.92
federal withholding

$$\$677.92 - (6 \times \$52.88) = \$360.64$$
$$(\$360.64 - \$124) \times 15\% = \$35.50$$

federal tax	= $ 35.50
state withholding	= $ 22.18
FICA (6.2% × $677.92)	= $ 42.03
medicare (1.45% × $677.92)	= $ 9.83
SDI (1% × $677.92)	= $ 6.78
union dues	= $ 14.00
charitable contributions	= $ 15.00
total deductions	$145.32

Net pay

$$\$677.92 - \$145.32 = \$532.60$$

30. $12,720.15 × 6.2% = $788.65
$12,720.15 × 1.45% = $184.44

$788.65	employee FICA
788.65	employer FICA
184.44	employee medicare
184.44	employer medicare
$1946.18	total due

31. $29,185.17 × 6.2% = $1809.48
$29,185.17 × 1.45% = $423.18

$1809.48	employee FICA
1809.48	employer FICA
423.18	employee medicare
423.18	employer medicare
$4921.00	withholding tax
$9386.32	total due

32. Gross earnings

$452 + ($712 + $523 + $1002
+ $391 + $609 − $114) × 2%
$452 + ($3123 × 0.02)
  = $514.46

(a) FICA

$$\$514.46 \times 6.2\% = \$31.90$$

(b) medicare

$$\$514.46 \times 1.45\% = \$7.46$$

(c) SDI

$$\$514.46 \times 1\% = \$5.14$$

33. $78,325.18 + $1956.44 = $80,281.62
$80,281.62 − $80,000.00 = $281.62

(a) FICA tax

$$\$1956.44 - \$281.62 = \$1674.82$$
$$\$1674.82 \times 6.2\% = \$103.84$$

(b) medicare

$$\$1956.44 \times 1.45\% = \$28.37$$

34.

$1496.11	employees FICA
1496.11	employer FICA
345.30	employees medicare
345.30	employer medicare
+ 1768.43	federal withholding tax
$5451.25	total due

35. **(a)** $38,795.22 × 12.4% = $4810.61

(b) $38,795.22 × 2.9% = $1125.06

36. **(a)** $27,618.53 × 12.4% = $3424.70

(b) $27,618.53 × 2.9% = $800.94

37. **(a)** $2875 + $3212 = $6087
$7000 − $6087 = $913

(b) FUTA tax

$$\$913 \times 6.2\% = \$56.61$$

38. **(a)** $810 × 6 = $4860
$7000 − $4860 = $2140

(b) FUTA tax

$$\$2140 \times 6.2\% = \$132.68$$

Chapter 5 Summary Exercise

(a) Regular weekly earnings

$32,240 \div 52 = 620

(b) Overtime earnings

$620 \div 40 = 15.50 (hourly rate)

$1\frac{1}{2} \times 15.50×12 hours $= 279

(c) Total gross earnings

$620 + $279 = 899

(d) FICA

$899 \times 6.2\% = 55.74

(e) Medicare

$899 \times 1.45\% = 13.04

(f) Federal withholding

$899 - $52.88 = 846.12

$71.10 + ($846.12 - $525) \times 28\% = 161.01

(g) State disability

$899 \times 1\% = 8.99

(h) State withholding

$899 \times 4.4\% = 39.56

(i) Net pay

FICA	=	$ 55.74
medicare	=	$ 13.04
federal withholding	=	$161.01
SDI	=	$ 8.99
state withholding	=	$ 39.56
credit union payment	=	$125.00
retirement	=	$ 75.00
association dues	=	$ 12.00
Diabetes Association	=	$ 15.00
total deductions		$505.34

Net pay

$899 - $505.34 = 393.66

TAXES

6.1 Sales Tax

For Exercise 1-9, use *amount of sale + sales tax + excise tax = total sales price.*

1. **(a)** Sales tax
   $76.20 \times 3\% = \$2.29$

   (b) Excise tax
   $76.20 \times 11\% = \$8.38$

   (c) Total sale price
   $76.20 + \$2.29 + \$8.38 = \$86.87$

3. **(a)** Sales tax
   $47.70 \times 4.5\% = \$2.15$

   (b) Excise tax
   $47.70 \times 3\% = \$1.43$

   (c) Total sale price
   $47.70 + \$2.15 + \$1.43 = \$51.28$

5. **(a)** Sales tax
   $173.50 \times 5\% = \$8.68$

   (b) Excise tax
   $165 \times \$0.09 = \$14.85$

   (c) Total sale price
   $173.50 + \$8.68 + \$14.85 = \$197.03$

7. **(a)** Sales tax
   $822.18 \times 7\% = \$57.55$

   (b) Excise tax
   $822.18 \times 12\% = \$98.66$

   (c) Total sale price
   $822.18 + \$57.55 + \$98.66 = \$978.39$

9. **(a)** Sales tax
   $29,400 \times 6.25\% = \$1837.50$

   (b) Excise tax
   $168 \times \$3.00 = \$504$

   (c) Total sale price
   $29,400 + \$1837.50 + \$504 = \$31,741.50$

For Exercises 11-17, use the basic percent equation. *Sales tax = amount of sale × sales tax rate.*

11. $P = BR$
    $\$9.60 = B \times 6\%$
    $\$9.60 = 0.06B$

    $$\frac{\$9.60}{0.06} = \frac{0.06B}{0.06}$$

    $\$160 = B$

    Total price

    $\$160 + \$9.60 = \$169.60$

13. $P = BR$
    $\$6.30 = B \times 4\%$
    $\$6.30 = 0.04B$

    $$\frac{\$6.30}{0.04} = \frac{0.04B}{0.04}$$

    $\$157.50 = B$

    Total price

    $\$157.50 + \$6.30 = \$163.80$

15. $P = BR$
    $\$21.45 = B \times 6.5\%$
    $\$21.45 = 0.065B$

    $$\frac{\$21.45}{0.065} = \frac{0.065B}{0.065}$$

    $\$330 = B$

    Total price

    $\$330 + \$21.45 = \$351.45$

17. $P = BR$
    $\$63.84 = B \times 5\%$
    $\$63.84 = 0.05B$

    $$\frac{\$63.84}{0.05} = \frac{0.05B}{0.05}$$

    $\$1276.80 = B$

    Total price

    $\$1276.80 + \$63.84 = \$1340.64$

19. Amount of sale

$$B + (5\%)B = \$107.31$$
$$B + 0.05B = \$107.31$$
$$1.05B = \$107.31$$
$$B = \frac{\$107.31}{1.05}$$
$$B = \$102.20$$

Sales tax

$$\$107.31 - \$102.20 = \$5.11$$

21. Amount of sale

$$B + (6\%)B = \$551.52$$
$$B + 0.06B = \$551.52$$
$$1.06B = \$551.52$$
$$B = \frac{\$551.52}{1.06}$$
$$B = \$520.30$$

Sales tax

$$\$551.52 - \$520.30 = \$31.22$$

23. Amount of sale

$$B + (4.25\%)B = \$20.60$$
$$B + 0.0425B = \$20.60$$
$$1.0425B = \$20.60$$
$$B = \frac{\$20.60}{1.0425}$$
$$B = \$19.76$$

Sales tax

$$\$20.60 - \$19.76 = \$0.84$$

25. Amount of sale

$$B + (6\%)B = \$333.90$$
$$B + 0.06B = \$333.90$$
$$1.06B = \$333.90$$
$$B = \frac{\$333.90}{1.06}$$
$$B = \$315$$

Sales tax

$$\$333.90 - \$315.00 = \$18.90$$

27. $B + (7\%)B = \$2945.76$
$$B + 0.07B = \$2945.76$$
$$1.07B = \$2945.76$$
$$B = \frac{\$2945.76}{1.07}$$
$$B = \$2753.05$$

Sales tax

$$\$2945.76 - \$2753.05 = \$192.71$$

29. Answers will vary.

31. **(a)** Sales tax
$$\$119.80 \times 6\% = \$7.19$$

(b) Excise tax
$$\$119.80 \times 12.4\% = \$14.86$$

(c) Total price
$$\$119.80 + \$7.19 + \$14.86 = \$141.85$$

33. $P = BR$

$$\$17.10 = B \times 4\tfrac{1}{2}\%$$

$$\$17.10 = 0.045B$$

$$\frac{\$17.10}{0.045} = \frac{0.045B}{0.045}$$

$$\$380 = B$$

35. $B + (6\%)B = \$1979.55$
$$B + 0.06B = \$1979.55$$
$$1.06B = \$1979.55$$
$$B = \frac{\$1979.55}{1.06}$$
$$B = \$1867.50$$

37. $B + (4\%)B = \$1285.44$
$$B + 0.04B = \$1285.44$$
$$1.04B = \$1285.44$$
$$B = \frac{\$1285.44}{1.04}$$
$$B = \$1236$$

Sales tax

$$\$1285.44 - \$1236 = \$49.44$$

39. The procedure is not correct.

(a) $B + (6\%)B = \$1908$
$$B + 0.06B = \$1908$$
$$1.06B = \$1908$$
$$B = \frac{\$1908}{1.06}$$
$$B = \$1800$$

Sales tax

$$\$1908 - \$1800 = \$108$$

(b) $\$1908 \times 6\% = \$114.48$

The amount of error is

$$\$6.48 \ (\$114.48 - \$108).$$

41. Sales tax

$182.00 \times 7.5\% = \$13.65$

Excise tax

$4.50 + (74 - 70)(\$0.30)$

$\quad = \$4.50 + (4)(\$0.30)$

$\quad = \$4.50 + \$1.20$

$\quad = \$5.70$

Total cost

$182.00 + \$13.65 + \$5.70 = \$201.35$

43. Sales tax

$33,850 \times 7.5\% = \$2538.75$

Excise tax

$240 \times \$12.20 = \$2928$

Total cost

$33,850 + \$2538.75 + \$2928 = \$39,316.75$

6.2   Property Tax

1. Assessed valuation

$64,000 \times 0.4 = \$25,600$

3. Assessed valuation

$173,800 \times 0.35 = \$60,830$

5. Assessed valuation

$1,300,500 \times 0.25 = \$325,125$

7. Tax rate

$625,000 \div \$5,200,000 = 0.120 = 12\%$

9. Tax rate

$1,580,000 \div \$19,750,000 = 0.08 = 8\%$

11. Tax rate

$1,224,000 \div \$40,800,000 = 0.03 = 3\%$

13. 28   **(a)** 2.8%   **(b)** \$2.80   **(c)** \$28

15. 2.41%   **(a)** \$2.41   **(b)** \$24.10   **(c)** 24.1

17. \$7.08   **(a)** 7.08%   **(b)** \$70.80   **(c)** 70.8

19. Answers will vary.

21. $Tax = tax\ rate \times assessed\ valuation$

$\quad = \$6.80 \times 862$

$\quad = \$5861.60$

23. $Tax = tax\ rate \times assessed\ valuation$

$\quad = 6.93\% \times \$685,400$

$\quad = 0.0693 \times \$685,400$

$\quad = \$47,498.22$

25. $Tax = tax\ rate \times assessed\ valuation$

$\quad = \$1.80 \times 582$

$\quad = \$1047.60$

27. Tax rate

$2856.50 \div \$49,250 = 0.058 = 5.8\%$

29. Tax

$73,800 \times 0.085 = \$6273$

31. Tax rate

$8015.70 \div \$152,680$

$\quad = 0.0525 = \$5.25$ per \$100

33. Assessed valuation

$\dfrac{\$10,182.40}{4.3\%} = \$10,182.40 \div 0.043$

$\quad\quad\quad = \$236,800$

35. Assessed valuation

$378,000 \times 0.3 = \$113,400$

Property tax

$0.0428 \times \$113,400 = \$4853.52$

37. Assessed valuation

$334,400 \times 0.25 = \$83,600$

Property tax

$(\$83,600 \div \$1000) \times \$75.30 = \$6295.08$

39. Assessed valuation

$5,700,000 \times 0.25 = \$1,425,000$

Property tax

$(\$1,425,000 \div \$100) \times \$14.10 = \$200,925$

41. (a) The first county:

Assessed valuation

$95,000 \times 0.40 = \$38,000$

Property tax

$38,000 \times 0.0321 = \$1219.80$

The second county:

Assessed valuation

$95,000 \times 0.24 = \$22,800$

Property tax

$$\$22{,}800 \times 0.0502 = \$1144.56$$

The second county would charge the lower property tax.

(b) The annual amount saved is

$$\$1219.80 - \$1144.56 = \$75.24.$$

43. Use proportion to find the original assessed valuation.

$$\frac{\text{dollars}}{\text{thousand}} = \frac{\text{tax}}{\text{assessed valuation (AV)}}$$

$$\frac{12.50}{1000} = \frac{\$4625}{AV}$$

$$12.50AV = (1000)(4625)$$
$$12.50AV = 4{,}625{,}000$$
$$AV = \$370{,}000$$

New assessed valuation

$$\$370{,}000 + \$25{,}000 = \$395{,}000$$

Use proportion to find new tax rate.

$$\frac{\text{dollars (D)}}{\text{thousand}} = \frac{\text{tax}}{\text{new AV}}$$

$$\frac{D}{1000} = \frac{\$5350}{\$395{,}000}$$

$$395{,}000D = (1000)(5350)$$
$$395{,}000D = 5{,}350{,}000$$
$$D = \$13.5443038$$

Percent increase

$$\frac{\text{new rate} - \text{old rate}}{\text{old rate}} = \frac{\$13.5443038 - \$12.50}{\$12.50}$$

$$= 0.083544304$$
$$= 8.4\%$$

45. Use proportion to find the original tax rate.

$$\frac{\text{dollars (D)}}{\text{hundred}} = \frac{\text{tax}}{\text{assessed valuation (AV)}}$$

$$\frac{D}{100} = \frac{\$1327.50}{\$45{,}000}$$

$$45{,}000D = (100)(1327.50)$$
$$= 45{,}000D = 132{,}750$$
$$D = \$2.95$$

New tax rate $= \$2.95 - \$0.10$
$$= \$2.85 \text{ per } \$100$$

$$\frac{\text{dollars (D)}}{\text{hundred}} = \frac{\text{tax}}{\text{assessed valuation (AV)}}$$

$$\frac{2.85}{100} = \frac{\$1353.75}{AV}$$

$$2.85AV = (100)(1353.75)$$
$$AV = \$47{,}500$$

Increase in assessed valuation

$$\$47{,}500 - \$45{,}000 = \$2500$$

6.3   Personal Income Tax

For Exercises 1-5, use *Adjusted gross income = Total income − Total adjustments*.

1. Adjusted gross income

$$\$18{,}610 + \$74 + \$1936 + \$115 - \$135 = \$20{,}600$$

3. Adjusted gross income

$$\$21{,}380 + \$625 + \$139 + \$184 - \$618 = \$21{,}710$$

5. Adjusted gross income

$$\$38{,}643 + \$95 + \$188 + \$105 - \$0 = \$39{,}031$$

7. Taxable income

$$\$24{,}200 - \$4300 \text{ (standard deduction)}$$
$$- \$2750 \text{ (one exemption)}$$
$$= \$17{,}150$$

Tax

$$\$17{,}150 \times 15\% = \$2572.50$$

9. Taxable income

$$\$38{,}751 - \$7200 \text{ (standard deduction)}$$
$$- \$8250 \text{ (three exemptions)}$$
$$= \$23{,}301$$

Tax

$$\$23{,}301 \times 15\% = \$3495.15$$

11. Taxable income

$$\$71{,}800 - \$8851 \text{ (standard deduction)}$$
$$- \$13{,}750 \text{ (five exemptions)}$$
$$= \$49{,}199$$

Tax

$$\$49{,}199 - \$43{,}050 = \$6149$$
$$\$6457.50 + (\$6149 \times 28\%) = \$8179.22$$

13. Taxable income

$40,350 − $4300 (standard deduction)
 − $2750 (one exemption)
   = $33,300

Tax

$33,300 − $25,750 = $7550
$3862.50 + ($7550 × 28%) = $5976.50

15. Taxable income

$68,574 − $4300 (standard deduction)
−$2750 (one exemption)
   = $61,524

Tax

$61,524 − $25,750 = $35,774
$3862.50 + ($35,774 × 28%)
   = $13,879.22

17. Taxable income

$62,613 − $7681 (itemized deductions)
−$5500 (two exemptions)
   = $49,432

Tax

$49,432 − $43,050 = $6382
$6457.50 + ($6382 × 28%)
   = $8244.46

19. Tax

$12,947 × 15% = $1942.05

Tax withheld

$243.10 × 12 = $2917.20

Refund

$2917.20 − $1942.05 = $975.15

21. Tax

$22,988 × 15% = $3448.20

Tax withheld

$72.18 × 52 = $3753.36

Refund

$3753.36 − $3448.20 = $305.16

23. Tax

$31,786 × 15% = $4767.90

Tax withheld

$358.44 × 12 = $4301.28

Tax due

$4767.90 − $4301.28 = $466.62

25. Answers will vary.

27. Total deductions

$682 + $187 + $472 + $4260 + $785 = $6386

Use the standard deduction.
Personal exemptions

$2750 × 4 = $11,000

Taxable income

$54,378 − $7200 − $11,000 = $36,178

Tax

$36,178 × 15% = $5426.70

29. Use the standard deduction.
Taxable income

$36,998 − $4300 − $2750 = $29,948

Tax

$29,948 − $25,750 = $4198
$3862.50 + ($4198 × 28%) = $5037.94

31. Total income

$4108 + $2653 + $1838 + $137 = $8736

Taxable income

$8736 − $4300 = $4436

Tax

$4436 × 15% = $665.40

33. Total income

$68,645 + $385 + $672 − $1058 = $68,644

Total deductions

$877 + $342 + $786 + $8180 + $186 = $10,371

Personal exemptions

$2750 × 5 = $13,750

Taxable income

$68,644 - $10,371 - $13,750 = $44,523

Tax

$44,523 - $43,050 = $1473
$6457.50 + ($1473 \times 28\%) = $6869.94

35. Total income

$45,428 + $5283 + $324 + $668 - 2484
$= $49,219

Total deductions

$7615 + $729 + $1185 + $1219 = $10,748

Personal exemptions

$2750 \times 6 = $16,500

Taxable income

$49,219 - $10,748 - $16,500 = $21,971

Tax

$21,971 \times 15\% = $3295.65

Chapter 6 Review Exercises

1. Sales tax

$852.15 \times 6\% = $51.13

Excise tax

$852.15 \times 10\% = $85.22

Total sale price

$852.15 + $51.13 + $85.22 = $988.50

2. Sales tax

$86.15 \times 4\% = $3.45

Excise tax

$86.15 \times 11\% = $9.48

Total sale price

$86.15 + $3.45 + $9.48 = $99.08

3. Sales tax

$16,500 \times 5\% = $825

Excise tax

$110 \times $12.20 = $1342

Total sale price

$16,500 + $825 + $1342 = $18,667

4. Sales tax

$345.96 \times 7\% = $24.22

Excise tax

$285 \times $0.183 = $52.16

Total sale price

$345.96 + $24.22 + $52.16 = $422.34

5.
$$P = BR$$
$$\$68.04 = B \times 6\%$$
$$\$68.04 = 0.06B$$
$$\frac{\$68.04}{0.06} = \frac{0.06B}{0.06}$$
$$\$1134 = B$$

6.
$$P = BR$$
$$\$14.20 = B \times 5\%$$
$$\$14.20 = 0.05B$$
$$\frac{\$14.20}{0.05} = \frac{0.05B}{0.05}$$
$$\$284 = B$$

7.
$$P = BR$$
$$\$19.60 = B \times 7\%$$
$$\$19.60 = 0.07B$$
$$\frac{\$19.60}{0.07} = \frac{0.07B}{0.07}$$
$$\$280 = B$$

8.
$$P = BR$$
$$\$15.75 = B \times 4\tfrac{1}{2}\%$$
$$\$15.75 = 0.045B$$
$$\frac{\$15.75}{0.045} = \frac{0.045B}{0.045}$$
$$\$350 = B$$

9.
$$B + (6\%)B = \$447.32$$
$$B + 0.06B = \$447.32$$
$$1.06B = \$447.32$$
$$B = \frac{\$447.32}{1.06}$$
$$B = \$422$$

10. $B + (7\%)B = \$133.75$
$B + 0.07B = \$133.75$
$1.07B = \$133.75$

$$B = \frac{\$133.75}{1.07}$$

$B = \$125$

11. $B + (5\%)B = \$292.95$
$B + 0.05B = \$292.95$
$1.05B = \$292.95$

$$B = \frac{\$292.95}{1.05}$$

$B = \$279$

12. $B + (4\%)B = \$430.56$
$B + 0.04B = \$430.56$
$1.04B = \$430.56$

$$B = \frac{\$430.56}{1.04}$$
$B = \$414$

13. $4.06 (a) 4.06\% (b) 40.60 (c) 40.6$

14. $27 (a) 2.7\% (b) \$2.70 (c) \$27$

15. $1.27\% (a) \$1.27 (b) \$12.70 (c) 12.7$

16. $19.50 (a) 1.95\% (b) \$1.95 (c) 19.5$

17. $Tax = tax\ rate \times assessed\ valuation$
$= 0.032 \times \$426,000$
$= \$13,632$

18. Tax rate

$\$1816.70 \div \$98,200 = 0.0185$
$= \$18.50$ per $\$1000$

19. Assessed valuation
$$\frac{\$1627.50}{3.5\%} = \$1627.50 \div 0.035$$
$$= \$46,500$$

20. Tax rate

$\$3934 \div \$140,500 = 0.028 = 2.8\%$

21. Assessed valuation
$$\frac{\$3655.60}{3.8\%} = \$3655.6 \div 0.038$$
$$= \$96,200$$

22. $Tax = tax\ rate \times assessed\ valuation$
$= 0.027 \times \$103,600$
$= \$2797.20$

23. Taxable income

$\$38,415 - \$4516$ (itemized deductions)
$- \$2750$ (one exemption)
$= \$31,149$

Tax

$\$31,149 - \$25,750 = \$5399$
$\$3862.50 + (5399 \times 28\%) = \$5374.22$

24. Taxable income

$\$78,628 - \$7634$ (itemized deductions)
$- \$13,750$ (five exemptions)
$= \$57,244$

Tax

$\$57,244 - \$43,050 = \$14,194$
$\$6457.50 + (\$14,194 \times 28\%) = \$10,431.82$

25. Taxable income

$\$54,110 - \$7200$ (standard deduction)
$- \$8250$ (three exemptions)
$= \$38,660$

Tax

$\$38,660 \times 15\% = \$5799$

26. Taxable income

$\$48,752 - \$4300$ (standard deduction)
$- \$2750$ (one exemption)
$= \$41,702$

Tax

$\$41,702 - \$25,750 = \$15,952$
$\$3862.50 + (\$15,952 \times 28\%) = \$8329.06$

27. Tax rate

$\$1,978,000 \div \$90,550,000 = 0.0218 = 2.2\%$

28. $B + (6\%)B = \$3442.88$
$B + 0.06B = \$3442.88$
$1.06B = \$3442.88$

$$\frac{1.06B}{1.06} = \frac{\$3442.88}{1.06}$$

$B = \$3248$

Sales tax

$\$3442.88 - \$3248 = \$194.88$

29. Assessed valuation

$2,608,300 \times 0.28 = \$730,324$

Property tax

$(\$730,324 \div \$1000) \times \$21.50 = \$15,701.97$

30. Use the standard deduction.

Personal exemptions

$\$2750 \times 6 = \$16,500$

Taxable income

$\$63,280 - \$7200 - \$16,500 = \$39,580$
$\$39,580 \times 15\% = \$5937$

31. Total deductions

$\$817 + \$875 + \$1495 + \$343 = \$3530$

Use the standard deduction.

Taxable income

$\$27,760 - \$4300 - \$2750 = \$20,710$

Tax

$\$20,710 \times 15\% = \$3106.50$

32. Total income

$\$69,750 + \$852 + \$2880 - \$2450$
$= \$71,032$

Total deductions

$\$7218 + \$471 + \$1040 = \$8729$

Personal exemptions

$\$2750 \times 4 = \$11,000$

Taxable income

$\$71,032 - \$8729 - \$11,000$
$= \$51,303$

Tax

$\$51,303 - \$43,050 = \$8253$
$\$6457.50 + (\$8253 \times 28\%) = \$8768.34$

33. Tax

$\$62,450 - \$43,050 = \$19,400$
$\$6457.50 + (\$19,400 \times 28\%) = \$11,889.50$

Tax withheld

$\$198.50 \times 52 = \$10,322$

Tax due

$\$11,889.50 - \$10,322 = \$1567.50$

34. Tax

$\$33,825 - \$25,750 = \$8075$
$\$3862.50 + (\$8075 \times 28\%) = \$6123.50$

Tax withheld

$\$533.20 \times 12 = \$6398.40$

Refund

$\$6398.40 - \$6123.50 = \$274.9$

35. Tax

$\$28,315 - \$25,750 = \$2565$
$\$3862.50 + (\$2565 \times 28\%) = \$4580.70$

Tax withheld

$\$105.40 \times 52 = \$5480.80$

Refund

$\$5480.80 - \$4580.70 = \$900.10$

36. Tax

$\$40,180 \times 15\% = \$6027$

Tax withheld

$\$537.30 \times 12 = \$6447.60$

Refund

$\$6447.60 - \$6027 = \$420.60$

Chapter 6 Summary Exercise

(a) Anderson land cost

$(11)(43,560)(\$0.40) = \$191,664$

Anderson building and improvements

$(90,000)(\$32.80) = \$2,952,000$

Total cost at Anderson
$=$ Land $+$ Building and improvements

$\$191,664 + \$2,952,000 = \$3,143,664$

Bentonville land cost

$(11)(43,560)(\$0.35) = \$167,706$

Bentonville building and improvements

$(90,000)(\$36.90) = \$3,321,000$

Total cost at Bentonville
$=$ Land $+$ Building and improvements

$\$167,706 + \$3,321,000 = \$3,488,706$

(b) Assessed valuation at Anderson
= ($3,143,664)(0.25) = $785,916

Assessed valuation at Bentonville
= ($3,488,706)(0.20) = $697,741.20

(c) Property tax at Anderson

$785,916 × 0.032 = $25,149.31

Property tax at Bentonville

($697,741.20 ÷ $100) × $2.95 = $20,583.37

(d) Total cost at Anderson for 10 years

$3,143,664 + (10 × $25,149.31)
= $3,395,157.10

Total cost at Bentonville for 10 years

$3,488,706 + (10 × $20,583.37)
= $3,694,539.70

(e) Anderson

RISK MANAGEMENT

7.1   Business Insurance

For Exercises 1-7, use
Premium = Value (per $100) × Rate.

1. $140,000 = 1400 hundreds
1400 × $0.84 = $1176
$75,000 = 750 hundreds
750 × $0.90 = $675

Total annual premium

$1176 + $675 = $1851

3. $596,400 = 5964 hundreds
5964 × $0.37 = $2206.68
$206,700 = 2067 hundreds
2067 × $0.46 = $950.82

Total annual premium

$2206.68 + $950.82 = $3157.50
$$= \$3158 \text{ (rounded)}$$

5. $782,600 = 7826 hundreds
7826 × $0.92 = $7199.92
$212,000 = 2120 hundreds
2120 × $0.99 = $2098.80

Total annual premium

$7199.92 + $2098.80 = $9298.72
$$= \$9299 \text{ (rounded)}$$

7. $583,200 = 5832 hundreds
5832 × $1.14 = $6648.48
$221,400 = 2214 hundreds
2214 × $1.05 = $2324.70

Total annual premium

$6648.48 + $2324.70 = $8973.18
$$= \$8973 \text{ (rounded)}$$

9. Amount of refund

$2162 × 0.80 = $1729.60
$2162 − 1729.60 = $432.40

11. Amount of refund

$807 × 0.18 = $145.26
$807 − $145.26 = $661.74

13. Amount of refund

$1507 × 0.45 = $678.15
$1507 − $678.15 = $828.85

15. Amount of refund

$4860 × 0.95 = $4617
$4860 − $4617 = $243

17. **(a)** Amount of premium retained

$$\$2680 \times \frac{9}{12} = \$2010$$

(b) Amount of refund

$2680 − $2010 = $670

19. **(a)** Amount of premium retained

$$\$4375 \times \frac{7}{12} = \$2552.08$$

(b) Amount of refund

$4375 − $2552.08 = $1822.92

21. **(a)** Amount of premium retained

$$\$5308 \times \frac{6}{12} = \$2654$$

(b) Amount of refund

$5308 − $2654 = $2654

23. $277,000 × 0.80 = $221,600

The face value is greater than $221,600, so the insurance company will pay $19,850.

25. $78,500 × 0.80 = $62,800

$$\$1500 \times \frac{\$47,500}{\$62,800} = \$1134.55$$

27. $218,500 × 0.80 = $174,800

The face value is greater than $174,800, so the insurance company will pay $36,500.

29. $147,850 × 0.80 = $118,280

$$\$14,850 \times \frac{\$100,000}{\$118,280} = \$12,554.95$$

31. $750,000 + $250,000 = $1,000,000

Company A: $\dfrac{\$750,000}{\$1,000,000} \times \$80,000 = \$60,000$

Company B: $\dfrac{\$250,000}{\$1,000,000} \times \$80,000 = \$20,000$

33. $1,350,000 + $1,200,000 + $450,000 = $3,000,000

Company 1: $\dfrac{\$1,350,000}{\$3,000,000} \times \$650,000 = \$292,500$

Company 2: $\dfrac{\$1,200,000}{\$3,000,000} \times \$650,000 = \$260,000$

Company 3: $\dfrac{\$450,000}{\$3,000,000} \times \$650,000 = \$97,500$

35. $90,000 \times 0.80 = $72,000

Coverage:

$35,000 + $25,000 = $60,000

(a) $\$36,000 \times \dfrac{\$60,000}{\$72,000} = \$30,000$

(b) Company A:

$\dfrac{\$35,000}{\$60,000} \times \$30,000 = \$17,500$

Company B:

$\dfrac{\$25,000}{\$60,000} \times \$30,000 = \$12,500$

(c) The insured will pay

$36,000 − $17,500 − $12,500 = $6000.

37. $250,000 \times 0.80 = $200,000

Coverage

$75,000 + $50,000 = $125,000

(a) $\$20,000 \times \dfrac{\$125,000}{\$200,000} = \$12,500$

(b) Company 1:

$\dfrac{\$75,000}{\$125,000} \times \$12,500 = \$7500$

Company 2:

$\dfrac{\$50,000}{\$125,000} \times \$12,500 = \$5000$

(c) The insured will pay

$20,000 − $7500 − $5000 = $7500.

39. $165,400 = 1654 hundreds
1654 × $0.54 = $893.16
$128,000 = 1280 hundreds
1280 × $0.60 = $768

Total annual premium

$893.16 + $768 = $1661.16
           = $1661 (rounded)

41. $84,000 = 840 hundreds
840 × $0.36 = $302.40
$18,500 = 185 hundreds
185 × $0.49 = $90.65

Total annual premium

$302.40 + $90.65 = $393.05
           = $393 (rounded)

43. Amount of refund

$2350 × 0.55 = $1292.50
$2350 − $1292.50 = $1057.50

45. Amount of refund

$2750 × 0.85 = $2337.50
$2750 − $2337.50 = $412.50

47. (a) Amount of premium retained

$\$2670 \times \dfrac{5}{12} = \$1112.50$

(b) Amount of refund

$2670 − $1112.50 = $1557.50

49. (a) Amount of premium retained

$\$1944 \times \dfrac{7}{12} = \$1134$

(b) Amount of refund

$1944 − $1134 = $810

51. Answers will vary.

53. (a) $395,000 × 0.80 = $316,000

$\$22,500 \times \dfrac{\$280,000}{\$316,000} = \$19,936.71$

(b) The insured will pay

$22,500 − $19,936.71 = $2563.29.

55. (a) $550,000 × 0.80 = $440,000

$\$45,000 \times \dfrac{\$300,000}{\$440,000} = \$30,681.82$

(b) The insured will pay

$$\$45,000 - \$30,681.82 = \$14,318.18.$$

57. Answers will vary.

59. $600,000 + $400,000 + $200,000 = $1,200,000

Company A:

$$\frac{\$600,000}{\$1,200,000} \times \$548,000 = \$274,000$$

Company B:

$$\frac{\$400,000}{\$1,200,000} \times \$548,000 = \$182,666.67$$

$$= \$182,667 \text{ (rounded)}$$

Company C:

$$\frac{\$200,000}{\$1,200,000} \times \$548,000 = \$91,333.33$$

$$= \$91,333 \text{ (rounded)}$$

61. $4,000,000 \times 0.80 = $3,200,000
$1,800,000 + $600,000 = $2,400,000

(a) $\$500,000 \times \dfrac{\$2,400,000}{\$3,200,000} = \$375,000$

(b) Company A:

$$\frac{\$1,800,000}{\$2,400,000} \times \$375,000 = \$281,250$$

Company B:

$$\frac{\$600,000}{\$2,400,000} \times \$375,000 = \$93,750$$

63. $360,000 \times 0.80 = $288,000
$100,000 + $50,000 + $30,000 = $180,000

(a) $\$120,000 \times \dfrac{\$180,000}{\$288,000} = \$75,000$

(b) Company 1:

$$\frac{\$100,000}{\$180,000} \times \$75,000 = \$41,666.67$$

$$= \$41,667 \text{ (rounded)}$$

Company 2:

$$\frac{\$50,000}{\$180,000} \times \$75,000 = \$20,833.33$$

$$= \$20,833 \text{ (rounded)}$$

Company 3:

$$\frac{\$30,000}{\$180,000} \times \$75,000 = \$12,500$$

7.2 Motor Vehicle Insurance

1. Annual premium

$$\$284 + \$112 + \$58 + \$107 + \$70 = \$631$$

3. Annual premium

$$\$459 + \$145 + \$90 + \$162 + \$76 = \$932$$

5. Annual premium

$$\$282 + \$93 + \$44 + \$116 = \$535$$

7. Annual premium

$$\$310 + \$112 + \$66 + \$124 + \$70 = \$682$$

9. Annual premium

$$\$375 + \$134 + \$52 + \$111 + \$76 = \$748$$

11. Annual premium

$$\$196 + \$84 + \$16 + \$60 = \$356$$

13. Answers will vary.

15. Annual premium

$$\$253 + \$93 + \$44 + \$116 + \$66 = \$572$$
$$\$572 \times 1.15 = \$657.80$$

17. Annual premium

$$\$459 + \$145 + \$78 + \$157 + \$76 = \$915$$

19. The coverage limit is $25,000

(a) The company will pay $25,000.

(b) You will pay $11,500 ($36,500 − $25,000).

21. (a) To repair the car the insurance company will pay the entire cost except the $250 deductible.

$$\$1878 - \$250 = \$1628$$

(b) The insurance company will pay the entire bill of $6936.

(c) The insurance company will pay up to the policy limit of $100,000.

(d) Silva must pay the amount in excess of $100,000. The court awarded a total of $115,000 ($60,000 + $55,000). So, Silva will pay $15,250

($115,000 − $100,000 + $250).

23. (a) Since there were two injuries, the insurance company will pay $50,000 plus the $10,000 property damage for a total of $60,000.

(b) The business owner was liable for

$250 + ($20,000 + $40,000 − $50,000) = $10,250

25. Answers will vary.

27. (a) Life, medical, auto, homeowner's insurance

(b) Check with employer, avoid specialty insurance, check prices with different insurance agents

7.3 Life Insurance

1. $30,000 = 30 thousands

annual: 30 × $31.75 = $952.50
semiannual: $952.50 × 0.51 = $485.78
quarterly: $952.50 × 0.26 = $247.65
monthly: $952.50 × 0.0908 = $86.49

3. $35,000 = 35 thousands

annual: 35 × $24.26 = $849.10
semiannual: $849.10 × 0.51 = $433.04
quarterly: $849.10 × 0.26 = $220.77
monthly: $849.10 × 0.0908 = $77.10

5. $85,000 = 85 thousands

annual: 85 × $6.08 = $516.80
semiannual: $516.80 × 0.51 = $263.57
quarterly: $516.80 × 0.26 = $134.37
monthly: $516.80 × 0.0908 = $46.93

7. $100,000 = 100 thousands

annual: 100 × $297 = $297
semiannual: $297 × 0.51 = $151.47
quarterly: $297 × 0.26 = $77.22
monthly: $297 × 0.0908 = $26.97

9. $50,000 = 50 thousands

annual: 50 × $10.62 = $531
semiannual: $531 × 0.51 = $270.81
quarterly: $531 × 0.26 = $138.06
monthly: $531 × 0.0908 = $48.21

11. $70,000 = 70 thousands

annual: 70 × $26.23 = $1836.10
semiannual: $1836.10 × 0.51 = $936.41
quarterly: $1836.10 × 0.26 = $477.39
monthly: $1836.10 × 0.0908 = $166.72

13. $100,000 = 100 thousands

annual: 100 × $31.75 = $3175
semiannual: $3175 × 0.51 = $1619.25
quarterly: $3175 × 0.26 = $825.50
monthly: $3175 × 0.0908 = $288.29

15. Answers will vary.

17. Cash value

$310 × 50 (thousands) = $15,500

19. Paid-up insurance

$1000 × 30 (thousands) = $30,000

21. Cash value

$1968 × 100 (thousands) = $196,800

23. Extended term

23 years and 315 days

25. Monthly payment

$5.78 × 50 (thousands) = $289

27. $\dfrac{\$60.70}{10 \text{ (thousands)}} = \$6.07$ per $1000 of face value

Read down the Amount column of Table 7.13. Find $6.07. Payment will continue for 18 years.

29. Monthly payment

$5.34 × 30 (thousands) = $160.20

31. (a) Use Table 7.10

Annual premium

$7.68 × 50 (thousands) = $384

(b) Since Matthew is the beneficiary, he will receive the face value of the policy, which is $50,000.

33. Semiannual premium

$4.15 × 90 (thousands) = $373.50
$373.50 × 0.51 = $190.49

35. Monthly premium

$32.59 × 75 (thousands) = $2444.25
$2444.25 × 0.0908 = $221.94

37. (a) Semiannual premium

$872 \times 0.51 = \$444.72$

(b) Quarterly premium

$872 \times 0.26 = \$226.72$

(c) Monthly premium

$872 \times 0.0908 = \$79.18$

(d) Total cost

$444.72 \times 2 = \$889.44$
$226.72 \times 4 = \$906.88$
$79.18 \times 12 = \$950.16$

39. (a) Cash value

$283 \times 20 \text{ (thousands)} = \$5660$

(b) Amount of paid up insurance

$579 \times 20 \text{ (thousands)} = \$11{,}580$

(c) The time period for paid up term insurance is 23 years and 315 days.

41. Cash value

$1495 \times 100 \text{ (thousands)} = \$149{,}500$

43. (a) Monthly payment for 12 years

$8.46 \times 25 \text{ (thousands)} = \$211.50$

(b) $\dfrac{\$145}{25 \text{ (thousands)}} = \$5.80$

Use Table 7.13 to find that the payment, $5.78, is for approximately 20 years.

(c) Monthly amount from life annuity

$4.63 \times 25 \text{ (thousands)} = \$115.75$

(d) Monthly amount from a life annuity with 15 years certain

$4.38 \times 25 \text{ (thousands)} = \$109.50$

45. (a) Monthly payment

$6.91 \times 50 \text{ (thousands)} = \$345.50$

(b) $\dfrac{\$305}{50 \text{ (thousands)}} = \$6.10$

Use Table 7.13 to find that the payment, $6.10, is for approximately 18 years.

(c) Monthly amount from life annuity

$5.94 \times 50 \text{ (thousands)} = \$297$

(d) Monthly amount from life annuity with 10 years certain

$5.73 \times 50 \text{ (thousands)} = \$286.50$

47. Answers will vary.

49. Bankruptcy, foreclosure, the court appointed a guardian for children.

Chapter 7 Review Exercises

1. $640{,}000 = 6400$ hundreds
$6400 \times \$0.50 = \$3200$
$275{,}000 = 2750$ hundreds
$2750 \times \$0.52 = \$1430$

Total annual premium

$3200 + \$1430 = \$4630$

2. $375{,}000 = 3750$ hundreds
$3750 \times \$0.54 = \$2025$
$198{,}000 = 1980$ hundreds
$1980 \times \$0.75 = \$1485$

Total annual premium

$2025 + \$1485 = \$3510$

3. $80{,}000 = 800$ hundreds
$800 \times \$0.37 = \$296$
$30{,}000 = 300$ hundreds
$300 \times \$0.46 = \$138$

Total annual premium

$296 + \$138 = \$434$

4. $193{,}000 = 1930$ hundreds
$1930 \times \$0.36 = \$694.80$
$68{,}000 = 680$ hundreds
$680 \times \$0.49 = \$333.20$

Total annual premium

$694.80 + \$333.20 = \$1028$

5. Amount of refund

$1773 \times 0.35 = \$620.55$
$1773 - \$620.55 = \$1152.45$

6. Amount of refund

$1078 \times 0.70 = \$754.60$
$1078 - \$754.60 = \$323.40$

7. Amount of refund

$1486 × 0.85 = $1263.10
$1486 − $1263.10 = $222.90

8. Amount of refund

$2878 × 0.65 = $1870.70
$2878 − $1870.70 = $1007.30

9. (a) Amount of premium retained

$$\$3150 \times \frac{5}{12} = \$1312.50$$

(b) Amount of refund

$3150 − $1312.50 = $1837.50

10. (a) Amount of premium retained

$$\$1975 \times \frac{9}{12} = \$1481.25$$

(b) Amount of refund

$1975 − $1481.25 = $493.75

11. (a) Amount of premium retained

$$\$1476 \times \frac{10}{12} = \$1230$$

(b) Amount of refund

$1476 − $1230 = $246

12. (a) Amount of premium retained

$$\$2784 \times \frac{2}{12} = \$464$$

(b) Amount of refund

$2784 − $464 = $2320

13. $456,000 × 0.80 = $364,800

$$\$45,000 \times \frac{\$320,000}{\$364,800} = \$39,473.68$$

14. $277,500 × 0.80 = $222,000

$$\$97,800 \times \frac{\$165,000}{\$222,000} = \$72,689.19$$

15. $186,700 × 0.80 = $149,360

$$\$3400 \times \frac{\$120,000}{\$149,360} = \$2731.66$$

16. $325,000 × 0.80 = $260,000

$$\$42,200 \times \frac{\$220,000}{\$260,000} = \$35,707.69$$

17. Annual premium

$253 + $103 + $90 + $184 + $66 = $696

18. Annual premium

$310 + $129 + $64 + $122 + $76 = $701
$701 × 1.55 = $1086.55

19. Annual premium

$222 + $93 + $82 + $171 + $66 = $634
$634 × 1.40 = $887.60

20. Annual premium

$310 + $124 + $42 + $97 + $70 = $643

21. Annual premium

$32.59 × 70 (thousands) = $2281.30

22. Annual premium

$4.92 × 50 (thousands) = $246

23. Annual premium

$4.45 × 30 (thousands) = $133.50

24. Annual premium

$4.80 × 40 (thousands) = $192

25. Total coverage

$250,000 + $150,000 + $100,000
  = $500,000

Company A:

$$\frac{\$250,000}{\$500,000} \times \$72,000 = \$36,000$$

Company B:

$$\frac{\$150,000}{\$500,000} \times \$72,000 = \$21,600$$

Company C:

$$\frac{\$100,000}{\$500,000} \times \$72,000 = \$14,400$$

26. $820,000 × 0.80 = $656,000

Total coverage

$350,000 + $200,000 = $550,000

(a) Insurance payment

$$\frac{\$550,000}{\$656,000} \times \$150,000 = \$125,762$$

(b) Company 1:

$$\frac{\$350,000}{\$550,000} \times \$125,762 = \$80,030$$

Company 2:

$$\frac{\$200,000}{\$550,000} \times \$125,762 = \$45,732$$

27. (a) With 25/50 coverage, the insurance company will pay $25,000.

(b) The driver must pay
$9000 ($34,000 − $25,000).

28. (a) With 50/100 coverage, the insurance company will pay $50,000 per person up to a maximum of $100,000. Each person will receive $15,000.

(b) Payment does not exceed the limits, so the driver pays nothing.

29. (a) With 15/30 coverage, $10,000 property damage and two people injured, the insurance company will pay $40,000 ($10,000 + $30,000).

(b) Total damage
$16,800 + $25,000 + $35,000 = $76,800

Driver liability
$76,800 − $40,000 = $36,800

30. (a) Semiannual premium
$970 × 0.51 = $494.70

(b) Quarterly premium
$970 × 0.26 = $252.20

(c) Monthly premium
$970 × 0.0908 = $88.08

(d) Total cost: semiannual
$494.70 × 2 = $989.40

Total cost: quarterly
$252.20 × 4 = $1008.80

Total cost: monthly
$88.08 × 12 = $1056.96

31. Cash value

$1125 × 60 (thousands) = $67,500

32. (a) Cash value

$283 × 40 (thousands) = $11,320

(b) Paid-up insurance

$579 × 40 (thousands) = $23,160

(c) The time for which he could have paid-up term insurance is 23 years 315 days.

33. (a) Monthly fixed payment

$6.91 × 80 (thousands) = $552.80

(b) $\dfrac{\$675}{80 \text{ (thousands)}} = \$8.44$

Jim can receive $675 per month for about 12 years.

(c) Monthly amount of life annuity

$5.94 × 80 (thousands) = $475.20

(d) The monthly amount that he could receive as a life annuity with 15 years certain is

$4.91 × 80 (thousands) = $392.80

34. (a) Monthly payment

$5.78 × 40 (thousands) = $231.20

(b) $\dfrac{\$245}{40 \text{ (thousands)}} = \$6.13$

She can receive $245 a month for about 18 years.

(c) Monthly amount of life annuity

$4.63 × 40 (thousands) = $185.20

(d) The monthly amount she could receive as a life annuity with 10 years certain is

$4.49 × 40 (thousands) = $179.60

Chapter 7 Summary Exercise

(a) Value of building

$1,730,000 = 17,300 hundreds

Premium for building

17,300 × $0.75 = $12,975

Value of contents

$3,502,000 = 35,020 hundreds
35,020 × $0.77 = $26,965.40

Annual insurance premium

$12,975 + $26,965.40 = $39,940.40

(b) Use Table 7.10.

$4.15 \times \$175$ (thousands) $= \$726.25$
$726.25 \times 0.51 = \$370.39$

(c) Total premium

$39,940.40 + \$370.39 = \$40,310.79$

(d) $41,700 - \$40,310.79 = \$1389.21$

Cumulative Review Exercises (Chapters 4-7)

1. Total sales

$93.50 + \$117.75 + \$173.05 + \$315.26$
$+ \$38.00 + \$92.18 + \$22.51 + \$162.15$
$= \$1014.40$

2. Total credits

$99.84 + \$72.68 + \$35.63 = \$208.15$

3. Total deposit

$1014.40 - \$208.15 = \$806.25$

4. Discount charge

$806.25 \times 0.0225(2\frac{1}{4}\%) = \$18.14$

5. Credit

$806.25 - \$18.14 = \$788.11$

6. Bank statement balance $16,298.60

Add: Deposits not recorded

$3631.28	
+ 7136.60	+ 10,767.88
	$27,066.48

Less: Outstanding checks

$2570.20	
4331.92	
293.00	
+ 1887.32	− 9082.44
Adjusted balance	$17,984.04

Checkbook balance $18,645.24

Add: Interest credit   + 40.72
                                     18,685.96

Less: Bank charges

$673.60	
+ 28.32	− 701.92
Adjusted balance	$17,984.04

7. Regular hours

$7 + 8 + 8 + 8 + 8 = 39$ hours

Overtime hours

$2 + 1 + 2 = 5$ hours

Overtime rate

$1\frac{1}{2} \times \$12.80 = \$19.20$

Regular earnings

39 hours $\times \$12.80 = \$499.20$

Overtime earnings

5 hours $\times \$19.20 = \$96$

Total gross earnings

$499.20 + \$96 = \$595.20$

8. (a) Total commission

$28,400	(total sales)		
− 5000	5000×0.05	=	$ 250
$23,400			
− 15,000	$15,000 \times 0.08$	=	$1200
$8400	8400×0.15	=	$1260
			$2710

(b) Gross earnings
= commissions − draw
= $2710 − $1200
$1510

9. Piecework earnings

$268 \times \$2.18 =$	$584.24
$1\frac{1}{2} \times 32 \times \$2.18 =$	+ 104.64
	$688.88

Gross earnings
= $688.88 − (9 \times \$1.05)$
= $688.88 − \$9.45$
= $679.43

10. Total earnings

$650 + (\$33,482 \times 0.03) = \$1654.46$

(a) Social security tax

$1654.46 \times 0.062 = \$102.58$

(b) Medicare tax

$1654.46 \times 0.0145 = \$23.99$

(c) State disability

$1654.46 \times 0.01 = \$16.54$

11. federal withholding

$722 - (3 \times \$52.88) = \$563.36$
$563.36 - \$525 = \$38.36$
$71.10 + (\$38.36 \times 28\%) = \$81.84$

federal tax	=	$ 81.84
state tax (2.8% × $722)	=	$ 20.22
FICA tax (6.2% × $722)	=	$ 44.76
medicare (1.45% × $722)	=	$ 10.47
SDI (1% × $722)	=	$ 7.22
United Way	=	$ 10.00
Savings bond	=	$ 50.00
total deductions	=	$224.51

Net pay

$722 - \$224.51 = \$497.49$

12.

$ 968.50	employee FICA
968.50	employer FICA
223.50	employee medicare
223.50	employer medicare
1975.38	withholding tax
$4359.38	total due

13. $B + (5.5\%) = \$286.96$
$B + 0.055B = \$286.96$
$1.055B = \$286.96$
$B = \dfrac{\$286.96}{1.055}$
$B = \$272$

14. $2.68 **(a)** 2.68% **(b)** $26.80 **(c)** 26.8

15. 4.62% **(a)** $4.62 **(b)** $46.20 **(c)** 46.2

16. Adjusted gross income

$38,514.75 + \$675.18 - \$1800 = \$37,389.93$

17. $179,480 = 1794.8$ hundreds
$1794.8 \times \$0.50 = \$897.40$
$83,300 = 833$ hundreds
$833 \times \$0.52 = \$433.16$
Total annual premium

$897.40 + \$433.16 = \$1330.56$

18. $720,000 \times 0.80 = \$576,000$

Coverage

$200,000 + \$100,000 + \$60,000 = \$360,000$

(a) $\dfrac{\$360,000}{\$576,000} \times \$240,000 = \$150,000$

(b) Company A:

$\dfrac{\$200,000}{\$360,000} \times \$150,000 = \$83,333.33$
$= \$83,333$ (rounded)

Company B:

$\dfrac{\$100,000}{\$360,000} \times \$150,000 = \$41,666.67$
$= \$41,667$ (rounded)

Company C:

$\dfrac{\$60,000}{\$360,000} \times \$150,000 = \$25,000$

19. (a) The insurance company will pay $6090 ($6340 − $250).

(b) The insurance company will pay the coverage limit, $25,000.

(c) Since there were two injuries, the insurance company will pay $25,000 for each personal injury for a total of $50,000.

(d) Nelson must pay $125,050

($100,000 + \$65,000 + \$34,800 + \$250 - \$25,000 - \$50,000$).

20. Use Table 7.10

32.59×100 (thousands) $= \$3259$

(a) Semiannual premium

$3259 \times 0.51 = \$1662.09$

(b) Quarterly premium

$3259 \times 0.26 = \$847.34$

(c) Monthly premium

$3259 \times 0.0908 = \$295.92$

(d) Total cost

$1662.09 \times 2 = \$3324.18$
$847.34 \times 4 = \$3389.36$
$295.92 \times 12 = \$3551.04$

Chapter 8

MATHEMATICS OF BUYING

8.1 Invoices and Trade Discounts

1. foot

3. sack

5. great gross

7. case

9. drum

11. cost per thousand

13. gallon

15. cash on delivery

17. Extension totals

$24 \times \$2.25 = \$54.00$

$12 \times \$4.75 = \$57.00$

$6 \times \$10.80 = \$64.80$

$2 \times \$14.20 = \$28.40$

$18 \times \$16.50 = \$297.00$

Invoice total $= \$501.20$

Total amount due

$\$501.20 + \$23.75 = \$524.95$

19. Answers will vary.

21. 10/20

$0.9 \times 0.8 = 0.72$

23. 20/20/20

$0.8 \times 0.8 \times 0.8 = 0.512$

25. 10/20/25

$0.9 \times 0.8 \times 0.75 = 0.54$

27. 30/42$\frac{1}{2}$ 42$\frac{1}{2}$% = 0.425

$0.7 \times 0.575 = 0.4025$

29. 20/30/5

$0.8 \times 0.7 \times 0.95 = 0.532$

31. 50/10/20/5

$0.5 \times 0.9 \times 0.8 \times 0.95 = 0.342$

33. $418 less 20/20

Net cost

$0.8 \times 0.8 \times \$418 = \$267.52$

35. $8.20 less 5/10

Net cost

$0.95 \times 0.9 \times \$8.20 = \$7.01$

37. $9.80 less 10/10/10

Net cost

$0.9 \times 0.9 \times 0.9 \times \$9.80 = \$7.14$

39. $1630 less 10/5/10

Net cost

$0.9 \times 0.95 \times 0.9 \times \$1630 = \$1254.29$

41. $25 less 30/32$\frac{1}{2}$; 32$\frac{1}{2}$% = 0.325

Net cost

$0.7 \times 0.675 \times \$25 = \$11.81$

43. $1250 less 20/20/20

Net cost

$0.8 \times 0.8 \times 0.8 \times \$1250 = \$640$

45. Answers will vary.

47. Answers will vary.

49. $399.99 less 10/10/25

Net amount

$0.9 \times 0.9 \times 0.75 \times \$399.99 = \$242.99$

51. (a) $120 less 10/15/10

$0.9 \times 0.85 \times 0.9 \times \$120 = \$82.62$

$120 less 20/15

$0.8 \times 0.85 \times \$120 = \$81.60$

The 20/15 discount gives the lower price.

(b) The difference is

$1.02 ($82.62−$81.60).

53. $28,500 less 20/20/10

Net cost

$0.8 \times 0.8 \times 0.9 \times \$28,500 = \$16,416$

55. $47 less 20/10

Net cost

$0.8 \times 0.9 \times \$47 = \$33.84$

57. $3500 = 3.5$ thousands
$3.5 \times \$135 = \$472.50$

$472.50 less 40/33\frac{1}{3}; 33\frac{1}{3}\% = \frac{1}{3}$

Net cost

$0.60 \times \frac{2}{3} \times \$472.50 = \$189$

59. $\$9.95 \times 48 \ (4 \text{ dozen}) = \$477.60$

(a) $477.60 less 10/25/15

$0.9 \times 0.75 \times 0.85 \times \$477.60 = \$274.02$

$477.60 less 20/15/10

$0.8 \times 0.85 \times 0.9 \times \$477.60 = \$292.29$

The total cost of the less expensive supplier is $274.02.

(b) The amount saved by selecting the lower price is $18.27($292.29 − $274.02).

8.2 Single Discount Equivalents

1. Net cost equivalent

10/20
$0.9 \times 0.8 = 0.72$

Percent form of single discount

$1 - 0.72 = 0.28 = 28\%$

3. Net cost equivalent

15/35
$0.85 \times 0.65 = 0.5525$

Percent form of single discount

$1 - 0.5525 = 0.4475 = 44.75\%$

5. Net cost equivalent

20/20
$0.8 \times 0.8 = 0.64$

Percent form of single discount

$1 - 0.64 = 0.36 = 36\%$

7. Net cost equivalent

20/20/10
$0.8 \times 0.8 \times 0.9 = 0.576$

Percent form of single discount

$1 - 0.576 = 0.424 = 42.4\%$

9. Net cost equivalent

25/10
$0.75 \times 0.9 = 0.675$

Percent form of single discount

$1 - 0.675 = 0.325 = 32.5\%$

11. Net cost equivalent

$16\frac{2}{3}/10; \ 16\frac{2}{3}\% = \frac{1}{6}$

$\frac{5}{6} \times 0.9 = 0.75$

Percent form of single discount

$1 - 0.75 = 0.25 = 25\%$

13. Net cost equivalent

10/10/20
$0.9 \times 0.9 \times 0.8 = 0.648$

Percent form of single discount

$1 - 0.648 = 0.352 = 35.2\%$

15. Net cost equivalent

55/40/10
$0.45 \times 0.6 \times 0.9 = 0.243$

Percent form of single discount

$1 - 0.243 = 0.757 = 75.7\%$

17. Net cost equivalent

40/25
$0.6 \times 0.75 = 0.45$

Percent form of single discount

$1 - 0.45 = 0.55 = 55\%$

19. Net cost equivalent

$20/12\frac{1}{2}; \ 12\frac{1}{2}\% = 0.125$

$0.8 \times 0.875 = 0.7$

Percent form of single discount

$1 - 0.7 = 0.3 = 30\%$

21. Net cost equivalent

20/10/20/10
$0.8 \times 0.9 \times 0.8 \times 0.9 = 0.5184$

Percent form of single discount

$1 - 0.5184 = 0.4816 = 48.16\%$

23. Net cost equivalent

5/20/30/5
$0.95 \times 0.8 \times 0.7 \times 0.95 = 0.5054$

Percent form of single discount

$1 - 0.5054 = 0.4946 = 49.46\%$

25. Answers will vary.

27. Net cost equivalent

20/10
$0.8 \times 0.9 = 0.72$

List price

$0.72 \times B = \$518.40$

$$\frac{0.72B}{0.72} = \frac{\$518.40}{0.72}$$

$$B = \$720$$

29. Net cost equivalent

40/5/30
$0.6 \times 0.95 \times 0.7 = 0.399$

List price

$0.399 \times B = \$279.30$

$$\frac{0.399B}{0.399} = \frac{\$279.30}{0.399}$$

$$B = \$700$$

31. Net cost equivalent

5/10/20
$0.95 \times 0.9 \times 0.8 = 0.684$

List price

$0.684 \times B = \$1313.28$

$$\frac{0.684B}{0.684} = \frac{\$1313.28}{0.684}$$

$$B = \$1920$$

33. (a) Wholesaler/s price

$89.95 less 20/10/10
$0.8 \times 0.9 \times 0.9 \times \$89.95 = \$58.29$

(b) Retailer's price

$89.95 less 20/10
$0.8 \times 0.9 \times \$89.95 = \$64.76$

(c) The difference between the two prices is $6.47
($64.76 − $58.29)

35. Net cost equivalent

30/20
$0.7 \times 0.8 = 0.56$

List price

$0.56 \times B = \$414.40$

$$\frac{0.56B}{0.56} = \frac{\$414.40}{0.56}$$

$$B = \$740$$

37. (a) Wholesaler's price

10/20/10
$0.9 \times 0.8 \times 0.9 \times \$61.90 = \$40.11$

(b) Retailer's price

10/20
$0.9 \times 0.8 \times \$61.90 = \$44.57$

(c) The difference between the two prices is $4.46 ($44.57 − $40.11).

39. The amount of discount is $74 ($295.95 − $221.95).

Single trade discount

$R \times \$295.95 = \$74$

$$\frac{\$295.95R}{\$295.95} = \frac{\$74}{\$295.95}$$

$$R = 0.2500 = 25.0\%$$

41. Net cost equivalent

$10/10/12\frac{1}{2}$; $12\frac{1}{2}\% = 0.125$
$0.9 \times 0.9 \times 0.875 = 0.70875$

List price

$0.70875 \times B = \$2733.75$

$$\frac{0.70875B}{0.70875} = \frac{\$2733.75}{0.70875}$$

$$B = \$3857.14$$

43. $27.60 less 10%

$0.9 \times \$27.60 = \$24.84$

Single trade discount

$R \times \$24.84 = \$23.60$

$$\frac{\$24.84R}{\$24.84} = \frac{\$23.60}{\$24.84}$$

$$R = 0.9500 = 95.0\%$$

An additional trade discount of

5% (100% − 95) must be given to meet the competitor's price.

8.3   Cash Discounts: Ordinary Dating Method

1. Final discount date

Oct. 8 + 10 days = Oct. 18

The net payment date is Nov. 7

(23 days in Oct. + 7 days in Nov.).

3. Final discount date

Mar. 10 + 15 days = Mar. 25

The net payment date is Mar. 30

(20 days from Mar. 10).

5. Final discount date

Sept. 11 + 10 days = Sept. 21

The net payment date is Nov. 10

(19 days in Sept. + 31 days in Oct. + 10 days in Nov.).

7. Final discount date

Jan. 14 + 10 days = Jan. 24

The net payment date is Mar. 15

(17 days in Jan. + 28 days in Feb. + 15 days in March).

9. Final discount date

Jan. 5 + 15 days = Jan. 20

The net payment date is Mar. 6

(26 days in Jan. + 28 days in Feb. + 6 days in March).

11. Amount of discount

$\$151.35 \times 2\% = \$3.03$

Amount due

$\$151.35 − \$3.03 + \$12.58 = \$160.90$

13. Amount of discount
There is no discount. $0
Amount due

$\$96.06 + \$5.22 = \$101.28$

15. Amount of discount

$\$724 \times 2\% = \$14.48$

Amount due

$\$724 − \$14.48 + \$38.14 = \$747.66$

17. Amount of discount

$\$780.70 − \$125 = \$655.70$

There is no discount. $0

Amount due

$\$655.70 + \$3.80 = \$659.50$

19. Amount of discount

$\$635 − \$52 = \$583$
$\$583 \times 5\% = \$29.15$

Amount due

$\$583 − \$29.15 + \$53.18 = \$607.03$

21. Answers will vary.

23. Amount of discount

$\$2010.70 \times 1\% = \$20.11$

Amount due

$\$2010.70 − \$20.11 = \$1990.59$

25. Amount due

$0.9 \times 0.8 \times 0.95 \times \$986 = \$674.42$

27. Amount due

$0.8 \times 0.8 \times \$215.80 = \$138.11$
$\$138.11 \times 4\% = \$5.52$
$\$138.11 − \$5.52 = \$132.59$

29. Amount due

$0.9 \times 0.8 \times 0.7 \times \$5,190 = \$2615.76$
$\$2615.76 \times 4\% = \$104.63$
$\$2615.76 − \$104.63 = \$2511.13$

31. Amount due

$0.9 \times 0.8 \times 0.9 \times \$3215.80 = \$2083.84$
$\$2083.84 \times 3\% = \$62.52$
$\$2083.84 - \$62.52 = \$2021.32$

33. **(a)** 6/10: April 24
(10 days from April 14)

4/20: May 4
(16 days in April + 4 days in May)

1/30: May 14
(16 days in April + 14 days in May)

(b) Net payment date: June 3
(16 days in April + 31 days in May + 3 days in June)

35. **(a)** Final discount date: Apr. 25
(20 days from April 5)

(b) Net payment date: May 5
(25 days in April + 5 days in May)

37. Amount of discount

$\$3724.40 - \$104.50 = \$3619.90$
$\$3619.90 \times 2\% = \$72.40$

Amount due

$\$3619.90 - \$72.40 = \$3547.50$

39. $\$3322.80 - \$152.80 = \$3170$
$\$3170 \times 2\% = \$63.40$

Amount due

$\$3170 - \$63.40 = \$3106.60$

41. Answers will vary.

8.4 Cash Discounts: Other Dating Methods

1. Discount date: Mar. 10
(10 days from the end of the month, Feb. 28)
The net payment is due Mar. 30 (20 days after the final discount date).

3. Discount date: Dec. 22
(10 + 20 or 30 days from Nov. 22)
The net payment is due Jan. 11 (20 days after the final discount date).

5. Discount date: June 16
(18 days in April + 31 days in May + 16 days in June or 65 days)
The net payment is due July 6 (20 days after the final discount date).

7. Discount date: Aug. 10
(10 days from the end of the month, July 31)
The net payment is due Aug. 30 (20 days after the final discount date).

9. Discount date: Aug. 24
(20 days from the receipt of goods)
The net payment is due Sept. 13 (7 days in Aug. + 13 days in Sept. which is 20 days after the final discount date).

11. Amount of discount

$\$682.28 \times 0.03 = \$20.47$

Amount due

$\$682.28 - \$20.47 = \$661.81$

13. The last day to take the 5% discount was Oct. 31, 5 days before the invoice was paid.

Amount due: $194.04

15. Final discount date: Dec. 10
(10 days from the end of the month, Nov. 30)

Amount of discount

$\$2960 \times 0.02 = \$59.20$

Amount due

$\$2960 - \$59.20 = \$2900.80$

17. Final discount date: Nov. 15
(15 days from the end of the month, Oct. 31)

Amount of discount

$\$4220 \times 0.04 = \$168.80$

Amount due

$\$4220 - \$168.80 = \$4051.20$

19. Final discount date: April 30
(15 days from the receipt of goods)

Amount of discount

$\$12.38 \times 0.02 = \$0.25$

Amount due

$\$12.38 - \$0.25 = \$12.13$

21. Final discount date: Nov. 21
(35 days from Oct. 17)

Amount of discount

$3250.60 \times 0.03 = \$97.52$

Amount due

$3250.60 - \$97.52 = \$3153.08$

23. Final discount date: Dec. 10

Amount of discount

$1708.18 \times 0.04 = \$68.33$

Amount due

$1708.18 - \$68.33 = \$1639.85$

25. Answers will vary.

27. Credit given

$100\% - 2\% = 98\%$

$1862 = 0.98 \times C$

$\dfrac{\$1862}{0.98} = \dfrac{0.98C}{0.98}$

$1900 = C$

Balance due

$3150 - \$1900 = \$1250$

29. Credit given

$100\% - 5\% = 95\%$

$684 = 0.95 \times C$

$\dfrac{\$684}{0.95} = \dfrac{0.95C}{0.95}$

$720 = C$

Balance due

$1750 - \$720 = \$1030$

31. Credit given

$100\% - 3\% = 97\%$

$97 = 0.97 \times C$

$\dfrac{\$97}{0.97} = \dfrac{0.97C}{0.97}$

$100 = C$

Balance due

$160 - \$100 = \$60$

33. Answers will vary.

35. (a) Final discount date: Sept. 15
(15 days from the end of the month)

(b) The net payment date is Oct. 5
(20 days after the final discount date).

37. (a) Final discount date: Dec. 23
(20 days after the receipt of goods)

(b) The net payment date is Jan. 12
(20 days after the final discount date,
8 days in Dec. + 12 days in Jan.).

39. Amount of discount

$1525 \times 0.01 = \$15.25$

Amount due

$1525 - \$15.25 = \$1509.75$

41. (a) Final discount date: Dec. 13
(27 days in Nov. + 13 days in Dec.)

(b) Amount of discount

$970.68 \times 0.02 = \$19.41$

Amount due

$970.68 - \$19.41 = \$951.27$

43. Final discount date: Apr. 9
(11 days in Mar. + 9 days in Apr.)

Amount of discount

$4358.50 \times 0.02 = \$87.17$

Amount due

$4358.50 - \$87.17 = \$4271.33$

45. (a) Final discount date: Dec. 20
Credit given

$100\% - 6\% = 94\%$

$940 = 0.94C$

$\dfrac{\$940}{0.94} = \dfrac{0.94C}{0.94}$

$1000 = C$

(b) Balance due

$1920 - \$1000 = \$920$

47. Final discount date: May 3
(15 days after receipt of goods)

Balance due with discount

$0.75 \times 0.9 \times 0.9 \times 0.97 \times \$2538 = \$1495.58$

49. Final discount date: July 19

Balance due with discount

$0.85 \times 0.9 \times 0.95 \times \$128 = \$93.02$

51. (a) Final discount date: Nov. 20
(20 days from the end of the month)

Credit given

$100\% - 2\% = 98\%$

$\$300 = 0.98 \times C$

$\dfrac{\$300}{0.98} = \dfrac{0.98C}{0.98}$

$\$306.12 = C$

(b) Balance due

$\$526.80 - \$306.12 = \$220.68$

53. (a) Final discount date: July 4

Credit given

$100\% - 3\% = 97\%$

$\$580 = 0.97 \times C$

$\dfrac{\$580}{0.97} = \dfrac{0.97C}{0.97}$

$\$597.94 = C$

(b) Balance due

$\$1120.15 - \$597.94 = \$522.21$

Chapter 8 Review Exercises

1. $480 less 20/10

Net cost

$0.8 \times 0.9 \times \$480 = \$345.60$

2. $276 less $10/12\frac{1}{2}/10$; $12\frac{1}{2}\% = 0.125$

Net cost

$0.9 \times 0.875 \times 0.9 \times \$276 = \$195.62$

3. $2830 less 5/15/20

Net cost

$0.95 \times 0.85 \times 0.8 \times \$2830 = \$1828.18$

4. $1620 less 20/25/15

Net cost

$0.8 \times 0.75 \times 0.85 \times \$1620 = \$826.20$

5. Net cost equivalent

25/15
$0.75 \times 0.85 = 0.6375$

Percent form of a single discount

$1 - 0.6375 = 0.3625 = 36.25\%$

6. Net cost equivalent

20/10/20
$0.8 \times 0.9 \times 0.8 = 0.576$

Percent form of a single discount

$1 - 0.576 = 0.424 = 42.4\%$

7. Net cost equivalent

$20/32\frac{1}{2}$; $32\frac{1}{2}\% = 0.325$

$0.8 \times 0.675 = 0.54$

Percent form of single discount

$1 - 0.54 = 0.46 = 46\%$

8. Net cost equivalent

10/20/10/30
$0.9 \times 0.8 \times 0.9 \times 0.7 = 0.4536$

Percent form of single discount

$1 - 0.4536 = 0.5464 = 54.64\%$

9. Net cost equivalent

10/20
$0.9 \times 0.8 = 0.72$

List price

$0.72 \times B = \$361.50$

$\dfrac{0.72B}{0.72} = \dfrac{\$361.50}{0.72}$

$B = \$502.08$

10. Net cost equivalent

15/20
$0.85 \times 0.8 = 0.68$

List price

$0.68 \times B = \$1050.74$

$$\frac{0.68B}{0.68} = \frac{\$1050.74}{0.68}$$

$$B = \$1545.21$$

11. Net cost equivalent

10/20/15
$0.9 \times 0.8 \times 0.85 = 0.612$

List price

$0.612 \times B = \$328.70$

$$\frac{0.612B}{0.612} = \frac{\$328.70}{0.612}$$

$$B = \$537.09$$

12. Net cost equivalent

5/20/25
$0.95 \times 0.8 \times 0.75 = 0.57$

List price

$0.57 \times B = \$1289.40$

$$\frac{0.57B}{0.57} = \frac{\$1289.40}{0.57}$$

$$B = \$2262.11$$

13. Final discount date: Mar. 15
(15 days from the end of the month)

Net payment date: Apr. 4
(20 days after the final discount date)

14. Final discount date: May 30
(10 days from the receipt of goods)

Net payment date: June 19
(20 days after the final discount date)

15. Final discount date: Jan. 10
(10 days from the end of the month, Dec. 31)

Net payment date: Jan. 30
(20 days after the final discount date)

16. Final discount date: Dec. 19
(11 days in Oct. + 30 days in Nov. + 19 days in Dec.)

Net payment date: Jan 8
(20 days after the final discount date—12 days in Dec. + 8 days in Jan.)

17. Amount of discount

$\$1280.40 \times 0.03 = \$38.41$

Amount due

$\$1280.40 - \$38.41 + \$76.18 = \$1318.17$

18. Amount of discount

$\$945.60 \times 0.03 = \$28.37$

Amount due

$\$945.60 - \$28.37 = \$917.23$

19. Amount of discount

$\$875.50 \times 0.04 = \$35.02$

Amount due

$\$875.50 - \$35.02 + \$67.18 = \$907.66$

20. Amount of discount

$\$2210.60 \times 0.02 = \$44.21$

Amount due

$\$2210.60 - \$44.21 = \$2166.39$

21. Credit given

$100\% - 2\% = 98\%$

$\$300 = 0.98 \times C$

$$\frac{\$300}{0.98} = \frac{0.98C}{0.98}$$

$\$306.12 = C$

Balance due

$\$660 - \$306.12 = \$353.88$

22. Credit given

$$100\% - 3\% = 97\%$$

$$\$2520 = 0.97 \times C$$

$$\frac{\$2520}{0.97} = \frac{0.97C}{0.97}$$

$$\$2597.94 = C$$

Balance due

$$\$5310 - \$2597.94 = \$2712.06$$

23. Credit given

$$100\% - 1\% = 99\%$$

$$\$500 = 0.99 \times C$$

$$\frac{\$500}{0.99} = \frac{0.99C}{0.99}$$

$$\$505.05 = C$$

Balance due

$$\$860 - \$505.05 = \$354.95$$

24. Credit given

$$100\% - 3\% = 97\%$$

$$\$2050 = 0.97 \times C$$

$$\frac{\$2050}{0.97} = \frac{0.97C}{0.97}$$

$$\$2113.40 = C$$

Balance due

$$\$3850 - \$2113.40 = \$1736.60$$

25. Extension totals

$$16 \times \$17.50 = \$280.00$$
$$8 \times \$3.25 = \$26.00$$
$$4 \times \$12.65 = \$50.60$$
$$12 \times \$3.15 = \$37.80$$

(a) Invoice total: $394.40

(b) Amount due after cash discount

$$\$394.40 \times 0.02 = \$7.89$$
$$\$394.40 - \$7.89 = \$386.51$$

(c) Total amount due

$$\$386.51 + \$11.55 = \$398.06$$

26. (a) Fireside Shop

$$0.75 \times 0.9 \times \$120 = \$81$$

Builder's Supply

$$0.75 \times 0.95 \times \$111 = \$79.09$$

Builder's Supply offers the lowest price.

(b) The difference in price is $1.91

($81 − $79.09).

27. Net cost equivalent

$$0.8 \times 0.8 \times 0.9 = 0.576$$

List price

$$0.576 \times B = \$36,458$$

$$\frac{0.576B}{0.576} = \frac{\$36,458}{0.576}$$

$$B = \$63,295.14$$

28. Restaurant Distributing

$$\$980 \times 0.75 = \$735$$
$$\$735 \times R = \$661.50$$

$$\frac{\$735R}{\$735} = \frac{\$661.50}{\$735}$$

$$R = 0.9 = 90\%$$

Trade discount rate is

10% (100% − 90%)

29. Final discount date: Oct. 20
$2018 − $183 (goods returned)
    = $1835

Amount of discount

$$\$1835 \times 0.03 = \$55.05$$

Amount due

$$\$1835 - \$55.05 = \$1779.95$$

30. Final discount date: May 15

Amount of discount

$$\$1854 \times 0.03 = \$55.62$$

Amount due

$$\$1854 - \$55.62 = \$1798.38$$

31. (a) Final discount date: April 15
(15 days from the end of the month)

(b) Amount of discount

$838 \times 0.02 = \$16.76$

Amount due

$\$838 - \$16.76 = \$821.24$

32. Final discount date: Dec. 15

(a) Credit given

$100\% - 4\% = 96\%$

$\$1800 = 0.96 \times C$

$\dfrac{\$1800}{0.96} = \dfrac{0.96C}{0.96}$

$\$1875 = C$

(b) Balance due

$\$5280 - \$1875 = \$3405$

(c) Cash discount

$\$1875 - \$1800 = \$75$

(e) Credit given

$100\% - 3\% = 97\%$

$\$2500 = 0.97 \times C$

$\dfrac{\$2500}{0.97} = \dfrac{0.97C}{0.97}$

$\$2577.32 = C$

Balance due

$\$3844.76 - \$2577.32 + \$175.14 = \$1442.58$

Chapter 8 Summary Exercise

(a) Total price

$\$2893 + \$3138 = \$6031$

Total amount of invoice

$0.75 \times 0.85 \times \$6031 = \$3844.76$

(b) Final discount date: June 15

(15 days after the end of the month May 31)

(c) Net payment date: July 5
(20 days after the final discount
date. 15 days in June + 5 days
in July)

(d) Amount of discount

$\$3844.76 \times 0.03 = \$115.34$

Total amount due

$\$3844.76 - \$115.34 + \$175.14 = \$3904.56$

MARKUP

9.1 Markup on Cost

1. Markup

$$M = 0.40(\$12.40)$$
$$M = \$4.96$$

Selling price

$$C + M = S$$
$$\$12.40 + \$4.96 = S$$
$$\$17.36 = S$$

3. Percent of markup

$$R \times B = P$$
$$R \times \$23.50 = \$11.75$$
$$\$23.50R = \$11.75$$

$$\frac{\$23.50R}{\$23.50} = \frac{\$11.75}{\$23.50}$$
$$R = 0.5 = 50\%$$

Selling price

$$C + M = S$$
$$\$23.50 + \$11.75 = S$$
$$\$35.25 = S$$

5. Markup

$$C + M = S$$
$$\$158.70 + M = \$198.50$$
$$M = \$198.50 - \$158.70$$
$$M = \$39.80$$

Percent of markup

$$R \times B = P$$
$$R \times \$158.70 = \$39.80$$
$$\$158.70R = \$39.80$$

$$\frac{\$158.70R}{\$158.70} = \frac{\$39.80}{\$158.70}$$
$$R = 0.2507 = 25.1\%$$

7. Cost price

$$C + M = S$$
$$C + \$13.50 = \$81$$
$$C = \$81 - \$13.50$$
$$C = \$67.50$$

Percent of markup

$$R \times B = P$$
$$R \times \$67.50 = \$13.50$$
$$\$67.50R = \$13.50$$

$$\frac{\$67.50R}{\$67.50} = \frac{\$13.50}{\$67.50}$$
$$R = 0.2 = 20\%$$

9. Markup

$$C + M = S$$
$$\$210 + M = \$328$$
$$M = \$328 - \$210$$
$$M = \$118$$

Percent of markup

$$R \times B = S$$
$$R \times \$210 = \$118$$
$$\$210R = \$118$$

$$\frac{\$210R}{\$210} = \frac{\$118}{\$210}$$
$$R = 0.5619 = 56.2\%$$

11. Markup

$$R \times B = P$$
$$0.27 \times \$495 = M$$
$$\$133.65 = M$$

Selling price

$$C + M = S$$
$$\$495 + \$133.65 = S$$
$$\$628.65 = S$$

13. Answers will vary.

15. Cost price

$$C + M = S$$
$$C + 0.35C = \$138$$
$$1.35C = \$138$$

$$\frac{1.35C}{1.35} = \frac{\$138}{1.35}$$
$$C = \$102.22$$

17. Cost price

$$C + 0.38C = \$18.95$$
$$1.38C = \$18.95$$
$$\frac{1.38C}{1.38} = \frac{\$18.95}{1.38}$$
$$C = \$13.73$$

19. Selling price

$$C + M = S$$
$$\$11.96 + (0.25 \times \$11.96) = S$$
$$\$11.96 + \$2.99 = S$$
$$\$14.95 = S$$

21. Markup

$$C + M = S$$
$$\$1740 + M = \$2049.20$$
$$M = \$2049.20 - \$1740$$
$$M = \$309.20$$

Percent of markup

$$R \times B = P$$
$$R \times \$1740 = \$309.20$$
$$\frac{\$1740R}{\$1740} = \frac{\$309.20}{\$1740}$$
$$R = 0.1777 = 17.8\%$$

23. **(a)** Cost price

$$C + M = S$$
$$C + \$23.99 = \$119.95$$
$$C = \$119.95 - \$23.99$$
$$C = \$95.96$$

(b) Percent of markup

$$R \times B = P$$
$$R \times \$95.96 = \$23.99$$
$$\frac{95.96R}{95.96} = \frac{23.99}{95.96}$$
$$R = 0.25 = 25\%$$

(c) Selling price as a percent of cost

$$R \times B = P$$
$$R \times \$95.96 = \$119.95$$
$$\frac{95.96R}{95.96} = \frac{119.95}{95.96}$$
$$R = 1.25 = 125\%$$

25. **(a)** Selling price as a percent of cost

$$100\% + 32\% = 132\%$$

(b) Selling price

$$C + M = S$$
$$\$73.50 + (0.32 \times \$73.50) = \$97.02$$

(c) Markup

$$C + M = S$$
$$\$73.50 + M = \$97.02$$
$$M = \$97.02 - \$73.50$$
$$M = \$23.52$$

27. **(a)** Cost price

$$R \times B = P$$
$$1.272 \times C = \$44.52$$
$$\frac{1.272C}{1.272} = \frac{\$44.52}{1.272}$$
$$C = \$35$$

(b) Markup as a percent of cost

$$127.2\% - 100\% = 27.2\%$$

(c) Markup

$$C + M = S$$
$$\$35 + M = \$44.52$$
$$M = \$44.52 - \$35$$
$$M = \$9.52$$

29. Selling price

$$\$3860 + (\$3860 \times 0.22) + (\$3860 \times 0.12)$$
$$= \$3860 + \$849.20 + \$463.20$$
$$= \$5172.40$$

31. Cost price

$$C + 0.18C + 0.17C = \$8.95$$
$$1.35C = \$8.95$$
$$\frac{1.35C}{1.35} = \frac{\$8.95}{1.35}$$
$$C = \$6.63$$

9.2   Markup on Selling Price

1. Selling price

$$C + M = S$$
$$\$21 + 0.25S = S$$
$$\$21 + 0.25S - 0.25S = S - 0.25S$$
$$\$21 = 0.75S$$
$$\frac{\$21}{0.75} = \frac{0.75S}{0.75}$$
$$\$28 = S$$

Markup

$$C + M = S$$
$$\$21 + M = \$28$$
$$M = \$28 - \$21$$
$$M = \$7$$

3. Selling price

$$R \times B = P$$
$$46\% \times S = \$112$$
$$0.46S = \$112$$
$$S = \frac{\$112}{0.46}$$
$$S = \$243.48$$

Cost price

$$C + M = S$$
$$C + \$112 = \$243.48$$
$$C + \$112 - \$112 = \$243.48 - \$112$$
$$C = \$131.48$$

5. Selling price

$$66\frac{2}{3}\% = \frac{2}{3}$$
$$C + M = S$$
$$\$18.60 + \frac{2}{3}S = S$$
$$\$18.60 + \frac{2}{3}S - \frac{2}{3}S = S - \frac{2}{3}S$$
$$\$18.60 = \frac{1}{3}S$$
$$3 \times \$18.60 = 3 \cdot \frac{1}{3}S$$
$$\$55.80 = S$$

Markup

$$C + M = S$$
$$\$18.60 + M = \$55.80$$
$$M = \$55.80 - \$18.60$$
$$M = \$37.20$$

7. Markup

$$R \times B = P$$
$$0.35 \times \$71.32 = M$$
$$\$24.96 = M$$

Cost price

$$C + M = S$$
$$C + \$24.96 = \$71.32$$
$$C = \$71.32 - \$24.96$$
$$C = \$46.36$$

9. Cost price

$$C + M = S$$
$$C + \$42.18 = \$120$$
$$C = \$120 - \$42.18$$
$$C = \$77.82$$

Percent of markup on selling price

$$R \times B = P$$
$$R \times \$120 = \$42.18$$
$$\$120R = \$42.18$$
$$R = \frac{\$42.18}{\$120}$$
$$R = 0.3515 = 35.2\%$$

11. Cost

$$R \times B = P$$
$$25\% \times C = \$57.50$$
$$0.25C = \$57.50$$
$$C = \frac{\$57.50}{0.25}$$
$$C = \$230$$

Selling price

$$R \times B = P$$
$$20\% \times S = \$57.50$$
$$0.2S = \$57.50$$
$$S = \frac{\$57.50}{0.2}$$
$$S = \$287.50$$

13. Selling price

$$C + M = S$$
$$\$13.80 + 0.38S = S$$
$$\$13.80 + 0.38S - 0.38S = S - 0.38S$$
$$\$13.80 = 0.62S$$
$$\frac{\$13.80}{0.62} = S$$
$$\$22.26 = S$$

Markup

$$C + M = S$$
$$\$13.80 + M = \$22.26$$
$$M = \$22.26 - \$13.80$$
$$M = \$8.46$$

Percent markup on cost

$$R \times B = P$$
$$R \times \$13.80 = \$8.46$$
$$\$13.80R = \$8.46$$
$$R = \frac{\$8.46}{\$13.80}$$
$$R = 0.6130 = 61.3\%$$

15. Cost

$$R \times B = P$$
$$40\% \times C = \$300$$
$$0.40C = \$300$$
$$C = \frac{\$300}{0.40}$$
$$C = \$750$$

Selling price

$$C + M = S$$
$$\$750 + \$300 = S$$
$$\$1050 = S$$

Percent markup on selling price

$$R \times B = P$$
$$R \times \$1050 = \$300$$
$$\$1050R = \$300$$
$$R = \frac{\$300}{\$1050}$$
$$R = 0.2857 = 28.6\%$$

17. Cost

$$C + M = S$$
$$C + \$78.48 = \$436$$
$$C = \$436 - \$78.48$$
$$C = \$357.52$$

Percent markup on cost

$$R \times B = P$$
$$R \times \$357.52 = \$78.48$$
$$\$357.52R = \$78.48$$
$$R = \frac{\$78.48}{\$357.52}$$
$$R = 0.2195 = 22.0\%$$

19. $M_C = \dfrac{M_S}{100\% - M_S}$

$$= \frac{20\%}{100\% - 20\%}$$
$$= \frac{20\%}{80\%} = \frac{0.2}{0.8} = \frac{1}{4} = 25\%$$

21. $M_C = \dfrac{M_S}{100\% - M_S}$

$$= \frac{26\%}{100\% - 26\%}$$
$$= \frac{26\%}{74\%} = \frac{0.26}{0.74}$$
$$= 0.3513 = 35.1\%$$

23. $M_S = \dfrac{M_C}{100\% + M_C}$

$$= \frac{50\%}{100\% + 50\%}$$
$$= \frac{50\%}{150\%} = \frac{0.5}{1.5}$$
$$= \frac{1}{3} = 33.3\% \text{ or } 33\tfrac{1}{3}\%$$

25. $M_C = \dfrac{M_S}{100\% - M_S}$

$$= \frac{40\%}{100\% - 40\%}$$
$$= \frac{40\%}{60\%} = \frac{0.4}{0.6}$$
$$= \frac{2}{3} = 66.7\% \text{ or } 66\tfrac{2}{3}\%$$

27. Answers will vary.

29. Selling price

$$C + M = S$$
$$\$41.40 + 0.28S = S$$
$$\$41.40 + 0.28S - 0.28S = S - 0.28S$$
$$\$41.40 = 0.72S$$
$$\frac{\$41.40}{0.72} = S$$
$$\$57.50 = S$$

31.
$$C + M = S$$
$$C + (0.35 \times \$595) = \$595$$
$$C + \$208.25 = \$595$$
$$C = \$595 - \$208.25$$
$$C = \$386.75$$

33.
$$C + M = S$$
$$\$92.82 + 0.22S = S$$
$$\$92.82 + 0.22S - 0.22S = S - 0.22S$$
$$\$92.82 = 0.78S$$
$$\frac{\$92.82}{0.78} = S$$
$$\$119 = S$$

35. (a) Selling price
$$R \times B = P$$
$$36\% \times S = \$1.62$$
$$0.36S = \$1.62$$
$$S = \frac{\$1.62}{0.36}$$
$$S = \$4.50$$

(b) Cost
$$C + M = S$$
$$C + \$1.62 = \$4.50$$
$$C = \$4.50 - \$1.62$$
$$C = \$2.88$$

(c) Cost as a percent of selling price
$$R \times B = P$$
$$R \times \$4.50 = \$2.88$$
$$\$4.50R = \$2.88$$
$$R = \frac{\$2.88}{\$4.50}$$
$$R = 0.64 = 64\%$$

37. Cost per item
$$\$258 \div 12 = \$21.50$$

Selling price
$$C + M = S$$
$$\$21.50 + \$7.74 = S$$
$$\$29.24 = S$$

Percent of gross profit
$$R \times B = P$$
$$R \times \$29.24 = \$7.74$$
$$\$29.24R = \$7.74$$
$$R = \frac{\$7.74}{\$29.24}$$
$$R = 0.2647 = 26.5\%$$

39. (a) Cost
$$R \times B = P$$
$$72\% \times \$38.50 = C$$
$$0.72 \times \$38.50 = C$$
$$\$27.72 = C$$

(b) Markup
$$C + M = S$$
$$\$27.72 + M = \$38.50$$
$$M = \$38.50 - \$27.72$$
$$M = \$10.78$$

(c) Markup as a percent of selling price
$$R \times B = P$$
$$R \times \$38.50 = \$10.78$$
$$\$38.50R = \$10.78$$
$$R = \frac{\$10.78}{\$38.50}$$
$$R = 0.28 = 28\%$$

41. (a) Markup
$$R \times B = P$$
$$0.50 \times \$2880 = P$$
$$\$1440 = P$$
$$\$1440 \div 2000 = \$0.72$$

(b) Equivalent markup percent on cost
$$\$2,880 \div 2,000 = \$1.44$$
$$C + M = S$$
$$C + \$0.72 = \$1.44$$
$$C = \$0.72$$
$$R \times B = P$$
$$R \times \$0.72 = \$0.72$$
$$R = \frac{\$0.72}{\$0.72}$$
$$R = 1 = 100\%$$

43. (a) Total amount received

$$(\$45 \times 158) + (\$35 \times 74) + (\$30 \times 56) + (\$25 \times 92)$$
$$= \$7110 + \$2590 + \$1680 + \$2300$$
$$= \$13,680$$

(b) Total markup

$$C + M = S$$
$$\$7600 + M = \$13,680$$
$$M = \$13,680 - \$7600$$
$$M = \$6080$$

(c)   Markup percent on selling price

$$R \times B = P$$
$$R \times \$13,680 = \$6080$$
$$\$13,680R = \$6080$$
$$R = \frac{\$6080}{\$13,680}$$
$$R = 0.4444 = 44.4\%$$

(d) Equivalent markup percent on cost

$$R \times B = P$$
$$R \times \$7600 = \$6080$$
$$\$7600R = \$6080$$
$$R = \frac{\$6080}{\$7600}$$
$$R = 0.8 = 80\%$$

45. (a) Cost

$$R \times B = P$$
$$0.50 \times \$27.90 = C$$
$$\$13.95 = C$$

(b) Markup

$$\$27.90 - \$13.95 = \$13.95$$

Percent of markup on cost

$$R \times \$13.95 = \$13.95$$
$$\$13.95R = \$13.95$$
$$R = \frac{\$13.95}{\$13.95}$$
$$R = 1 = 100\%$$

47. Markup

$$C + M = S$$
$$\$4.80 + M = \$6.90$$
$$M = \$6.90 - \$4.80$$
$$M = \$2.10$$

(a) Percent of markup on cost

$$R \times B = P$$
$$R \times \$4.80 = \$2.10$$
$$\$4.80R = \$2.10$$
$$R = \frac{\$2.10}{\$4.80}$$
$$R = 0.4375 = 43.8\%$$

(b) Equivalent percent of markup on selling price

$$R \times B = P$$
$$R \times \$6.90 = \$2.10$$
$$\$6.90R = \$2.10$$
$$R = \frac{\$2.10}{\$6.90}$$
$$R = 0.3043 = 30.4\%$$

49. Cost

$$\$288 \div 12 = \$24$$

(a) Selling price

$$\$24 + 0.35C = S$$
$$\$24 + (0.35 \times \$24) = S$$
$$\$24 + \$8.40 = S$$
$$\$32.40 = S$$

(b) Markup

$$C + M = S$$
$$\$24 + M = \$32.40$$
$$M = \$32.40 - \$24.00$$
$$M = \$8.40$$

Percent of markup on selling price

$$R \times B = P$$
$$R \times \$32.40 = \$8.40$$
$$\$32.40R = \$8.40$$
$$R = \frac{\$8.40}{\$32.40}$$
$$R = 0.2592 = 25.9\%$$

51. Cost

$$\$2100 \div 12 = \$175$$

(a) Markup

$$C + M = S$$
$$\$175 + M = \$199.90$$
$$M = \$199.90 - \$175$$
$$M = \$24.90$$

(b) Percent of markup on selling price

$$R \times B = P$$
$$R \times \$199.90 = \$24.90$$
$$\$199.90R = \$24.90$$
$$R = \frac{\$24.90}{\$199.90}$$
$$R = 0.1245 = 12.5\%$$

(c) Percent of markup on cost

$$R \times B = P$$
$$R \times \$175 = \$24.90$$
$$\$175R = \$24.90$$
$$R = \frac{\$24.90}{\$175}$$
$$R = 0.1422 = 14.2\%$$

9.3   Markup with Spoilage

1. Total selling price

$$C + M = S$$
$$C + 10\%S = S$$
$$\$81 + 0.1S = S$$
$$\$81 = 1.00S - 0.1S$$
$$\$81 = 0.9S$$
$$\frac{\$81}{0.9} = S$$
$$\$90 = S$$

Selling price per item

$$\frac{\$90}{90 - 9} = \frac{\$90}{81} = \$1.11$$

3. Total selling price

$$C + M = S$$
$$C + 20\%S = S$$
$$\$340 + 0.2S = S$$
$$\$340 = 1.00S - 0.2S$$
$$\$340 = 0.8S$$
$$\frac{\$340}{0.8} = S$$
$$\$425 = S$$

Selling price per item

$$\frac{\$425}{144 - 8} = \frac{\$425}{136} = \$3.13$$

5. $33\frac{1}{3}\% = \frac{1}{3}$

Total selling price

$$C + M = S$$
$$C + \frac{1}{3}S = S$$
$$\$120 + \frac{1}{3}S = S$$
$$\$120 = 1S - \frac{1}{3}S$$
$$\$120 = \frac{2}{3}S$$
$$\frac{3}{2} \cdot \$120 = \frac{3}{2} \cdot \frac{2}{3}S$$
$$\$180 = S$$

Selling price per item

$$\frac{\$180}{25 - 5} = \frac{\$180}{20} = \$9$$

7. Total selling price

$$C + M = S$$
$$C + 0.15S = S$$
$$\$126 + 0.15S = S$$
$$\$126 = 1.00S - 0.15S$$
$$\$126 = 0.85S$$
$$\frac{\$126}{0.85} = S$$
$$\$148.24 = S$$

Selling price per item

$$\frac{\$148.24}{(8 \times 12) - 6} = \frac{\$148.24}{90} = \$1.65$$

9. Number to sell

$$100\% - 4\% = 96\%$$
$$0.96 \times 25 = 24$$

Total selling price

$$C + M = S$$
$$\$161 + 0.25S = S$$
$$\$161 = 0.75S$$
$$\frac{\$161}{0.75} = S$$
$$\$214.67 = S$$

Selling price per item

$$\frac{\$214.67}{24} = \$8.94$$

11. Number to sell

$100\% - 25\% = 75\%$

$0.75 \times 24 = 18$

Total selling price

$$C + M = S$$
$$\$198 + 0.5S = S$$
$$\$198 = 0.5S$$
$$\frac{\$198}{0.5} = S$$
$$\$396 = S$$

Selling price per item

$$\frac{\$396}{18} = \$22$$

13. Number to sell

$100\% - 5\% = 95\%$

$0.95 \times 80 \text{ pr.} = 76 \text{ pr.}$

Total selling price

$$33\frac{1}{3}\% = \frac{1}{3}$$
$$C + M = S$$
$$\$190 + \frac{1}{3}S = S$$
$$\$190 = \frac{2}{3}S$$
$$\frac{3}{2} \cdot \$190 = \frac{3}{2} \cdot \frac{2}{3}S$$
$$\$285 = S$$

Selling price per item

$$\frac{\$285}{76} = \$3.75$$

15. Number to sell

$100\% - 5\% = 95\%$

$0.95 \times 2000 \text{ gal.} = 1900 \text{ gal.}$

Total selling price

$$C + M = S$$
$$\$25,200 + 0.2S = S$$
$$\$25,200 = 0.8S$$
$$\frac{\$25,200}{0.8} = S$$
$$\$31,500 = S$$

Selling price per item

$$\frac{\$31,500}{1900} = \$16.58$$

17. Number at regular price

$0.8 \times 200 = 160$

Number at reduced price

$0.2 \times 200 = 40$

Total selling price

$$C + M = S$$
$$C + 25\%C = S$$
$$1.25C = S$$
$$1.25 \times \$500 = \$625$$

Amount at reduced price

$40 \times \$2.20 = \$88$

Amount from regular sales

$\$625 - \$88 = \$537$

Regular selling price per item

$$\frac{\$537}{160} = \$3.36$$

19. Number at regular price

$0.85 \times 40 = 34$

Number at reduced price

$0.15 \times 40 = 6$

Total selling price

$$C + M = S$$
$$C + 35\%M = S$$
$$1.35C = S$$
$$1.35 \times \$2200 = \$2970$$

Amount at reduced price

$6 \times \$50.00 = \$300.00$

Amount from regular sales

$\$2970 - \$300 = \$2670$

Regular selling price

$$\frac{\$2670}{34} = \$78.53$$

21. Number at regular price

$0.75 \times 144 = 108$

Number at reduced price

$0.25 \times 144 = 36$

Total selling price

$$C + M = S$$
$$C + 25\%C = S$$
$$1.25C = S$$
$$1.25 \times \$432 = \$540$$

Amount at reduced price

$$36 \times \$2.50 = \$90$$

Amount from regular sales

$$\$540 - \$90 = \$450$$

Regular selling price

$$\frac{\$450}{108} = \$4.17$$

23. Number at regular price

$$0.7 \times 1000 \text{ pr.} = 700 \text{ pr.}$$

Number at reduced price

$$0.3 \times 1000 \text{ pr.} = 300 \text{ pr.}$$

Total selling price

$$C + M = S$$
$$C + 50\%C = S$$
$$1.5C = S$$
$$1.5 \times \$2800 = \$4200$$

Amount at reduced price

$$300 \times \$3.00 = \$900$$

Amount at regular price

$$\$4200 - \$900 = \$3300$$

Regular selling price
$$\frac{\$3300}{700} = \$4.71$$

25. Answers will vary.

27. Number to sell

$$100\% - 15\% = 85\%$$
$$0.85 \times 100 = 85$$

Total cost

$$100 \times \$2.15 = \$215$$

Total selling price

$$C + M = S$$
$$\$215 + 0.4S = S$$
$$\$215 = 0.6S$$

$$\frac{\$215}{0.6} = S$$

$$\$358.33 = S$$

Selling price per item

$$\frac{\$358.33}{85} = \$4.22$$

29. Number to sell

$$75 - 2 = 73 \text{ trees}$$

Total selling price

$$33\frac{1}{3}\% = \frac{1}{3}$$
$$C + M = S$$
$$\$511 + \frac{1}{3}S = S$$
$$\$511 = \frac{2}{3}S$$
$$\frac{3}{2} \cdot \$511 = \frac{3}{2} \cdot \frac{2}{3}S$$
$$\$766.50 = S$$

Selling price per tree

$$\frac{\$766.50}{73} = \$10.50$$

31. 12 gross = 144 dozen
    Number at regular price

$$0.75 \times 144 = 108 \text{ dozen}$$

Number at reduced price

$$0.25 \times 144 = 36 \text{ dozen}$$

Total selling price

$$C + M = S$$
$$C + 100\%C = S$$
$$2C = S$$
$$2 \times \$945 = \$1890$$

Amount at reduced price

$$36 \times \$7.50 = \$270$$

Amount at regular price

$$\$1890 - \$270 = \$1620$$

Regular selling price per dozen

$$\frac{\$1620}{108} = \$15$$

33. Number unsaleable

$$0.2 \times 55 = 11$$

Number sold at regular price

$$0.8 \times 55 = 44$$

Total selling price

$$C + M = S$$
$$C + 30\%C = S$$
$$1.3C = S$$
$$1.3 \times \$10,450 = \$13,585$$

Selling price per stove

$$\frac{\$13,585}{44} = \$308.75$$

35. Total cost

$$500 \times \$80.50 = \$40,250$$

Number at reduced price

$$0.12 \times 500 = 60$$

Number at regular price

$$0.88 \times 500 = 440 \quad .$$

Total selling price

$$C + M = S$$
$$C + 110\%C = S$$
$$2.1C = S$$
$$2.1 \times \$40,250 = \$84,525$$

Amount sold as blemishes

$$60 \times \$105 = \$6300$$

Amount at regular price

$$\$84,525 - \$6300 = \$78,225$$

Regular price per tire

$$\frac{\$78,225}{440} = \$177.78$$

37. Total cost

$$50,000 \times \$10.80 = \$540,000$$

Number at discount

$$0.2 \times 50,000 = 10,000$$

Number at regular price

$$0.8 \times 50,000 = 40,000$$

Amount at reduced price

$$10,000 \times \$11.50 = \$115,000$$

Total selling price

$$C + M = S$$
$$\$540,000 + 0.5S = S$$
$$\$540,000 = 0.5S$$
$$\frac{\$540,000}{0.5} = S$$
$$\$1,080,000 = S$$

Amount at regular price

$$\$1,080,000 - \$115,000 = \$965,000$$

Regular price per book

$$\frac{\$965,000}{40,000} = \$24.13$$

Chapter 9 Review Exercises

1. Markup

$$M = 0.20(\$32)$$
$$M = \$6.40$$

Selling price

$$C + M = S$$
$$\$32 + \$6.40 = S$$
$$\$38.40 = S$$

2. Cost price

$$R \times B = P$$
$$0.25 \times C = \$6.15$$
$$0.25C = \$6.15$$
$$\frac{0.25C}{0.25} = \frac{\$6.15}{0.25}$$
$$C = \$24.60$$

Selling price

$$C + M = S$$
$$\$24.60 + \$6.15 = S$$
$$\$30.75 = S$$

3. Cost price

$$C + M = S$$
$$C + \$73.50 = \$220.50$$
$$C = \$220.50 - \$73.50$$
$$C = \$147$$

Percent of markup on cost

$$R \times B = P$$
$$R \times \$147 = \$73.50$$
$$\$147R = \$73.50$$

$$\frac{\$147R}{\$147} = \frac{\$73.50}{\$147}$$

$$R = 0.5 = 50\%$$

4. Markup

$$C + M = S$$
$$\$108 + M = \$153.90$$
$$M = \$153.90 - \$108$$
$$M = \$45.90$$

Percent of markup on cost

$$R \times B = P$$
$$R \times \$108 = \$45.90$$
$$\$108R = \$45.90$$

$$\frac{108R}{108} = \frac{45.90}{108}$$

$$R = 0.425 = 42.5\%$$

5. Selling price

$$C + M = S$$
$$\$72.32 + 0.2S = S$$
$$\$72.32 + 0.2S - 0.2S = S - 0.2S$$
$$\$72.32 = 0.8S$$

$$\frac{\$72.32}{0.8} = \frac{0.8S}{0.8}$$

$$\$90.40 = S$$

Markup

$$C + M = S$$
$$\$72.32 + M = \$90.40$$
$$M = \$90.40 - \$72.32$$
$$M = \$18.08$$

6. Cost price

$$C + M = S$$
$$C + \$35 = \$140$$
$$C = \$140 - \$35$$
$$C = \$105$$

Percent markup on selling price

$$R \times B = P$$
$$R \times \$140 = \$35$$
$$\$140R = \$35$$

$$\frac{\$140R}{\$140} = \frac{\$35}{\$140}$$

$$R = 0.25 = 25\%$$

7. Selling price

$$33\tfrac{1}{3}\% = \frac{1}{3}$$

$$R \times B = P$$

$$\frac{1}{3} \times S = \$17.35$$

$$\frac{3}{1} \cdot \frac{1}{3}S = 3 \cdot \$17.35$$

$$S = \$52.05$$

Cost price

$$C + M = S$$
$$C + \$17.35 = \$52.05$$
$$C = \$52.05 - \$17.35$$
$$C = \$34.70$$

8. Selling price

$$C + M = S$$
$$\$283.02 + \$177.18 = S$$
$$\$460.20 = S$$

Percent markup on selling price

$$R \times B = P$$
$$R \times \$460.20 = \$177.18$$
$$\$460.20R = \$177.18$$

$$R = \frac{\$177.18}{\$460.20}$$

$$R = 0.3850 = 38.5\%$$

9. Cost price

$$R \times B = P$$
$$25\% \times C = \$480$$
$$0.25C = \$480$$

$$C = \frac{\$480}{0.25}$$

$$C = \$1920$$

Selling price

$$C + M = S$$
$$\$1920 + \$480 = S$$
$$\$2400 = S$$

10. Markup

$$C + M = S$$
$$\$64.50 + M = \$129$$
$$M = \$129 - \$64.50$$
$$M = \$64.50$$

Percent markup on cost

$$R \times B = P$$
$$R \times \$64.50 = \$64.50$$
$$\$64.50R = \$64.50$$
$$R = \frac{\$64.50}{\$64.50}$$
$$R = 1 = 100\%$$

Percent markup on selling price

$$R \times B = P$$
$$R \times \$129 = \$64.50$$
$$\$129R = \$64.50$$
$$R = \frac{\$64.50}{\$129}$$
$$R = 0.5 = 50\%$$

11. Cost price

$$C + M = S$$
$$C + \$3.68 = \$11.68$$
$$C = \$11.68 - \$3.68$$
$$C = \$8$$

Percent markup on cost

$$R \times B = P$$
$$R \times \$8 = \$3.68$$
$$\$8R = \$3.68$$
$$R = \frac{\$3.68}{\$8}$$
$$R = 0.46 = 46\%$$

Percent markup on selling price

$$R \times B = P$$
$$R \times \$11.68 = \$3.68$$
$$\$11.68R = \$3.68$$
$$R = \frac{\$3.68}{\$11.68}$$
$$R = 0.3150 = 31.5\%$$

12. Cost price

$$R \times B = P$$
$$100\% \times C = \$474.28$$
$$C = \$474.28$$

Selling price

$$C + M = S$$
$$\$474.28 + \$474.28 = S$$
$$\$948.56 = S$$

Percent markup on selling price

$$R \times B = P$$
$$R \times \$948.56 = \$474.28$$
$$\$948.56R = \$474.28$$
$$R = \frac{\$474.28}{\$948.56}$$
$$R = 0.5 = 50\%$$

13. $M_C = \dfrac{M_S}{100\% - M_s}$

$$= \frac{20\%}{100\% - 20\%}$$
$$= \frac{20\%}{80\%} = \frac{0.2}{0.8} = \frac{1}{4} = 25\%$$

14. $M_S = \dfrac{M_C}{100\% + M_C}$

$$= \frac{100\%}{100\% + 100\%}$$
$$= \frac{100\%}{200\%} = \frac{1}{2} = 50\%$$

15. $M_C = \dfrac{M_S}{100\% - M_S}$

$$= \frac{15.3\%}{100\% - 15.3\%}$$
$$= \frac{15.3\%}{84.7\%} = \frac{0.153}{0.847}$$
$$= 0.1806 - 18.1\%$$

16. $M_S = \dfrac{M_C}{100\% + M_C}$

$$M_S = \frac{20\%}{100\% + 20\%}$$
$$= \frac{20\%}{120\%} = \frac{0.2}{1.2} = \frac{1}{6} = 16\frac{2}{3}\%$$

17. Total selling price

$$C + M = S$$
$$C + 20\%S = S$$
$$\$324 + 0.2S = S$$
$$\$324 = 0.8S$$
$$\frac{\$324}{0.8} = S$$
$$\$405 = S$$

Selling price per item

$$\frac{\$405}{360 - 36} = \frac{\$405}{324} = \$1.25$$

18. Total selling price

$$C + M = S$$
$$C + 40\%S = S$$
$$\$780 + 0.4S = S$$
$$\$780 = 0.6S$$
$$\frac{\$780}{0.6} = S$$
$$\$1300 = S$$

Selling price per item

$$\frac{\$1300}{52 - 12} = \frac{\$1300}{40} = \$32.50$$

19. Total selling price

$$C + M = S$$
$$C + 30\%S = S$$
$$\$970 + 0.3S = S$$
$$\$970 = 0.7S$$
$$\frac{\$970}{0.7} = S$$
$$\$1385.71 = S$$

Selling price per item

$$\frac{\$1385.71}{(9 \times 12) - 6} = \frac{\$1385.71}{102} = \$13.59$$

20. Total selling price

$$C + M = S$$
$$C + 45\%S = S$$
$$\$12,650 + 0.45S = S$$
$$\$12,650 = 0.55S$$
$$\frac{\$12,650}{0.55} = S$$
$$\$23,000 = S$$

Selling price per item

$$\frac{\$23,000}{1500 - 150} = \frac{\$23,000}{1350} = \$17.04$$

21. Number at regular price

$$0.9 \times 150 = 135$$

Number at reduced price

$$0.1 \times 150 = 15$$

Total selling price

$$C + M = S$$
$$C + 25\%C = S$$
$$1.25C = S$$
$$1.25 \times \$750 = \$937.50$$

Amount at reduced price

$$15 \times \$2.50 = \$37.50$$

Amount from regular sales

$$\$937.50 - \$37.50 = \$900$$

Regular selling price per item

$$\frac{\$900}{135} = \$6.67$$

22. Number at regular price

$$0.8 \times 90 \text{ pr.} = 72 \text{ pr.}$$

Number at reduced price

$$0.2 \times 90 \text{ pr.} = 18 \text{ pr.}$$

Total selling price

$$C + M = S$$
$$C + 40\%C = S$$
$$1.4C = S$$
$$1.4 \times \$135 = \$189$$

Amount at reduced price

$$18 \times \$1.00 = \$18$$

Amount from regular sales

$$\$189 - \$18 = \$171$$

Regular selling price per item

$$\frac{\$171}{72} = \$2.38$$

23. Number at regular price

$$0.75 \times 2 \times 144 = 216$$

Number at reduced price

$$0.25 \times 2 \times 144 = 72$$

Total selling price

$$C + M = S$$
$$C + 20\%C = S$$
$$1.2C = S$$
$$1.2 \times \$1728 = \$2073.60$$

Amount at reduced price

$72 \times \$4.00 = \$288$

Amount from regular sales

$\$2073.60 - \$288 = \$1785.60$

Regular selling price per item

$\dfrac{\$1785.60}{216} = \$8.27$

24. Number at regular price

$0.7 \times 1000 \text{ pr.} = 700 \text{ pr.}$

Number at reduced price

$0.3 \times 1000 \text{ pr.} = 300 \text{ pr.}$

Total selling price

$$C + M = S$$
$$C + 50\%C = S$$
$$1.5C = S$$
$$1.5 \times \$2800 = \$4200$$

Amount at reduced price

$300 \times \$3.00 = \$900$

Amount from regular sales

$\$4200 - \$900 = \$3300$

Regular selling price per item

$\dfrac{\$3300}{700} = \$4.71$

25. Selling price

$$C + M = S$$
$$\$97.50 + 0.35S = S$$
$$\$97.50 = 0.65S$$
$$\$150 = S$$

Selling price per item

$\$150 \div 12 = \$12.50$

26. Markup

$$C + M = S$$
$$\$334.75 + M = \$395$$
$$M = \$395 - \$334.75$$
$$M = \$60.25$$

Markup as a percent of cost

$$R \times B = P$$
$$R \times \$334.75 = \$60.25$$
$$\$334.75R = \$60.25$$

$$R = \dfrac{\$60.25}{\$334.75}$$
$$R = 0.1799 = 18.0\%$$

27. Markup

$$C + M = S$$
$$\$11.25 + M = \$18.75$$
$$M = \$18.75 - \$11.25$$
$$M = \$7.50$$

Percent markup on selling price

$$R \times B = P$$
$$R \times \$18.75 = \$7.50$$
$$\$18.75R = \$7.50$$

$$R = \dfrac{\$7.50}{\$18.75}$$
$$R = 0.4 = 40\%$$

28. Cost per boat

$\$1943.52 \div 12 = \$161.96$

(a) Markup

$$C + M = S$$
$$\$161.96 + M = \$199.95$$
$$M = \$199.95 - \$161.96$$
$$M = \$37.99$$

(b) Percent markup on selling price

$$R \times B = P$$
$$R \times \$199.95 = \$37.99$$
$$\$199.95R = \$37.99$$

$$R = \dfrac{\$37.99}{\$199.95}$$
$$R = 0.1899 = 19.0\%$$

(c) Percent markup on cost

$$R \times B = P$$
$$R \times \$161.96 = \$37.99$$
$$\$161.96R = \$37.99$$

$$R = \dfrac{\$37.99}{\$161.96}$$
$$R = 0.2345 = 23.5\%$$

29. (a) Total selling price

$$(580 \times \$13.95) + (635 \times \$9.95)$$
$$+ (318 \times \$8.95) + (122 \times \$7.95)$$
$$+ (165 \times \$5.00)$$
$$= \$19{,}050.25$$

(b) Markup

$$C + M = S$$
$$\$10{,}010 + M = \$19{,}050.25$$
$$M = \$19{,}050.25 - \$10{,}010$$
$$M = \$9040.25$$

Percent Markup on selling price

$$R \times B = P$$
$$R \times \$19{,}050.25 = \$9040.25$$
$$\$19{,}050.25R = \$9040.25$$
$$R = \frac{\$9040.25}{\$19{,}050.25}$$
$$R = 0.474 = 47\%$$

30. Number at reduced price

$$0.2 \times 200 = 40$$

Number at regular price

$$0.8 \times 200 = 160$$

Total selling price

$$C + M = S$$
$$C + 100\%C = S$$
$$2C = S$$
$$2 \times \$360 = \$720$$

Amount from reduced sales

$$40 \times \$2 = \$80$$

Amount from regular sales

$$\$720 - \$80 = \$640$$

Regular selling price per poster

$$\frac{\$640}{160} = \$4$$

31. (a) Markup

$$C + M = S$$
$$\$3.96 + M = \$4.95$$
$$M = \$4.95 - \$3.96$$
$$M = \$0.99$$

Percent markup on selling price

$$R \times B = P$$
$$R \times \$4.95 = \$0.99$$
$$\$4.95R = \$0.99$$
$$R = \frac{\$0.99}{\$4.95}$$
$$R = 0.2 = 20\%$$

(b) Percent markup on cost

$$R \times B = P$$
$$R \times \$3.96 = \$0.99$$
$$\$3.96R = \$0.99$$
$$\frac{\$3.96R}{\$3.96} = \frac{\$0.99}{\$3.96}$$
$$R = 0.25 = 25\%$$

32. Number sold at reduced price

$$0.15 \times 180 = 27 \text{ dozen}$$

Number sold at regular price

$$0.85 \times 180 = 153 \text{ dozen}$$

Total cost

$$180 \times \$1.93 = \$347.40$$

Total selling price

$$C + M = S$$
$$C + 100\%C = S$$
$$2C = S$$
$$2 \times \$347.40 = \$694.80$$

Amount at reduced price

$$27 \times \$2.40 = \$64.80$$

Amount from regular sales

$$\$694.80 - \$64.80 = \$630$$

Regular price per dozen

$$\frac{\$630}{153} = \$4.12$$

Chapter 9 Summary Exercise

(a) Cost

$$3400 \times \$4.95 = \$16{,}830$$

Trade discount: 30/10/20

$$0.7 \times 0.9 \times 0.8 \times \$16{,}830 = \$8482.32$$

Cash discount

$$0.05 \times \$8482.32 = \$424.12$$
$$\$8482.32 - \$424.12 = \$8058.20$$

(b)
$$
\begin{array}{rcl}
1080 \text{ at } \$3.95 &=& \$ \quad 4266.00 \\
1250 \text{ at } \$2.95 &=& \$ \quad 3687.50 \\
660 \text{ at } \$2.50 &=& \$ \quad 1650.00 \\
230 \text{ at } \$2.00 &=& \$ \quad 460.00 \\
\underline{180 \text{ unsold}} &=& \$ \qquad 0.00 \\
3400 && = \$10,063.50
\end{array}
$$

(c) Net profit

$$\$10,063.50 - \$8058.20 = \$2005.30$$

(d) Percent markup on selling price

$$R \times \$10,063.50 = \$2005.30$$

$$R = \frac{\$2005.30}{\$10,063.50}$$

$$R = 0.1992 = 20\%$$

(e) Percent markup on cost

$$R \times \$8058.20 = \$2005.30$$

$$R = \frac{\$2005.30}{\$8058.20}$$

$$R = 0.2488 = 25\%$$

MARKDOWN AND INVENTORY CONTROL

10.1 Markdown

1. Percent of markdown

$$R \times B = P$$
$$R \times \$860 = \$215$$
$$\$860R = \$215$$
$$R = \frac{\$215}{\$860}$$
$$R = 0.25 = 25\%$$

Reduced price

$$\$860 - \$215 = \$645$$

3. Markdown

$$\$30.80 - \$16.94 = \$13.86$$

Percent of markdown

$$R \times B = P$$
$$R \times \$30.80 = \$13.86$$
$$\$30.80R = \$13.86$$
$$R = \frac{\$13.86}{\$30.80}$$
$$R = 0.45 = 45\%$$

5. Original price

$$R \times B = P$$
$$0.80 \times B = \$5.20$$
$$0.8B = \$5.20$$
$$B = \frac{\$5.20}{0.8}$$
$$B \quad \$6.50$$

Markdown

$$\$6.50 - \$5.20 = \$1.30$$

7. Original price

$$R \times B = P$$
$$0.4 \times B = \$1.08$$
$$0.4B = \$1.08$$
$$B = \frac{\$1.08}{0.4}$$
$$B = \$2.70$$

Reduced price

$$\$2.70 - \$1.08 = \$1.62$$

9. Markdown

$$0.5 \times \$43.50 = \$21.75$$

Reduced price

$$\$43.50 - \$21.75 = \$21.75$$

11. Original price

$$\$175 + \$682 = \$857$$

Percent of markdown

$$R \times B = P$$
$$R \times \$857 = \$175$$
$$\$857R = \$175$$
$$R = \frac{\$175}{\$857}$$
$$R = 0.2042 = 20\%$$

13. Breakeven point

$$\$48 + \$12 = \$60$$

Operating loss

$$\$60 - \$50 = \$10$$

Absolute loss: none
Cost is less than the reduced price.

15. Operating expense

$$\$66 - \$50 = \$16$$

Operating loss

$$\$66 - \$44 = \$22$$

Absolute loss

$$\$50 - \$44 = \$6$$

17. Breakeven point

$$\$310 + \$75 = \$385$$

Reduced price

$$\$385 - \$135 = \$250$$

Absolute loss

$$\$310 - \$250 = \$60$$

19. Cost

$$\$25 + \$4 = \$29$$

Breakeven point

$$\$14 + 25 = \$39$$

Operating expense

$$\$39 - \$29 = \$10$$

21. Answers will vary.

23. Original price

$$R \times B = P$$
$$32\% \times B = \$76.48$$
$$0.32B = \$76.48$$
$$B = \frac{\$76.48}{0.32}$$
$$B = \$239$$

25. Reduced price

$$\$1675 - 1425 = \$250$$

Percent of markdown

$$R \times B = P$$
$$R \times \$1675 = \$250$$
$$\$1675R = \$250$$
$$R = \frac{\$250}{\$1675}$$
$$R = 0.1492 = 15\%$$

27. Operating expenses

$$33\tfrac{1}{3}\% = \frac{1}{3}$$
$$\frac{1}{3} \times \$2211 = \$737$$

Breakeven point

$$\$2211 + \$737 = \$2948$$

Operating loss

$$\$2948 - \$2650 = \$298$$

29. Breakeven point

$$\$153.49 + (0.149 \times \$153.49) = \$176.36$$

Reduced price

$$0.54 \times \$208.78 = \$112.74$$

(a) Operating loss

$$\$176.36 - \$112.74 = \$63.62$$

(b) Absolute loss

$$\$153.49 - \$112.74 = \$40.75$$

(c) Percent of absolute loss

$$R \times B = P$$
$$R \times \$153.49 = \$40.75$$
$$\$153.49R = \$40.75$$
$$R = \frac{\$40.75}{\$153.49}$$
$$R = 0.2654 = 27\%$$

31. (a) Selling price

$$\$190 + (0.2 \times \$190) + (0.15 \times \$190)$$
$$= \$256.50$$

(b) Cost

$$\$190 + (0.2 \times \$190) = \$228$$

Markdown

$$\$256.50 - \$228 = \$28.50$$

Percent of markdown

$$R \times B = P$$
$$R \times \$256.50 = \$28.50$$
$$\$256.50R = \$28.50$$
$$R = \frac{\$28.50}{\$256.50}$$
$$R = 0.111 = 11\%$$

10.2   Average Inventory and Inventory Turnover

1. Average inventory

$$\frac{\$10,603 + \$12,757}{2} = \$11,680$$

3. Average inventory

$$\frac{\$18,300 + \$26,580 + \$23,139}{3} = \$22,673$$

5. Average inventory

$$(\$16,250 + \$20,780 + \$28,720 + \$24,630$$
$$+ \$23,550 + \$34,800 + \$22,770) \div 7$$
$$= \$24,500$$

7. Stock turnover at retail

$$\frac{\$149{,}175}{\$53{,}085} = 2.81$$

Stock turnover at cost

$$\frac{\$75{,}591}{\$26{,}745} = 2.83$$

9. Stock turnover at retail

$$\frac{\$259{,}876}{\$32{,}730} = 7.94$$

Stock turnover at cost

$$\frac{\$178{,}687}{\$22{,}390} = 7.98$$

11. Stock turnover at retail

$$\frac{\$437{,}260}{\$42{,}660} = 10.25$$

Stock turnover at cost

$$\frac{\$270{,}600}{\$26{,}400} = 10.25$$

13. Stock turnover at retail

$$\frac{\$1{,}196{,}222}{\$256{,}700} = 4.66$$

Stock turnover at cost

$$\frac{\$846{,}336}{\$180{,}600} = 4.69$$

15. Answers will vary.

17. Average yearly inventory

$$\frac{\$655{,}974 + \$52{,}476}{12 + 1} = \$54{,}496.15$$

19. Stock turnover at cost

$$\frac{\$85{,}412}{\$15{,}730} = 5.43$$

21. Inventory at cost

$$0.6 \times \$27{,}250 = \$16{,}350$$

Turnover at cost

$$\frac{\$103{,}400}{\$16{,}350} = 6.32$$

23. Average inventory

$$\frac{\$208{,}180 + \$247{,}660 + \$114{,}438}{3} = \$190{,}092.67$$

Turnover at cost

$$\frac{\$2{,}108{,}410}{\$190{,}092.67} = 11.09$$

25. Average inventory at retail

$$(\$33{,}820 + \$46{,}240 + \$39{,}830 + \$52{,}040 + \$48{,}700) \div 5 = \$44{,}126$$

Average inventory at cost

$$\frac{\$44{,}126}{1.35} = \$32{,}685.93$$

Turnover at cost

$$\frac{\$136{,}450}{\$32{,}685.93} = 4.17$$

10.3 Valuation of Inventory

1. Total value of inventory

$$\$208 + \$274 + \$345 = \$827$$

3. Total value of inventory

$$\$79 + \$186 + \$295 = \$560$$

5. (a) Total cost

$$(10 \times \$8) + (25 \times \$9) + (15 \times \$10) = \$455$$

Average cost

$$\frac{\$455}{50} = \$9.10$$

Inventory value (weighted average)

$$20 \times \$9.10 = \$182$$

(b) Inventory value (FIFO)

$$(15 \times \$10) + (5 \times \$9) = \$195$$

(c) Inventory value (LIFO)

$$(10 \times \$8) + (10 \times \$9) = \$170$$

7. (a) Total cost

$$(50 \times \$30.50) + (70 \times \$31.50) + (30 \times \$33.25) + (40 \times \$30.75) = \$5957.50$$

Average cost

$$\frac{\$5957.50}{190} = \$31.36$$

Inventory value (weighted average)

$$75 \times \$31.36 = \$2352$$

(b) Inventory value (FIFO)

$(40 \times \$30.75) + (30 \times \$33.25) + (5 \times \$31.50)$
$= \$2385$

(c) Inventory value (LIFO)

$(50 \times \$30.50) + (25 \times \$31.50) = \$2313$

9. Inventory value

$\$182 + \$210 + \$132 + \$921 + \$325 = \$1770$

11. (a) Total cost

$(24 \times \$1.50) + (40 \times \$1.35) + (48 \times \$1.25)$
$+ (18 \times \$1.60)$
$= \$178.80$

Average cost

$\dfrac{\$178.80}{130} = \$1.38$

Inventory value (weighted average)

$35 \times \$1.38 = \$48$

(b) Inventory value (FIFO)

$(18 \times \$1.60) + (17 \times \$1.25) = \$50$

(c) Inventory value (LIFO)

$(24 \times \$1.50) + (11 \times \$1.35) = \$51$

13. (a) Total cost

$(200 \times \$3.10) + (250 \times \$3.50) + (300 \times \$4.25)$
$+ (280 \times \$4.50)$
$= \$4030$

Average cost

$\dfrac{\$4030}{1030} = 3.91$

Inventory value (weighted average)

$320 \times 3.91 = \$1251$

(b) Inventory value (FIFO)

$(280 \times \$4.50) + (40 \times \$4.25) = \$1430$

(c) Inventory value (FIFO)

$(200 \times \$3.10) + (120 \times \$3.50) = \$1040$

15. (a) Total cost

$(350 \times \$8.25) + (300 \times \$9.50) + (360 \times \$11.45)$
$+ (240 \times \$10.10)$
$= \$12{,}283.50$

Average cost

$\dfrac{\$12{,}283.50}{1250} = \$9.83$

Inventory value (weighted average)

$625 \times \$9.83 = \$6144$

(b) Inventory value (FIFO)

$(240 \times \$10.10) + (360 \times \$11.45) + (25 \times \$9.50)$
$= \$6784$

(c) Inventory value (LIFO)

$(350 \times \$8.25) + (275 \times \$9.50) = \$5500$

17. Merchandise in stock

$\$136{,}000 + \$148{,}000 = \$284{,}000$

Inventory value

$\$284{,}000 - (0.65 \times \$236{,}000) = \$130{,}600$

19. Merchandise for sale at cost

$\$43{,}750 + \$51{,}600 = \$95{,}350$

Merchandise for sale at retail

$\$62{,}500 + \$73{,}800 = \$136{,}300$

Inventory at retail

$\$136{,}300 - \$92{,}500 = \$43{,}800$

Inventory value

$\$43{,}800 \times \dfrac{\$95{,}350}{\$136{,}300} = \$30{,}641$

21. Answers will vary.

Chapter 10 Review Exercises

1. Markdown

$0.3 \times \$96 = \$28.80$

Reduced price

$\$96 - \$28.80 = \$67.20$

2. Original price

$$R \times B = P$$

$$\frac{2}{3} \times B = \$10$$

$$\frac{3}{2} \times \frac{2}{3}B = \frac{3}{2} \times \$10$$

$$B = \$15$$

Markdown

$$\$15 - \$10 = \$5$$

3. Original price

$$R \times B = P$$
$$0.5 \times B = \$2.70$$

$$0.5B = \$2.70$$
$$B = \frac{\$2.70}{0.5}$$
$$B = \$5.40$$

Reduced price

$$\$5.40 - \$2.70 = \$2.70$$

4. Markdown

$$\$2340 - \$1755 = \$585$$

Percent of markdown

$$R \times B = P$$
$$R \times \$2340 = \$585$$
$$\$2340R = \$585$$

$$R = \frac{\$585}{\$2340}$$
$$R = 0.25 = 25\%$$

5. Operating expense

$$\$198 - \$150 = \$48$$

Operating loss

$$\$198 - \$132 = \$66$$

6. Breakeven point

$$\$80 + \$20 = \$100$$

Operating loss

$$\$100 = \$93 - \$7$$

Absolute loss: none

Cost is less than the reduced price.

7. Breakeven point

$$\$78 + \$22 = \$100$$

Reduced price

$$\$100 - \$30 = \$70$$

Absolute loss

$$\$78 - \$70 = \$8$$

8. Breakeven point

$$\$5 + \$1.25 = \$6.25$$

Operating loss

$$\$6.25 - \$5.50 = \$0.75$$

Absolute loss: none

Cost is less than the reduced price.

9. Average inventory

$$\frac{\$44,398 + \$37,704}{2} = \$41,051$$

10. Average inventory

$$\frac{\$316,481 + \$432,185 + \$296,738}{3} = \$348,468$$

11. $(\$77,159 + \$67,305 + \$80,664 + \$95,229$
$+ \$61,702) \div 5$
$= \$76,411.80$

12. $(\$36,502 + \$27,331 + \$28,709 + \$32,153$
$+ \$39,604) \div 5$
$= \$32,859.80$

13. Stock turnover at retail
$$\frac{\$146,528}{\$25,572} = 5.73$$

Stock turnover at cost
$$\frac{\$81,312}{\$14,120} = 5.76$$

14. Stock turnover at retail
$$\frac{\$129,938}{\$16,365} = 7.94$$

Stock turnover at cost
$$\frac{\$89,343}{\$11,195} = 7.98$$

15. Stock turnover at retail
$$\frac{\$598,111}{\$128,350} = 4.66$$

Stock turnover at cost
$$\frac{\$423,168}{\$90,300} = 4.69$$

16. Stock turnover at retail

$$\frac{\$889,884}{\$195,150} = 4.56$$

Stock turnover at cost

$$\frac{\$476,417}{\$102,895} = 4.63$$

17. Inventory value

$$\$122 + \$199 + \$235 = \$556$$

18. Inventory value

$$\$314 + \$422 + \$506 = \$1242$$

19. Inventory value

$$\$795 + \$850 + \$915 + \$1080 = \$3640$$

20. Inventory value

$$\$1283 + \$1398 + \$1564 + \$1772 = \$6017$$

21. Markdown

$$\$1850 - \$1332 = \$518$$

Percent of markdown

$$R \times B = P$$
$$R \times \$1850 = \$518$$
$$\$1850R = \$518$$
$$R = \frac{\$518}{\$1850}$$
$$R = 0.28 = 28\%$$

22. (a) Selling price
$$\$56 + (0.3 \times \$56) + (0.1 \times \$56) = \$78.40$$

(b) Breakeven point

$$\$56 + (0.3 \times \$56) = \$72.80$$

Markdown

$$\$78.40 - \$72.80 = \$5.60$$

$$R \times B = P$$
$$R \times \$78.40 = \$5.60$$
$$\$78.40R = \$5.60$$
$$R = \frac{\$5.60}{\$78.40}$$
$$R = 0.0714 = 7\%$$

23. Average inventory
$$\frac{\$29,332 + \$36,908 + \$31,464}{3} = \$32,568$$

24. Stock turnover at cost

$$\frac{\$91,125}{\$8460} = 10.77$$

25. Average inventory

$$(\$53,820 + \$49,510 + \$60,820 + \$56,380) \div 4$$
$$= \$55,132.50$$

Stock turnover at retail

$$\frac{\$252,077}{\$55,132.50} = 4.57$$

26. (a) Total cost

$$(25 \times \$135) + (40 \times \$165) + (15 \times \$108.50) + (30 \times \$142)$$
$$= \$15,862.50$$

Average cost

$$\frac{\$15,862.50}{110} = \$144.20$$

Inventory value (weighted average)

$$45 \times \$144.20 = \$6489$$

(b) Inventory value (FIFO)

$$(30 \times \$142) + (15 \times \$108.50) = \$5887.50$$

(c) Inventory value (LIFO)

$$(25 \times \$135) + (20 \times \$165) = \$6675$$

27. Selling price

$$C + M = S$$
$$C + 0.5S = S$$
$$C = 0.5S$$
$$C = (0.5)(\$378,000)$$
$$= \$189,000$$

$118,000	inventory (December 31)
+ $186,000	purchases
$304,000	merchandise
− $189,000	cost of goods sold
$115,000	inventory (March 31)

28.

At cost	At retail		
$ 54,000	$ 90,000	inventory	(June 30)
+ $216,000	$360,000	purchases	
$270,000	$450,000		
	− $324,000	net sales	
	$126,000	inventory	(Sept. 30)

September inventory at cost

$$\$126,000 \times \frac{\$270,000}{\$450,000} = \$75,600$$

Chapter 10 Summary Exercise

(a) Cost per pair of skates

$$\frac{\$1950}{24} = \$81.25$$

Selling price per pair of skates

$$C + M = S$$
$$\$81.25 + 0.35S = S$$
$$\$81.25 = 0.65S$$

$$\frac{\$81.25}{0.65} = \frac{0.65S}{0.65}$$

$$\$125 = S$$

(b) Total selling price

$$6 \times (1.0 \times \$125) = \$ \ 750.00$$
$$6 \times (0.75 \times \$125) = \$ \ 562.50$$
$$12 \times (0.5 \times \$125) = \underline{\$ \ 750.00}$$
$$\$2062.50$$

(c) Breakeven point

$$\$1950 + (0.25 \times \$1950) = \$2437.50$$

Operating loss

$$\$2437.50 - \$2062.50 = \$375$$

(d) Absolute loss: none
Selling price is greater than cost.

Cumulative Review Exercises (Chapters 8-10)

1. $280 less 10/20

$$0.9 \times 0.8 \times \$280 = \$201.60$$

2. $375 less 25/10/5

$$0.75 \times 0.9 \times 0.95 \times \$375 = \$240.47$$

3. Net cost equivalent
10/20

$$0.9 \times 0.8 = 0.72$$

Percent form of single discount

$$1 - 0.72 = 0.28 = 28\%$$

4. Net cost equivalent
20/20

$$0.8 \times 0.8 = 0.64$$

Percent form of single discount

$$1 - 0.64 = 0.36 = 36\%$$

5. Net cost equivalent
20/30/5

$$0.8 \times 0.7 \times 0.95 = 0.532$$

Percent form of single discount

$$1 - 0.532 = 0.468 = 46.8\%$$

6. Net cost equivalent
50/40/10

$$0.5 \times 0.6 \times 0.9 = 0.27$$

Percent form of single discount

$$1 - 0.27 = 0.73 = 73\%$$

7. Amount of discount

$$\$740.58 \times 0.02 = \$14.81$$

Amount due

$$\$740.58 - \$14.81 + \$36.80 = \$762.57$$

8. Amount of discount

$$\$874.22 \times 0.04 = \$34.97$$

Amount due

$$\$874.22 - \$34.97 = \$839.25$$

9. Amount of discount

$$\$3788.20 \times 0.03 = \$113.65$$

Amount due

$$\$3788.20 - \$113.65 + \$71.18 = \$3745.73$$

10. Amount of discount

$$\$4692.50 \times 0.02 = \$93.85$$

Amount due

$$\$4692.50 - \$93.85 = \$4598.65$$

11. Cost price

$$C + M = S$$
$$C + \$288.14 = \$576.28$$
$$C = \$576.28 - \$288.14$$
$$C = \$288.14$$

Percent markup on cost

$$R \times B = P$$
$$R \times \$288.14 = \$288.14$$
$$\$288.14R = \$288.14$$
$$R = \frac{\$288.14}{\$288.14}$$
$$R = 1 = 100\%$$

Percent markup on selling price

$$R \times B = P$$
$$R \times \$576.28 = \$288.14$$
$$\$576.28R = \$288.14$$
$$R = \frac{\$288.14}{\$576.28}$$
$$R = 0.5 = 50\%$$

12. Cost

$$33\tfrac{1}{3}\% = \frac{1}{3}$$
$$R \times B = P$$
$$\frac{1}{3} \times C = \$38.22$$
$$\frac{1}{3}C = \$38.22$$
$$3 \cdot \frac{1}{3}C = 3 \times \$38.22$$
$$C = \$114.66$$

Selling price

$$C + M = S$$
$$\$114.66 + \$38.22 = S$$
$$\$152.88 = S$$

Percent markup on selling price

$$R \times B = P$$
$$R \times \$152.88 = \$38.22$$
$$\$152.88R = \$38.22$$
$$R = \frac{\$38.22}{\$152.88}$$
$$R = 0.25 = 25\%$$

13. Number at reduced price

$$30\% \times 1000 = 300$$

Number at regular price

$$70\% \times 1000 = 700$$

Total selling price

$$C + M = S$$
$$C + 50\%C = S$$
$$1.5C = S$$
$$1.5 \times \$1400 = \$2100$$

Amount at reduced price

$$300 \times \$1.50 = \$450$$

Amount from regular sales

$$\$2100 - \$450 = \$1650$$

Regular selling price each

$$\frac{\$1650}{700} = \$2.36$$

14. Number at reduced price

$$10\% \times 150 = 15$$

Number at regular price

$$90\% \times 150 = 135$$

Total selling price

$$C + M = S$$
$$C + 25\%C = S$$
$$1.25C = S$$
$$1.25 \times \$2250 = \$2812.50$$

Amount at reduced price

$$15 \times \$7.50 = \$112.50$$

Amount from regular sales

$$\$2812.50 - \$112.50 = \$2700$$

Regular selling price each

$$\frac{\$2700}{135} = \$20$$

15. Operating expense

$198 - $150 = $48

Operating loss

$198 - $132 = $66

Absolute loss

$150 - $132 = $18

16. Breakeven point

$39 + $11 = $50

Reduced price

$50 - $15 = $35

Absolute loss

$39 - $35 = $4

17. Average inventory

$$\frac{\$74{,}422 + \$58{,}320 + \$61{,}889}{3}$$
$$= \$64{,}877$$

18. Average inventory

$(\$218{,}143 + \$186{,}326 + \$275{,}637$
$+ \$207{,}448 + \$172{,}351) \div 5$
$= \$211{,}981$

19. Stock turnover at retail

$$\frac{\$146{,}528}{\$25{,}572} = 5.73$$

Stock turnover at cost

$$\frac{\$81{,}312}{\$14{,}120} = 5.76$$

20. Stock turnover at retail

$$\frac{\$299{,}056}{\$64{,}175} = 4.66$$

Stock turnover at cost

$$\frac{\$211{,}584}{\$45{,}150} = 4.69$$

21. (a) Wholesaler's price
20/20/20

$0.8 \times 0.8 \times 0.8 \times \$2995 = \$1533.44$

(b) Retailer's price
20/20

$0.8 \times 0.8 \times \$2995 = \$1916.80$

(c) Difference

$1916.80 - $1533.44 = $383.36

22. Amount of discount

$3578 \times 0.02 = $71.56

Amount due

$3578 - $71.56 = $3506.44

23. Cost

$$C + M = S$$
$$C + 25\%C = S$$
$$1.25C = \$5250$$
$$C = \frac{\$5250}{1.25}$$
$$C = \$4200$$

24. Markup

$$C + M = S$$
$$\$15.96 + M = \$19.95$$
$$M = \$19.95 - \$15.96$$
$$M = \$3.99$$

Percent markup on selling price

$$R \times B = P$$
$$R \times \$19.95 = \$3.99$$
$$\$19.95R = \$3.99$$
$$R = \frac{\$3.99}{\$19.95}$$
$$R = 0.2 = 20\%$$

25. Cost per rug

$4499.55 \div 12 = $374.96

(a) Markup

$$C + M = S$$
$$\$374.96 + M = \$499.95$$
$$M = \$499.95 - \$374.96$$
$$M = \$124.99$$

(b) Percent markup on selling price

$$R \times B = P$$
$$R \times \$499.95 = \$124.99$$
$$\$499.95R = \$124.99$$
$$R = \frac{\$124.99}{\$499.95}$$
$$R = 0.2500 = 25\%$$

(c)   Percent markup on cost

$$R \times B = P$$
$$R \times \$374.96 = \$124.99$$
$$\$374.96R = \$124.99$$
$$R = \frac{\$124.99}{\$374.96}$$
$$R = 0.3333 = 33.3\%$$

26. Markdown

$$\$9250 - \$6660 = \$2590$$

Percent of markdown

_____% of $9250 is $2590

$$R \times \$9250 = \$2590$$
$$\$9250R = \$2590$$
$$R = \frac{\$2590}{\$9250}$$
$$R = 0.28 = 28\%$$

27. Reduced price

$$0.6 \times \$399 = \$239.40$$

Breakeven point

$$\$285 + (0.3 \times \$285) = \$370.50$$

(a)   Operating loss

$$\$370.50 - \$239.40 = \$131.10$$

(b)   Absolute loss

$$\$285 - \$239.40 = \$45.60$$

28. Average inventory

$$\frac{\$58,664 + \$73,815 + \$62,938}{3} = \$65,139$$

29. Total cost

$$(30 \times \$18.50) + (25 \times \$21.80)$$
$$+ (20 \times \$20.50) + (30 \times \$21.25)$$
$$= \$2147.50$$

Average cost

$$\frac{\$2147.50}{105} = \$20.45$$

Inventory value (weighted average)

$$55 \times \$20.45 = \$1125$$

30. (a)   Inventory value (FIFO)

$$(30 \times \$21.25) + (20 \times \$20.50)$$
$$+ (5 \times \$21.80) = \$1157$$

(b)   Inventory value (LIFO)

$$(30 \times \$18.50) + (25 \times \$21.80)$$
$$= \$1100$$

SIMPLE INTEREST

11.1 Basics of Simple Interest

For Exercises 1-9, use the formula $I = PRT$.

1. $I = \$6800 \times 0.10 \times \dfrac{5}{4} = \$850$

3. $I = \$12{,}400 \times 0.095 \times \dfrac{8}{12} = \$785.33$

5. $I = \$8250 \times 0.13 \times \dfrac{15}{12} = \$1340.63$

7. $I = \$9874 \times 0.07125 \times \dfrac{11}{12} = \$644.90$

9. $I = \$74{,}986.15 \times 0.1223 \times \dfrac{5}{12} = \$3821.17$

11. $I = PRT$

$I = \$10{,}000 \times 0.10 \times \dfrac{6}{12}$

$I = \$500$

13. $P = \dfrac{I}{RT}$

$P = \dfrac{\$162.50}{0.08 \times \frac{6}{12}}$

$P = \$4062.50$

15. $R = \dfrac{I}{PT}$

$R = \dfrac{\$199.50}{\$3800 \times \frac{9}{12}}$

$R = 0.07 = 7\%$

17. $T = \dfrac{I}{PR}$

$T = \dfrac{\$749}{\$5350 \times 0.12}$

$T = 1.1\overline{6}$ years

$(1.1\overline{6} \times 12 = 14 \text{ months})$

19. $T = \dfrac{I}{PR}$

$T = \dfrac{\$1353}{\$8200 \times 0.11}$

$T = 1.5$ years

21. $P = \dfrac{I}{RT}$

$P = \dfrac{\$245}{0.07 \times \frac{10}{12}}$

$P = \$4200$

23. $T = \dfrac{I}{PR}$

$T = \dfrac{\$77}{\$840 \times 0.10}$

$T = 0.91\overline{6}$ year

$(0.91\overline{6} \times 12 = 11 \text{ months})$

25. $R = \dfrac{I}{PT}$

$R = \dfrac{\$138.60}{\$1890 \times \frac{11}{12}}$

$R = 0.08 = 8\%$

27. $I = PRT$

$I = \$5850 \times 0.10 \times \dfrac{3}{12}$

$I = \$146.25$

29. $P = \dfrac{I}{RT}$

$P = \dfrac{\$114{,}375}{0.10 \times \frac{9}{12}}$

$P = \$1{,}525{,}000$

31. $T = \dfrac{I}{PR}$

$T = \dfrac{\$560}{\$7000 \times 0.12}$

$T = 0.\overline{6}$ year

$(0.\overline{6} \times 12 = 8 \text{ months})$

33. $I = \$5544 - \$5400 = \$144$

$R = \dfrac{I}{PT}$

$R = \dfrac{\$144}{\$5400 \times \frac{8}{12}}$

$R = 0.04 = 4\%$

35. Answers will vary.

37. Credit union

$I = PRT$

$I = \$3200 \times 0.10 \times \dfrac{8}{12}$

$I = \$213.33$

Loan company

$I = PRT$

$I = \$3200 \times 0.18 \times \dfrac{8}{12}$

$I = \$384$

Lupe will save $170.67 ($384 − $213.33).

39. $I = \$7276 - \$6800 = \$476$

$R = \dfrac{I}{PT}$

$R = \dfrac{\$476}{\$6800 \times \frac{6}{12}}$

$R = 0.14 = 14\%$

41. Cash discount

$\$2543 \times 0.02 = \$50.86$

Net price

$\$2543 - \$50.86 = \$2492.14$

$P = \$2492.14,\ R = 12\%,\ T = 60 - 15 = 45$ days

$I = PRT$

$I = \$2492.14 \times 0.12 \times \dfrac{45}{360}$

$I = \$37.38$

Net savings

$\$50.86 - \$37.38 = \$13.48$

11.2   Simple Interest for a Given Number of Days

1. May 7 is day 127;
Dec. 2 is day 336.

Number of days

$336 - 127 = 209$

3. October 27 is day 300;
December 2 is day 336.

Number of days

$336 - 300 = 36$

5. September 2 is day 245;
March 17 is day 76.

The number of days from September 2
to the end of the year is

$$\begin{array}{r} 365 \\ -\ 245 \\ \hline 120. \end{array}$$

Since March 17 is day 76, there are
$120 + 76 = 196$ days from September 2
to March 17.

7. June 14 is day 165.
$165 + 120 = 285$,
which is Oct. 12.

9. February 8 is day 39.
$39 + 30 = 69$,
which is March 10.

11. December 12 is day 346.
$346 + 120 = 466$ and $466 - 365 = 101$,
which is April 11.

13. $I = PRT$

$I = \$1800 \times 0.10 \times \dfrac{90}{360}$

$I = \$45$

15. $I = PRT$

$I = \$3250 \times 0.12 \times \dfrac{150}{360}$

$I = \$162.50$

17. May 9 is day 129;
August 25 is day 237.

Number of days:

$237 - 129 = 108$

$I = PRT$

$I = \$1250 \times 0.11 \times \dfrac{108}{360}$

$I = \$41.25$

19. February 27 is day 58;
August 5 is day 217.
Number of days: $217 - 58 = 159$

$I = PRT$

$I = \$1520 \times 0.10 \times \dfrac{159}{360}$

$I = \$67.13$

21. $I = PRT$

$I = \$3800 \times 0.10 \times \dfrac{100}{365}$

$I = \$104.11$

23. $I = PRT$

$I = \$4600 \times 0.13 \times \dfrac{60}{365}$

$I = \$98.30$

25. July 12 is day 193;
October 12 is day 285.

Number of days:

$285 - 193 = 92$

$I = PRT$

$I = \$6500 \times 0.07 \times \dfrac{92}{365}$

$I = \$114.68$

27. June 24 is day 175;
February 12 is day 43.

Number of days:

$(365 - 175) + 43 = 233$

$I = PRT$

$I = \$2050 \times 0.12 \times \dfrac{233}{365}$

$I = \$157.04$

29. Answers will vary.

31. $I = PRT$

$I = \$1800 \times 0.60 \times \dfrac{40}{365}$

$I = \$118.36$

33. October 23 is day 296;
March 15 is day 74.

Number of days:

$(365 - 296) + 74 = 143$

$I = PRT$

$I = \$56,000 \times 0.12 \times \dfrac{143}{365}$

$I = \$2632.77$

35. June 10 is day 161;
Dec. 25 is day 359.

Number of days:

$359 - 161 = 198$

$I = PRT$

$I = \$48,000 \times 0.1225 \times \dfrac{198}{360}$

$I = \$3234$

37. $I = PRT$

$I = \$80,000 \times 0.11 \times \dfrac{120}{360}$

$I = \$2933.33$

$I = \$80,000 \times 0.20 \times \dfrac{120}{360}$

$I = \$5333.33$

The difference is

$2400 ($5333.33 - $2933.33).

39. **(a)** Exact interest

$\$18,000 \times 0.09 \times \dfrac{120}{365} = \$532.60$

(b) Ordinary interest

$\$18,000 \times 0.09 \times \dfrac{120}{360} = \$540$

(c) Ordinary interest is larger by

$7.40 ($540 - $532.60).

41. **(a)** Exact interest

$\$145,000 \times 0.095 \times \dfrac{240}{365} = \$9057.53$

(b) Ordinary interest

$\$145,000 \times 0.095 \times \dfrac{240}{360} = \$9183.33$

(c) Ordinary interest is larger by

$125.80 ($9183.33 − $9057.53).

43. Exact interest

$$\$160{,}000 \times 0.09625 \times \frac{95}{365} = \$4008.22$$

Ordinary interest

$$\$160{,}000 \times 0.09625 \times \frac{95}{360} = \$4063.89$$

Ordinary interest is larger by

$55.67 ($4063.89 − $4008.22).

45. September 20 is day 263;
May 1 is day 121.
Number of days: $(365 − 263) + 121 = 223$

Exact interest at State Bank

$$\$880{,}000 \times 0.09 \times \frac{223}{365} = \$48{,}387.95$$

Ordinary interest at First Bank

$$\$880{,}000 \times 0.08875 \times \frac{223}{360} = \$48{,}378.61$$

Interest is less at First Bank by

$9.34 ($48,387.95 − $48,378.61).

11.3 Maturity Value

1. $I = PRT$

$$I = \$8500 \times 0.10 \times \frac{9}{12}$$

$$I = \$637.50$$

$M = P + I$
$M = \$8500 + \$637.50$
$M = \$9137.50$

3. $I = PRT$

$$I = \$5800 \times 0.085 \times \frac{140}{360}$$

$$I = \$191.72$$

$M = P + I$
$M = \$5800 + \$191.72$
$M = \$5991.72$

5. $I = PRT$
$I = \$8640 \times 0.10 \times 1.25$
$I = \$1080$

$M = P + I$
$M = \$8640 + \$1080$
$M = \$9720$

7. $I = PRT$

$$I = \$4800 \times 0.10 \times \frac{100}{360}$$

$$I = \$133.33$$

$M = P + I$
$M = \$4800 + \$133.33$
$M = \$4933.33$

9. $T = \dfrac{I}{PR} \times 360$

$$T = \frac{\$133.78}{\$8600 \times 0.08} \times 360$$

$T = 70$ days

$M = P + I$
$M = \$8600 + \$133.78$
$M = \$8733.78$

11. $R = \dfrac{I}{PT}$

$$R = \frac{\$490}{\$14{,}000 \times \frac{120}{360}}$$

$$R = 0.105 = 10.5\%$$

$M = P + I$
$M = \$14{,}000 + \$490$
$M = \$14{,}490$

13. $P = \dfrac{M}{1 + RT}$

$$P = \frac{\$17{,}117.50}{1 + (0.07875 \times \frac{200}{360})}$$

$$P = \$16{,}400$$

$I = M - P$
$I = \$17{,}117.50 - \$16{,}400$
$I = \$717.50$

15. $P = M - I$
$P = \$15,466 - \$666$
$P = \$14,800$

$T = \dfrac{I}{PR} \times 360$

$T = \dfrac{\$666}{\$14,800 \times 0.09} \times 360$

$T = 180$ days

17. $I = M - P$
$I = \$1926 - \$1800$
$I = \$126$

$T = \dfrac{I}{PR} \times 360$

$T = \dfrac{\$126}{\$1800 \times 0.14} \times 360$

$T = 180$ days

19. $P = M - I$
$P = \$45,732.36 - \$3272.92$
$P = \$42,459.44$

$R = \dfrac{I}{PT}$

$R = \dfrac{\$3272.92}{\$42,459.44 \times \frac{185}{360}}$

$R = 0.1500 = 15\%$

21. $I = PRT$

$I = \$12,000 \times 0.105 \times \dfrac{140}{360}$

$I = \$490$

$M = P + I$
$M = \$12,000 + \$490$
$M = \$12,490$

23. $I = M - P$
$I = \$12,912.36 - \$12,400$
$I = \$512.36$

$T = \dfrac{I}{PR} \times 360$

$T = \dfrac{\$512.36}{\$12,400 \times 0.10625} \times 360$

$T = 140$ days

25. $I = M - P$
$I = \$1000 - \$930$
$I = \$70$

$R = \dfrac{I}{PT}$

$R = \dfrac{\$70}{\$930 \times \frac{150}{360}}$

$R = 0.1806 = 18\%$

27. $P = \dfrac{M}{1 + RT}$

$P = \dfrac{\$854,166.67}{1 + (0.10 \times \frac{150}{360})}$

$P = \$820,000$

$I = M - P$
$I = \$854,166.67 - \$820,000$
$I = \$34,166.67$

29. Answers will vary.

11.4   Inflation and the Time Value of Money

1. Future value

$M = P(1 + RT)$

$M = \$4800(1 + 0.09 \times \dfrac{110}{360})$

$M = \$4932$

3. Future value

$M = P(1 + RT)$

$M = \$10,500(1 + 0.11 \times \dfrac{10}{12})$

$M = \$11,462.50$

5. Future value

$M = P(1 + RT)$

$M = \$4100(1 + 0.08375 \times \dfrac{5}{4})$

$M = \$4529.22$

7. Present value

$$P = \frac{M}{(1 + RT)}$$

$$P = \frac{\$2440}{(1 + 0.06 \times \frac{100}{360})}$$

$$P = \$2400$$

9. Present value

$$P = \frac{M}{(1 + RT)}$$

$$P = \frac{\$8867.25}{(1 + 0.11125 \times \frac{6}{12})}$$

$$P = \$8400$$

11. Present value

$$P = \frac{M}{(1 + RT)}$$

$$P = \frac{\$9275.02}{(1 + 0.05875 \times \frac{9}{8})}$$

$$P = \$8700$$

13. Answers will vary.

15. Future value
Time: Aug. 4 is day 216;
Dec. 31 is day 365
$365 - 216 = 149$

$$M = P(1 + RT)$$

$$M = \$6,500,000(1 + 0.095 \times \frac{149}{360})$$

$$M = \$6,755,576.39$$

17. Interest earned by Chase Bank

$$I = PRT$$

$$I = \$8,500,000 \times 0.095 \times \frac{90}{360}$$

$$I = \$201,875$$

Interest paid to the depositors

$$I = PRT$$

$$I = \$8,500,000 \times 0.0675 \times \frac{90}{360}$$

$$I = \$143,437.50$$

The difference is

$\$58,437.50$ ($\$201,875 - \$143,437.50$).

19. Maturity value of loan

$$M = P(1 + RT)$$

$$M = \$1200(1 + 0.09 \times \frac{120}{360})$$

$$M = \$1236$$

Present value

$$P = \frac{M}{(1 + RT)}$$

$$P = \frac{\$1236}{1 + (0.06 \times \frac{120}{360})}$$

$$P = \$1211.76$$

21. Maturity value of loan

$$M = P(1 + RT)$$

$$M = \$6980(1 + 0.11 \times \frac{214}{360})$$

$$M = \$7436.41$$

Present value on July 9
May 24 is day 144.
$144 + 214 = 358$ which is
December 24, the maturity date.
July 9 is day 190.

Time: $358 - 190 = 168$

$$P = \frac{M}{(1 + RT)}$$

$$P = \frac{\$7436.41}{1 + (0.10 \times \frac{168}{360})}$$

$$P = \$7104.85$$

23. Future value

$$M = P(1 + RT)$$

$$M = \$3200(1 + 0.0775 \times \frac{14}{12})$$

$$M = \$3489.33$$

25. Increase in wages of 3%

$0.03 \times \$26,500 = \$795$

Inflation of 4.5%

$0.045 \times \$26,500 = \$1192.50$

Net loss in purchasing power

$\$1192.50 - \$795 = \$397.50$

27. Present value of \$5600 today is \$5600.
Present value of \$5800 in 150 days is

$$\frac{\$5800}{1 + (0.12 \times \frac{150}{360})} = \$5523.81.$$

Present value of \$6000 in 210 days is

$$\frac{\$6000}{1 + (0.12 \times \frac{210}{360})} = \$5607.48.$$

The lowest bid is \$5800 in 150 days.

29. Answers will vary.

Chapter 11 Review Exercises

1. $I = PRT$
$I = \$8400 \times 0.09 \times 2$
$I = \$1512$

2. $I = PRT$

$I = \$9600 \times 0.105 \times \dfrac{10}{12}$

$I = \$840$

3. $P = \dfrac{I}{RT}$

$P = \dfrac{\$696.80}{0.12 \times \frac{8}{12}}$

$P = \$8710$

4. $P = \dfrac{I}{RT}$

$P = \dfrac{\$144}{0.08 \times \frac{9}{12}}$

$P = \$2400$

5. $R = \dfrac{I}{PT}$

$R = \dfrac{\$810}{\$12,000 \times \frac{9}{12}}$

$R = 0.09 = 9\%$

6. $R = \dfrac{I}{PT}$

$R = \dfrac{\$750}{\$8000 \times \frac{15}{12}}$

$R = 0.075 = 7.5\%$

7. $T = \dfrac{I}{PR}$

$T = \dfrac{\$540}{\$12,000 \times 0.06} \times 12$

$T = 9$ months

8. $T = \dfrac{I}{PR}$

$T = \dfrac{\$1600}{\$8000 \times 0.08} \times 12$

$T = 30$ months

9. April 15 is day 105;
August 7 is day 219.
Number of days:
$219 - 105 = 114$

10. July 12 is day 193:
November 4 is day 308.
Number of days:

$308 - 193 = 115$

11. $I = PRT$

$I = \$10,500 \times 0.075 \times \dfrac{120}{365}$

$I = \$258.90$

12. $I = PRT$

$I = \$8400 \times 0.0975 \times \dfrac{150}{365}$

$I = \$336.58$

13. July 12 is day 193;
November 30 is day 334.
Number of days:

$334 - 193 = 141$

$I = PRT$

$I = \$7200 \times 0.08 \times \dfrac{141}{365}$

$I = \$222.51$

14. February 4 is day 35;
May 9 is day 129.
Number of days:

$129 - 35 = 94$

$I = PRT$

$I = \$6800 \times 0.075 \times \dfrac{94}{365}$

$I = \$131.34$

15. $I = PRT$

$$I = \$7400 \times 0.07 \times \frac{30}{360}$$

$$I = \$43.17$$

16. $I = PRT$

$$I = \$52,000 \times 0.102 \times \frac{220}{360}$$

$$I = \$3241.33$$

17. $P = \dfrac{I}{RT}$

$$P = \frac{\$203.33}{0.075 \times \frac{80}{360}}$$

$$P = \$12,199.80$$

18. $P = \dfrac{I}{RT}$

$$P = \frac{\$340}{0.11 \times \frac{180}{360}}$$

$$P = \$6181.82$$

19. $R = \dfrac{I}{PT}$

$$R = \frac{\$70}{\$6000 \times \frac{60}{360}}$$

$$R = 0.07 = 7\%$$

20. $R = \dfrac{I}{PT}$

$$R = \frac{\$231}{\$8400 \times \frac{120}{360}}$$

$$R = 0.0825 = 8\tfrac{1}{4}\%$$

21. $T = \dfrac{I}{PR} \times 360$

$$T = \frac{\$78}{\$7800 \times 0.09} \times 360$$

$$T = 40 \text{ days}$$

22. $T = \dfrac{I}{PR} \times 360$

$$T = \frac{\$124.03}{\$4900 \times 0.10125} \times 360$$

$$T = 90 \text{ days}$$

23. $M = P(1 + RT)$

$$M = \$5500(1 + 0.08 \times 1.75)$$

$$M = \$6270$$

24. $M = P(1 + RT)$

$$M = \$6900(1 + 0.072 \times \frac{9}{12})$$

$$M = \$7272.60$$

25. $P = \dfrac{M}{1 + RT}$

$$P = \frac{\$12,180}{1 + (0.09 \times \frac{60}{360})}$$

$$P = \$12,000$$

26. $P = \dfrac{M}{1 + RT}$

$$P = \frac{\$6752.78}{1 + (0.10 \times \frac{140}{360})}$$

$$P = \$6500$$

27. $I = M - P$

$$I = \$8120 - \$8000 = \$120$$

$$R = \frac{I}{PT}$$

$$R = \frac{\$120}{\$8000 \times \frac{60}{360}}$$

$$R = 0.09 = 9\%$$

28. $I = M - P$

$$I = \$17,537.50 - \$15,250$$

$$I = \$2287.50$$

$$R = \frac{I}{PT}$$

$$R = \frac{\$2,287.50}{\$15,250 \times \frac{15}{12}}$$

$$R = 0.12 = 12\%$$

29. Present value

$$P = \frac{M}{1 + RT}$$

$$P = \frac{\$6600}{1 + (0.085 \times \frac{120}{360})}$$

$$P = \$6418.15$$

30. Present value

$$P = \frac{M}{1 + RT}$$

$$P = \frac{\$12,000}{1 + (0.12 \times \frac{100}{360})}$$

$$P = \$11,612.90$$

31. Maturity value

$$M = P(1 + RT)$$

$$M = \$8000(1 + 0.095 \times \frac{10}{12})$$

$$M = \$8633.33$$

Present value

$$P = \frac{M}{1 + RT}$$

$$P = \frac{\$8633.33}{1 + (0.08 \times \frac{10}{12})}$$

$$P = \$8093.75$$

32. Maturity value

$$M = P(1 + RT)$$

$$M = \$20,000(1 + 0.07 \times \frac{15}{12})$$

$$M = \$21,750$$

Present value

$$P = \frac{M}{1 + RT}$$

$$P = \frac{\$21,750}{1 + (0.09 \times \frac{15}{12})}$$

$$P = \$19,550.56$$

33. Maturity value

$$M = P(1 + RT)$$

$$M = \$800(1 + 0.10 \times \frac{60}{360})$$

$$M = \$813.33$$

Present value on March 15
February 1 is day 32.
$32 + 60 = 92$, which is April 2, the due date.
March 15 is day 74.
Time: $92 - 74 = 18$

$$P = \frac{M}{1 + RT}$$

$$P = \frac{\$813.33}{1 + (0.09 \times \frac{18}{360})}$$

$$P = \$809.69$$

34. Maturity value

$$M = P(1 + RT)$$

$$M = \$15,000(1 + 0.08 \times \frac{300}{360})$$

$$M = \$16,000$$

Present value on October 15
April 1 is day 91.
$91 + 300 = 391$, which is day 26 $(391 - 365)$ or
January 26.
October 15 is day 288.
Time: $(365 - 288) + 26 = 103$ days

$$P = \frac{M}{1 + RT}$$

$$P = \frac{\$16,000}{1 + (0.09 \times \frac{103}{360})}$$

$$P = \$15,598.34$$

35. $I = PRT$

$$I = \$4500 \times 0.075 \times \frac{90}{360}$$

$$I = \$84.38$$

36. $M = P(1 + RT)$

$$M = \$18,600 \left(1 + 0.09 \times \frac{150}{360}\right)$$

$$M = \$19,297.50$$

37. $I = M - P$
$I = \$15,567.71 - \$14,700$
$I = \$867.71$

$$T = \frac{I}{PR} \times 360$$

$$T = \frac{\$867.71}{\$14,700 \times 0.085} \times 360$$

$$T = 250 \text{ days}$$

38. $P = \dfrac{M}{1 + RT}$

$$P = \frac{\$3816}{1 + (0.09 \times \frac{8}{12})}$$

$$P = \$3600$$

39. $I = M - P$

$I = \$12,000 - \$11,200$

$I = \$800$

$R = \dfrac{I}{PT}$

$R = \dfrac{\$800}{\$11,200 \times \frac{270}{360}}$

$R = 0.0952 = 9.5\%$

40. $I = M - P$

$I = \$27,500 - \$26,000$

$I = \$1500$

$R = \dfrac{I}{PT}$

$R = \dfrac{\$1500}{\$26,000 \times \frac{120}{360}}$

$R = 0.1730 = 17.3\%$

41. $I = M - P$

$I = \$18,348 - \$17,600$

$I = \$748$

$T = \dfrac{I}{PR} \times 360$

$T = \dfrac{\$748}{\$17,600 \times 0.17} \times 360$

$T = 90$ days

42. $M = P(1 + RT)$

$M = \$9100(1 + 0.11 \times \dfrac{85}{360})$

$M = \$9336.35$

$P = \dfrac{M}{1 + RT}$

$P = \dfrac{\$9336.35}{1 + (0.097 \times \frac{85}{360})}$

$P = \$9127.31$

43. $M = P(1 + RT)$

$M = \$19,250(1 + 0.12 \times \dfrac{75}{360})$

$M = \$19,731.25$

Present value on November 27

October 15 is day 288.

$288 + 75 = 363$ which is December 29.

November 27 is day 331.

Time: $363 - 331 = 32$ days

$P = \dfrac{M}{1 + RT}$

$P = \dfrac{\$19,731.25}{1 + (0.117 \times \frac{32}{360})}$

$P = \$19,528.16$

44. $M = P(1 + RT)$

$M = \$11,800(1 + 0.09 \times \dfrac{153}{360})$

$M = \$12,251.35$

Present value on September 20

July 12 is day 193.

$193 + 153 = 346$ which is December 12, the maturity date.

September 20 is day 263.

Time: $346 - 263 = 83$ days

$P = \dfrac{M}{1 + RT}$

$P = \dfrac{\$12,251.35}{1 + (0.11 \times \frac{83}{360})}$

$P = \$11,948.33$

45. Increase in salary of 6%

$0.06 \times \$28,400 = \$1704$

Inflation of 3%

$0.03 \times \$28,400 = \$852$

Net gain in purchasing power

$\$1704 - \$852 = \$852$

46. Inflation of 6%

$0.06 \times \$46,850 = \$2811$

The net loss in purchasing power is $\$2811$.

47. Answers will vary.

48. Answers will vary.

Chapter 11 Summary Exercise

Loan Number 1

Maturity date

March 3 + 6 months = September 3

Maturity value

$$M = P(1 + RT)$$

$$M = \$12,000 \left(1 + 0.10 \times \frac{6}{12}\right)$$

$$M = \$12,600$$

Simple Interest

$$I = M - P$$
$$I = \$12,600 - \$12,000$$
$$I = \$600$$

Loan Number 2

Maturity date

April 17 is day 107;
107 + 90 = 197, which is July 16

Maturity value

$$M = P(1 + RT)$$

$$M = \$6200 \left(1 + 0.12 \times \frac{90}{360}\right)$$

$$M = \$6386$$

Simple Interest

$$I = M - P$$
$$I = \$6386 - \$6200$$
$$I = \$186$$

Loan Number 3

Maturity date

July 9 is day 190;
190 + 200 = 390;
390 − 365 = 25, which is Jan. 25
of the following year.

Maturity value

$$M = P(1 + RT)$$

$$M = \$18,400 \left(1 + 0.095 \times \frac{200}{360}\right)$$

$$M = \$19,371.11$$

Simple interest

$$I = M - P$$
$$I = \$19,371.11 - \$18,400$$
$$I = \$971.11$$

Totals

Maturity Value

$12,600.00
$ 6386.00
+ $19,371.11
$38,357.11

Simple Interest

$600.00
$186.00
+ $971.11
$1757.11

Chapter 12

NOTES AND BANK DISCOUNT

12.1   Simple Interest Notes

1. Helen Spence   **3.** Leta Clendenen

5. 90 days

7. Oct. 27 is day 300;
$300 + 90 = 390$
$390 - 365 = 25$ which is
Jan. 25 of the following year.

9. Six months from June 20 is December 20.

11. Six months from December 31 is June 30.
(June has only 30 days).

13. Seventy days from January 6 is day 76,
which is March 17.

15. 150 days from November 24
November 24 is day 328.
$365 - 328 = 37$
$150 - 37 = 113$, which is April 23 of the
following year.

17. June 12 is day 163.
$163 + 150 = 313$, which is November 9,
the due date.

Maturity value

$M = P(1 + RT)$

$M = \$6000 \left(1 + 0.09 \times \dfrac{150}{360} \right)$

$M = \$6225$

19. August 14 is day 226.
$(365 - 226) = 139$
$300 - 139 = 161$ which is June 10, the due date.

Maturity value

$M = P(1 + RT)$

$M = \$5000 \left(1 + 0.085 \times \dfrac{300}{360} \right)$

$M = \$5354.17$

21. $P = \dfrac{M}{1 + RT}$

$P = \dfrac{\$8820}{1 + 0.10 \times \frac{180}{360}}$

$P = \$8400$

23. $I = M - P$
$I = \$3696 - \$3600$
$I = \$96$

$T = \dfrac{I}{PR} \times 360$

$T = \dfrac{\$96}{\$3600 \times 0.08} \times 360$

$T = 120$ days

25. $I = M - P$
$I = \$8759.33 - \$8400$
$I = \$359.33$

$R = \dfrac{I}{PT}$

$R = \dfrac{\$359.33}{\$8400 \times \frac{140}{360}}$

$R = 0.1099 = 11.0\%$

27. $M = P(1 + RT)$

$M = \$12{,}240 \left(1 + 0.105 \times \dfrac{200}{360} \right)$

$M = \$12{,}954$

29. Answers will vary.

31. **(a)** October 1 is day 274.
$274 + 100 = 374$
$374 - 365 = 9$, which is January 9,
the due date.

(b) $M = P(1 + RT)$

$M = \$1125 \left(1 + 0.0825 \times \dfrac{100}{360} \right)$

$M = \$1150.78$

33. (a) Nine months from March 13 is December 13, which is the due date.

(b) $M = P(1 + RT)$

$$M = \$21,000 \left(1 + 0.11 \times \frac{9}{12}\right)$$

$$M = \$22,732.50$$

35. $P = \dfrac{M}{1 + RT}$

$$P = \frac{\$46,466.67}{1 + \left(0.105 \times \frac{320}{360}\right)}$$

$$P = \$42,500$$

37. (a) The loan was made 7 months before November 7 on April 7 of the prior year.

(b) Face value

$$P = \frac{M}{1 + RT}$$

$$P = \frac{\$30,222.33}{1 + \left(0.11 \times \frac{7}{12}\right)}$$

$$P = \$28,400$$

39. $T = \dfrac{I}{PR} \times 360$

$$T = \frac{\$120}{\$4000 \times 0.12} \times 360$$

$$T = 90 \text{ days}$$

41. $I = M - P$
$I = \$12,320 - \$12,000$
$I = \$320$

$$R = \frac{I}{PT}$$

$$R = \frac{\$320}{\$12,000 \times \frac{120}{360}}$$

$$R = 0.08 = 8\%$$

43. $M = P(1 + RT)$

$$M = \$6800\left(1 + 0.12 \times \frac{90}{360}\right)$$

$$M = \$7004$$

$$\$7004 - \$5000 = \$2004$$

12.2   Simple Discount Notes

1. $B = MDT$

$$B = \$4600 \times 0.09 \times \frac{90}{360}$$

$$B = \$103.50$$

$P = M - B$
$P = \$4600 - \$103.50$
$P = \$4496.50$

3. $B = MDT$

$$B = \$6200 \times 0.1425 \times \frac{180}{360}$$

$$B = \$441.75$$

$P = M - B$
$P = \$6200 - \$441.75$
$P = \$5758.25$

5. $B = MDT$

$$B = \$8400 \times 0.095 \times \frac{30}{360}$$

$$B = \$66.50$$

$P = M - B$
$P = \$8400 - \$66.50$
$P = \$8333.50$

7. Jan. 3 is day 3.
$3 + 100 = 103$,
which is April 13

$$B = MDT$$

$$B = \$8400 \times 0.11 \times \frac{100}{360}$$

$$B = \$256.67$$

$P = M - B$
$P = \$8400 - \$256.67$
$P = \$8143.33$

9. August 21 is day 233.
$365 - 233 = 132$
$180 - 132 = 48$, which is February 17, the due date.

$B = MDT$

$B = \$12,000 \times 0.095 \times \dfrac{180}{360}$

$B = \$570$

$P = M - B$
$P = \$12,000 - \$570$
$P = \$11,430$

11. February 4 is day 35.
$35 + 180 = 215$, which is August 3,
the due date.

$B = MDT$

$B = \$24,000 \times 0.095 \times \dfrac{180}{360}$

$B = \$1140$

$P = M - B$
$P = \$24,000 - \$1140$
$P = \$22,860$

13. $D = \dfrac{B}{MT}$

$D = \dfrac{\$660}{\$14,400 \times \frac{150}{360}}$

$D = 0.11 = 11\%$

January 4 is day 4.
$(365 - 150) + 4 = 219$, which is August 7,
the date made.

$P = M - B$
$P = \$14,400 - \$660$
$P = \$13,740$

15. $M = P + B$
$M = \$7092 + \$108$
$M = \$7200$

$D = \dfrac{B}{MT}$

$D = \dfrac{\$108}{\$7200 \times \frac{90}{360}}$

$D = 0.06 = 6\%$

November 12 is day 316.
$365 - 316 = 49$
$90 - 49 = 41$, which is February 10,
the due date.

17. $B = MDT$

$B = \$6000 \times 0.11 \times \dfrac{120}{360}$

$B = \$220$

$P = M - B$
$P = \$6000 - \$220$
$P = \$5780$

19. $T = \dfrac{B}{MD} \times 360$

$T = \dfrac{\$210}{\$6000 \times 0.105} \times 360$

$T = 120$ days

21. $R = \dfrac{I}{PT}$

$R = \dfrac{\$291.67}{\$14,000 \times \frac{60}{360}}$

$R = 0.1250 = 12.5\%$

23. $M = \dfrac{P}{1 - DT}$

$M = \dfrac{\$3200}{1 - (0.10 \times \frac{140}{360})}$

$M = \$3329.48$

25. $B = MDT$

$B = \$4200 \times 0.11 \times \dfrac{10}{12}$

$B = \$385$

$P = M - B$
$P = \$4200 - \$385$
$P = \$3815$

$R = \dfrac{I}{PT}$

$R = \dfrac{\$385}{\$3815 \times \frac{10}{12}}$

$R = 0.1211 = 12.1\%$

27. $M = \dfrac{P}{1 - DT}$

$M = \dfrac{\$165,000}{1 - (0.15 \times \frac{30}{360})}$

$M = \$167,088.61$

$B = M - P$
$B = \$167,088.61 - \$165,000$
$B = \$2088.61$

$R = \dfrac{I}{PT}$

$R = \dfrac{\$2088.61}{\$165,000 \times \frac{30}{360}}$

$R = 0.1518 = 15.2\%$

29. (a) $B = MDT$

$B = \$25,000,000 \times 0.06 \times \dfrac{13}{52}$

$B = \$375,000$

$P = M - B$
$P = \$25,000,000 - \$375,000$
$P = \$24,625,000$

(b) The maturity value is \$25,000,000.

(c) $I = PRT$

$I = \$25,000,000 \times 0.06 \times \dfrac{13}{52}$

$I = \$375,000$

(d) $R = \dfrac{I}{PT}$

$R = \dfrac{\$375,000}{\$24,625,000 \times \frac{13}{52}}$

$R = 0.06091 = 6.09\%$

31. Answers will vary.

33. Answers will vary.

12.3 Comparing Simple Interest and Simple Discount

1. Simple interest rate

$R = \dfrac{D}{1 - DT}$

$R = \dfrac{0.09}{1 - (0.09 \times \frac{120}{360})}$

$R = 0.09278 = 9.28\%$

3. Simple interest rate

$R = \dfrac{D}{1 - DT}$

$R = \dfrac{0.1425}{1 - (0.1425 \times \frac{200}{360})}$

$R = 0.15475 = 15.48\%$

5. $D = \dfrac{R}{1 + RT}$

$D = \dfrac{0.12}{1 + (0.12 \times \frac{220}{360})}$

$D = 0.11180 = 11.18\%$

7. $D = \dfrac{R}{1 + RT}$

$D = \dfrac{0.10}{1 + (0.10 \times \frac{100}{360})}$

$D = 0.09729 = 9.73\%$

9. Answers will vary.

11. Simple interest rate: 13%

$R = \dfrac{D}{1 - DT}$

$R = \dfrac{0.128}{1 - (0.128 \times \frac{90}{360})}$

$R = 0.13223 = 13.22\%$

Simple discount rate: 12.8%

$D = \dfrac{R}{1 + RT}$

$D = \dfrac{0.13}{1 + (0.13 \times \frac{90}{360})}$

$D = 0.12590 = 12.59\%$

Simple interest rate is better because 13% is less than 13.22%, while 12.8% is more than 12.59%.

13. Simple interest rate

$$M = P(1 + RT)$$

$$M = \$4500 \left(1 + 0.12 \times \frac{180}{360}\right)$$

$$M = \$4770$$

Simple discount note

$$M = \frac{P}{1 - DT}$$

$$M = \frac{\$4500}{1 - \left(0.12 \times \frac{180}{360}\right)}$$

$$M = \$4787.23$$

$17.23 ($4787.23−$4770) less interest is paid with a simple interest note.

15. Answers will vary.

12.4   Discounting a Note

1. June 28 is day 179.
179 + 120 = 299, which is October 26.
August 4 is day 216
Discount period

299 − 216 = 83 days

3. August 4 is day 216.
365 − 216 = 149
220 − 149 = 71, which is March 12.
January 12 is day 12.

Discount period

71 − 12 = 59 days

5. $B = MDT$

$$B = \$4800 \times 0.085 \times \frac{90}{360}$$

$$B = \$102$$

$$P = M - B$$
$$P = \$4800 - \$102$$
$$P = \$4698$$

7. $B = MDT$

$$B = \$15,000 \times 0.12 \times \frac{180}{360}$$

$$B = \$900$$

$$P = M - B$$
$$P = \$15,000 - \$900$$
$$P = \$14,100$$

9. January 9 is day 9; March 9 is day 68.
9 + 120 = 129
129 − 68 = 61 day discount period

$$M = P(1 + RT)$$

$$M = \$3500 \left(1 + 0.09 \times \frac{120}{360}\right)$$

$$M = \$3605$$

$$B = MDT$$

$$B = \$3605 \times 0.12 \times \frac{61}{360}$$

$$B = \$73.30 \quad Discount$$

$$P = M - B$$
$$P = \$3605 - \$73.30$$
$$P = \$3531.70 \quad Proceeds$$

11. May 5 is day 125;
July 6 is day 187.
125 + 130 = 255
255 − 187 = 68 day discount period

$$M = P(1 + RT)$$

$$M = \$6800 \left(1 + 0.105 \times \frac{130}{360}\right)$$

$$M = \$7057.83$$

$$B = MDT$$

$$B = \$7057.83 \times 0.12 \times \frac{68}{360}$$

$$B = \$159.98 \quad Discount$$

$$P = M - B$$
$$P = \$7057.83 - \$159.98$$
$$P = \$6897.85 \quad Proceeds$$

13. September 18 is day 261; February 4 is day 35.

$261 + 220 = 481$

$481 - (365 + 35) = 81$ day discount period

$M = P(1 + RT)$

$M = \$10{,}000 \left(1 + 0.10 \times \dfrac{220}{360}\right)$

$M = \$10{,}611.11$

$B = MDT$

$B = \$10{,}611.11 \times 0.12 \times \dfrac{81}{360}$

$B = \$286.50$ *Discount*

$P = M - B$

$P = \$10{,}611.11 - \$286.50$

$P = \$10{,}324.61$ *Proceeds*

15. (a) Discount period

May 10 is day 130; July 26 is day 207.

$130 + 190 = 320$

$320 - 207 = 113$ days

(b) $M = P(1 + RT)$

$M = \$12{,}000 \left(1 + 0.10 \times \dfrac{190}{360}\right)$

$M = \$12{,}633.33$

Discount

$B = MDT$

$B = \$12{,}633.33 \times 0.12 \times \dfrac{113}{360}$

$B = \$475.86$

(c) Proceeds

$P = M - B$

$P = \$12{,}633.33 - \$475.86$

$P = \$12{,}157.47$

17. $M = P(1 + RT)$

$M = \$4500 \left(1 + 0.10 \times \dfrac{150}{360}\right)$

$M = \$4687.50$

(a) Bank Discount

$B = MDT$

$B = \$4687.50 \times 0.14 \times \dfrac{120}{360}$

$B = \$218.75$

(b) Proceeds

$P = M - B$

$P = \$4687.50 - \$218.75$

$P = \$4468.75$

19. (a) Proceeds to customer

$B = MDT$

$B = \$24{,}000 \times 0.105 \times \dfrac{150}{360}$

$B = \$1050$

$P = M - B$

$P = \$24{,}000 - \$1050$

$P = \$22{,}950$

(b) Proceeds to Citizen's First Bank

$B = MDT$

$B = \$24{,}000 \times 0.11 \times \dfrac{90}{360}$

$B = \$660$

$P = M - B$

$P = \$24{,}000 - \$660$

$P = \$23{,}340$

(c) Interest earned by Citizen's First Bank

$\$23{,}340 - \$22{,}950 = \$390$

21. (a) December 3 is day 337;

May 4 is day 124.

$(365 - 337) + 124 = 152$ days

$B = MDT$

$B = \$36{,}500 \times 0.12 \times \dfrac{280}{360}$

$B = \$3406.67$

Proceeds to the maker of note

$P = M - B$

$P = \$36{,}500 - \$3406.67$

$P = \$33{,}093.33$

(b) $280 - 152 = 128$ day discount period

$$B = MDT$$

$$B = \$36,500 \times 0.10 \times \frac{128}{360}$$

$$B = \$1297.78$$

Proceeds to State Bank

$$P = M - B$$
$$P = \$36,500 - \$1297.78$$
$$P = \$35,202.22$$

(c) Actual interest earned

$$\$35,202.22 - \$33,093.33 = \$2108.89$$

23. (a) Maturity value

$$M = P(1 + RT)$$

$$M = \$78,000(1 + 0.12 \times \frac{150}{360})$$

$$M = \$81,900$$

(b) November 20 is day 324; 14 days after January 23 is February 6, which is day 37.

$$(365 - 324) + 37 = 78$$
$$150 - 78 = 72 \text{ day discount period}$$

$$B = MDT$$

$$B = \$81,900 \times 0.135 \times \frac{72}{360}$$

$$B = \$2211.30$$

Proceeds to Union State Bank

$$P = M - B$$
$$P = \$81,900 - \$2211.30$$
$$P = \$79,688.70$$

25. Answers will vary.

Chapter 12 Review Exercises

1. $I = PRT$

$$I = \$9800 \times 0.085 \times \frac{200}{360}$$

$$I = \$462.78$$

$$M = P + I$$
$$M = \$9800 + \$462.78$$
$$M = \$10,262.78$$

2. $R = \dfrac{I}{PT}$

$$R = \frac{\$78.75}{\$3000 \times \frac{90}{360}}$$

$$R = 0.105 = 10\frac{1}{2}\%$$

$$M = P + I$$
$$M = \$3000 + \$78.75$$
$$M = \$3078.75$$

3. $T = \dfrac{I}{PR} \times 360$

$$T = \frac{\$640}{\$8000 \times 0.12} \times 360$$

$$T = 240 \text{ days}$$

$$M = P + I$$
$$M = \$8000 + \$640$$
$$M = \$8640$$

4. $P = \dfrac{I}{RT}$

$$P = \frac{\$615}{0.10 \times \frac{180}{360}}$$

$$P = \$12,300$$

$$M = P + I$$
$$M = \$12,300 + \$615$$
$$M = \$12,915$$

5. Jan. 8 is day 8.
   $8 + 120 = 128$, which is May 8, the due date.

$$M = P(1 + RT)$$

$$M = \$12,000 \left(1 + 0.09 \times \frac{120}{360}\right)$$

$$M = \$12,360$$

6. June 19 is day 170.
   $170 + 200 = 370$
   $370 - 365 = 5$, which is January 5, the due date.

$$M = P(1 + RT)$$

$$M = \$6000 \left(1 + 0.125 \times \frac{200}{360}\right)$$

$$M = \$6416.67$$

7. $B = MDT$

$$B = \$18{,}000 \times 0.12 \times \frac{80}{360}$$

$$B = \$480$$

$P = M - B$
$P = \$18{,}000 - \$480$
$P = \$17{,}520$

8. $B = MDT$

$$B = \$26{,}000 \times 0.105 \times \frac{180}{360}$$

$$B = \$1365$$

$P = M - B$
$P = \$26{,}000 - \$1365$
$P = \$24{,}635$

9. (a) Discount period

September 4 is day 247;
October 25 is day 298.

$247 + 150 = 397$
$397 - 298 = 99$ days

(b) $M = P(1 + RT)$

$$M = \$12{,}000 \left(1 + 0.09 \times \frac{150}{360}\right)$$

$$M = \$12{,}450$$

Discount

$B = MDT$

$$B = \$12{,}450 \times 0.12 \times \frac{99}{360}$$

$$B = \$410.85$$

(c) Proceeds

$P = M - B$
$P = \$12{,}450 - \$410.85$
$P = \$12{,}039.15$

10. (a) Discount period

December 20 is day 354; February 28 is day 59.

$354 + 120 = 474$
$474 - (365 + 59) = 50$ days

(b) $M = P(1 + RT)$

$$M = \$8500 \left(1 + 0.11125 \times \frac{120}{360}\right)$$

$$M = \$8815.21$$

Discount

$B = MDT$

$$B = \$8815.21 \times 0.12 \times \frac{50}{360}$$

$$B = \$146.92$$

(c) Proceeds

$P = M - B$
$P = \$8815.21 - \$146.92$
$P = \$8668.29$

11. Simple interest rate

$$R = \frac{D}{1 - DT}$$

$$R = \frac{0.11}{1 - (0.11 \times \frac{150}{360})}$$

$$R = 0.11528 = 11.53\%$$

12. Simple interest rate

$$R = \frac{D}{1 - DT}$$

$$R = \frac{0.09}{1 - (0.09 \times \frac{170}{360})}$$

$$R = 0.09399 = 9.40\%$$

13. Simple discount rate

$$D = \frac{R}{1 + RT}$$

$$D = \frac{0.12}{1 + (0.12 \times \frac{180}{360})}$$

$$D = 0.11320 = 11.32\%$$

14. Maturity value

$M = P(1 + RT)$

$$M = \$38{,}000 \left(1 + 0.1175 \times \frac{120}{360}\right)$$

$$M = \$39{,}488.33$$

15. Discount

$B = MDT$

$$B = \$39{,}488.33 \times 0.125 \times \frac{65}{360}$$

$$B = \$891.23$$

16. $R = \dfrac{I}{PT}$

$R = \dfrac{\$2250}{\$45,000 \times \frac{200}{360}}$

$R = 0.09 = 9\%$

17. $T = \dfrac{I}{PR} \times 360$

$T = \dfrac{\$490}{\$9800 \times 0.10} \times 360$

$T = 180$ days

18. Discount

$\$50,000 - \$47,361.11 = \$2638.89$

$T = \dfrac{B}{MD} \times 360$

$T = \dfrac{\$2638.89}{\$50,000 \times 0.095} \times 360$

$T = 200$ days

19. Discount

$\$35,000 - \$34,319.44 = \$680.56$

$R = \dfrac{I}{PT}$

$R = \dfrac{\$680.56}{\$35,000 \times \frac{50}{360}}$

$R = 0.1400 = 14.00\%$

20. Interest

$\$12,330 - \$12,000 = \$330$

Simple interest rate

$R = \dfrac{I}{PT}$

$R = \dfrac{\$330}{\$12,000 \times \frac{90}{360}}$

$R = 0.11 = 11\%$

Note that the principal is the amount borrowed.

Simple discount rate

$R = \dfrac{I}{PT}$

$R = \dfrac{\$330}{\$12,330 \times \frac{90}{360}}$

$R = 0.10705 = 10.71\%$

Note that the principal is the maturity value.

21. Effective rate of interest

$R = \dfrac{D}{1 - DT}$

$R = \dfrac{0.12}{1 - (0.12 \times \frac{240}{360})}$

$R = 0.13043 = 13.04\%$

22. December 8 is day 342; February 12 is day 43.

$342 + 100 = 442$

$442 - (365 + 43) = 34$

For a discount note, the maturity value is the face value of \$14,000.

$B = MDT$

$B = \$14,000 \times 0.095 \times \dfrac{34}{360}$

$B = \$125.61 \quad Discount$

$P = M - B$

$P = \$14,000 - \$125.61$

$P = \$13,874.39$

23. $T = \dfrac{I}{PR} \times 360$

$T = \dfrac{\$410}{\$12,300 \times 0.08} \times 360$

$T = 150$ days

24. $I = PRT$

$I = \$42,000 \times 0.095 \times \frac{200}{360}$

$I = \$2216.67$

25. $I = M - P$

$I = \$83,187 - \$79,000$

$I = \$4187$

$R = \dfrac{I}{PT}$

$R = \dfrac{\$4187}{\$79,000 \times \frac{120}{360}}$

$R = 0.159 = 15.9\%$

26. $M = \dfrac{B}{DT}$

$M = \dfrac{\$1012}{0.11 \times \frac{180}{360}}$

$M = \$18,400$

27. May 25 is day 145; August 7 is day 219.

$145 + 270 = 415$
$415 - 219 = 196$ day discount period

$$M = P(1 + RT)$$

$$M = \$8000 \left(1 + 0.15 \times \frac{270}{360}\right)$$

$$M = \$8900$$

$$B = MDT$$

$$B = \$8900 \times 0.12 \times \frac{196}{360}$$

$$B = \$581.47 \quad \textit{Discount}$$

$P = M - B$
$P = \$8900 - \$581.47$
$P = \$8318.53 \;\; \textit{Proceeds}$

28. $T = \dfrac{B}{MD} \times 360$

$$T = \frac{\$788.40}{\$6570 \times 0.16} \times 360$$

$$T = 270 \text{ days}$$

29. $M = P + B$
$M = \$28,400 + \$1812$
$M = \$30,212$

30. Bank discount

$$B = MDT$$

$$B = \$26,000 \times 0.10 \times \frac{120}{360}$$

$$B = \$866.67$$

Proceeds

$P = M - B$
$P = \$26,000 - \$866.67$
$P = \$25,133.33$

31. March 3 is day 62;
May 26 is day 146.
$62 + 180 = 242$
$242 - 146 = 96$ day discount period

$$B = MDT$$

$$B = \$25,000 \times 0.095 \times \frac{96}{360}$$

$$B = \$633.33 \quad \textit{Discount}$$

$P = M - B$
$P = \$25,000 - \$633.33$
$P = \$24,366.67 \;\; \textit{Proceeds}$

32. $M = P(1 + RT)$

$$M = \$16,000 \left(1 + 0.09 \times \frac{120}{360}\right)$$

$$M = \$16,480$$

May 12 is day 132; July 20 is day 201.

$132 + 120 = 252$
$252 - 201 = 51$ day discount period

$$B = MDT$$

$$B = \$16,480 \times 0.11 \times \frac{51}{360}$$

$$B = \$256.81 \quad \textit{Discount}$$

$P = M - B$
$P = \$16,480 - \$256.81$
$P = \$16,223.19$

33. $M = P(1 + RT)$

$$M = \$83,000\left(1 + 0.10 \times \tfrac{200}{360}\right)$$

$$M = \$87,611.11$$

$$B = MDT$$

$$B = \$87,611.11 \times 0.10 \times \tfrac{90}{360}$$

$$B = \$2190.28$$

$P = M - B$
$P = \$87,611.11 - \$2190.28$
$P = \$85,420.83$

34. Face value

$$M = \frac{P}{1 - DT}$$

$$M = \frac{\$7580}{1 - \left(0.14 \times \tfrac{120}{360}\right)}$$

$$M = \$7951.05$$

35. $M = P(1 + RT)$

$$M = \$6420 \left(1 + 0.12 \times \frac{210}{360}\right)$$

$$M = \$6869.40$$

December 12 is day 346; January 19 is day 19.

$346 + 210 = 556$
$556 - (365 + 19) = 172$
day discount period

$$B = MDT$$

$$B = \$6869.40 \times 0.15 \times \frac{172}{360}$$

$$B = \$492.31 \quad Discount$$

$$P = M - B$$
$$P = \$6869.40 - \$492.31$$
$$P = \$6377.09 \quad Proceeds$$

36. $D = \dfrac{R}{1 + RT}$

$$D = \dfrac{0.15}{1 + (0.15 \times \frac{270}{360})}$$

$$D = 0.13483 = 13.48\%$$

37. Answers will vary.

38. Answers will vary.

Chapter 12 Summary Exercise

(a) Interest paid to Japanese investment house.

$$I = PRT$$

$$I = \$80,000,000 \times 0.09 \times \tfrac{180}{360}$$

$$I = \$3,600,000$$

Interest received from Canadian firm

$$I = PRT$$

$$I = \$38,000,000 \times 0.11 \times \tfrac{180}{360}$$

$$I = \$2,090,000$$

Interest received from French contractor

$$B = MDT$$

$$B = \$27,500,000 \times 0.128 \times \tfrac{180}{360}$$

$$B = \$1,760,000$$

Interest received from Louisiana company

$$B = MDT$$

$$B = \$14,500,000 \times 0.12 \times \tfrac{180}{360}$$

$$B = \$870,000$$

Interest received:	$2,090,000
	$1,760,000
	+ $ 870,000
	$4,720,000

Interest received:	$4,720,000
Interest paid:	− $3,600,000
Difference:	$1,120,000

(b) Total amount loaned out

$38,000,000	
$25,740,000	($27,500,000 − $1,760,000)
$13,630,000	($14,500,000 − $870,000)
$77,370,000	

COMPOUND INTEREST

13.1 Compound Interest

For the exercises in this section, use the interest tables in Appendix D and a scientific or finanical calculator.

1. $P = \$12,000;\ i = 8\%;\ n = 3 \times 1 = 3$

$$\begin{aligned} M &= P(1+i)^n \\ &= \$12,000(1 + 0.08)^3 \\ &= \$12,000(1.08)^3 \\ &= \$12,000 \times 1.08 \times 1.08 \times 1.08 \\ &= \$15,116.54 \end{aligned}$$

For calculator use, press

1.08 $\boxed{y^x}$ 3 $\boxed{=}$ $\boxed{\times}$ \$12,000 $\boxed{=}$.

$$\begin{aligned} I &= M - P \\ &= \$15,116.54 - \$12,000 \\ &= \$3116.54 \end{aligned}$$

3. $P = \$6000;\ i = \dfrac{12\%}{4} = 3\%;\ n = 2 \times 4 = 8$

$$\begin{aligned} M &= P(1+i)^n \\ &= \$6000(1 + 0.03)^8 \\ &= \$6000(1.03)^8 \\ &= \$6000 \times 1.03 \times 1.03 \times 1.03 \times 1.03 \times 1.03 \\ &\qquad \times 1.03 \times 1.03 \times 1.03 \\ &= \$7600.62 \end{aligned}$$

For calculator use, press

1.03 $\boxed{y^x}$ 8 $\boxed{\times}$ \$6,000 $\boxed{=}$.

$$\begin{aligned} I &= M - P \\ &= \$7600.62 - \$6000 \\ &= \$1600.62 \end{aligned}$$

5. $P = \$1000;\ i = 8\%;\ n = 40$

Look at row 40 in column A on the 8% page of the interest table and find 21.72452150.

$$\begin{aligned} M &= \$1000(21.72452150) \\ &= \$21,724.52 \end{aligned}$$

7. $P = \$470;\ i = \dfrac{12\%}{2} = 6\%;\ n = 9 \times 2 = 18$

Use row 18 of the 6% interest table.

$$\begin{aligned} M &= \$470(2.85433915) \\ &= \$1341.54 \end{aligned}$$

9. $P = \$8400;\ i = 7\%;\ n = 8$

Use row 8 of the 7% interest table.

$$\begin{aligned} M &= \$8400(1.71818618) \\ &= \$14,432.76 \end{aligned}$$

$$\begin{aligned} I &= M - P \\ &= \$14,432.76 - \$8400 \\ &= \$6032.76 \end{aligned}$$

11. $P = \$12,600;\ \dfrac{8\%}{4} = 2\%;\ n = 4\frac{3}{4} \times 4 = 19$

Use row 19 of the 2% interest table.

$$\begin{aligned} M &= \$12,600(1.45681117) \\ &= \$18,355.82 \end{aligned}$$

$$\begin{aligned} I &= M - P \\ &= \$18,355.82 - \$12,600 \\ &= \$5755.82 \end{aligned}$$

13. $P = \$1000;\ i = 6\% = 0.06;\ n = 5$

Simple interest

$$\begin{aligned} I &= PRT \\ &= \$1000 \times 0.06 \times 5 \\ &= \$300 \end{aligned}$$

Compound interest

$$\begin{aligned} M &= \$1000(1.33822558) \\ &= \$1338.23 \end{aligned}$$

$$\begin{aligned} I &= \$1338.23 - \$1000 \\ &= \$338.23 \end{aligned}$$

Compound interest is greater by $38.23 ($338.23 − $300).

15. $P = \$7908.42$; $i = 5\% = 0.05$; $n = 8$

Simple interest

$I = \$7908.42 \times 0.05 \times 8$
$\quad = \$3163.37$

$M = \$7908.42(1.47745544)$
$\quad = \$11,684.34$

$I = \$11,684.34 - \$7908.42$
$\quad = \$3775.92$

Compound interest is greater by \$612.55
(\$3775.92 − \$3163.37).

17. 8% compounded quarterly

$\dfrac{8\%}{4} = 2\%$ for 4 periods

Look in column A of the interest table for 2% and
4 periods.
Find: 1.08243216.
Subtract 1.

$1.08243216 - 1 = 0.08243216$

The effective rate is 8.24%.

19. 15% compounded monthly

$\dfrac{15\%}{12} = 1\frac{1}{4}\%$ for 12 periods

Look in column A of the interest table for $1\frac{1}{4}\%$
and 12 periods.
Find: 1.16075452.
Subtract 1.

$1.16075452 - 1 = 0.16075452$

The effective rate is 16.08%.

21. (a) Compounded quarterly

$P = \$2800$; $i = \dfrac{12\%}{4} = 3\%$; $i = 2\frac{1}{2} \times 4 = 10$

$M = P(1 + i)^n$
$\quad = \$2800(1.03)^{10}$
$\quad = \$2800(1.34391638)$
$\quad = \$3762.97$

(b) Compounded monthly

$P = \$2800$; $i = \dfrac{12\%}{12} = 1\%$; $n = 2\frac{1}{2} \times 12 = 30$

$M = P(1 + i)^n$
$\quad = \$2800(1.01)^{30}$
$\quad = \$2800(1.34784892)$
$\quad = \$3773.98$

23. (a) Yearly

$P = \$10,000$; $i = 6\%$; $n = 4$

$M = P(1 + i)^n$
$\quad = \$10,000(1.06)^4$
$\quad = \$10,000(1.26247696)$
$\quad = \$12,624.77$

$I = M - P$
$\quad = \$12,624.77 - \$10,000$
$\quad = \$2624.77$

(b) Semiannually

$P = \$10,000$; $i = \dfrac{6\%}{2} = 3\%$; $n = 4 \times 2 = 8$

$M = P(1 + i)^n$
$\quad = \$10,000(1.03)^8$
$\quad = \$10,000(1.26677008)$
$\quad = \$12,667.70$

$I = \$12,667.70 - \$10,000$
$\quad = \$2667.70$

(c) Quarterly

$P = \$10,000$; $i = \dfrac{6\%}{4} = 1\frac{1}{2}\%$; $n = 4 \times 4 = 16$

$M = P(1 + i)^n$
$\quad = \$10,000(1.015)^{16}$
$\quad = \$10,000(1.26898555)$
$\quad = \$12,689.86$

$I = \$12,689.86 - \$10,000$
$\quad = \$2689.86$

(d) Monthly

$P = \$10,000$; $i = \dfrac{6\%}{12} = \frac{1}{2}\%$; $n = 4 \times 12 = 48$

$M = P(1 + i)^n$
$\quad = \$10,000(1.005)^{48}$
$\quad = \$10,000(1.27048916)$
$\quad = \$12,704.89$

$I = \$12,704.89 - \$10,000$
$\quad = \$2704.89$

(e) Simple interest

$P = \$10,000$; $r = 6\%$; $t = 4$

$I = PRT$
$\quad = \$10,000 \times 0.06 \times 4$
$\quad = \$2400.00$

25. $P = \$8800$; $i = \dfrac{8\%}{2} = 4\%$; $n = 4 \times 2 = 8$

$M = P(1 + i)^n$
$= \$8800(1.04)^8$
$= \$8800(1.36856905)$
$= \$12{,}043.41$

27. (a) 4% compounded annually

$P = \$10{,}000$; $i = 4\%$; $n = 3$

$M = P(1 + 0.04)^3$
$= \$10{,}000(1.12486400)$
$= \$11{,}248.64$

4% compounded quarterly

$P = \$10{,}000$; $i = \dfrac{4\%}{4} = 1\%$; $n = 3 \times 4 = 12$

$M = P(1 + i)^n$
$= \$10{,}000(1.01)^{12}$
$= \$10{,}000(1.12682503)$
$= \$11{,}268.25$

(b) $\$11{,}268.25 - \$11{,}248.64 = \$19.61$

29. $P = \$80{,}000$; $i = \dfrac{12\%}{12} = 1\%$; $n = 3 \times 12 = 36$

$M = P(1 + i)^n$
$= \$80{,}000(1.01)^{36}$
$= \$80{,}000(1.43076878)$
$= \$114{,}461.50$

He will need to add $10,538.50

($125,000 − $114,461.50).

31. $i = \dfrac{6\%}{2} = 3\%$; $n = 2$

Use row 2 of the 3% interest table.

1.06090000
$1.06090000 - 1 = 0.06090000 = 6.09\%$

$i = \dfrac{5\%}{12} = \dfrac{5}{12}$; $n = 12$

Use row 12 of the $\frac{5}{12}\%$ interest table.

1.05116190
$1.05116190 - 1 = 0.05116190 = 5.12\%$

The first option would produce more income.

33. $P = \$12{,}800$; $i = 12.3\%$; $n = 3$

$M = P(1 + i)^n$
$= \$12{,}800(1.123)^3$
$= \$18{,}127.97$

35. $P = \$9500$; $i = \dfrac{10.3\%}{4} = 2.575\%$; $n = 2\frac{1}{2} \times 4 = 10$

$M = P(1 + i)^n$
$= \$9500(1.02575)^{10}$
$= \$12{,}250.08$

37. Answers will vary.

39. $P = \$25{,}000$; $r = 10\%$; $t = 1$

$I = PRT$
$= \$25{,}000 \times 0.10 \times 1$
$= \$2500$

$P = \$25{,}000$; $i = \dfrac{8\%}{4} = 2\%$; $n = 4$

$M = P(1 + i)^n$
$= \$25{,}000(1.02)^4$
$= \$25{,}000(1.08243216)$
$= \$27{,}060.80$

$I = \$27{,}060.80 - \$25{,}000$
$= \$2060.80$

Simple interest would generate

$\$2500 - \$2060.80 = \$439.20$

additional interest.

41. $P = \$12{,}000$; $i = \dfrac{8\%}{2} = 4\%$; $n = 25 \times 2 = 50$

$M = P(1 + i)^n$
$= \$12{,}000(1.04)^{50}$
$= \$12{,}000(7.10668335)$
$= \$85{,}280.20$

13.2   Daily and Continuous Compounding

1. October 7 is day 280;
December 10 is day 344.
$344 - 280 = 64$ days
Use Table 13.2.

$\$6200(1.006155560) = \$6238.16$

$I = \$6238.16 - \$6200$
$= \$38.16$

3. February 17 is day 48;
April 15 is day 105.
$105 - 48 = 57$ days
Use Table 13.2.

$6500(1.005480454) = \$6535.62$

$I = \$6535.62 - \$6500$
$\quad = \$35.62$

5. February 14 is day 45; April 1 is day 91.
$91 - 45 = 46$ days
Use Table 13.2.
Compound amount

$7235.82(1.004420489) = \$7267.81$

7. July 1 is day 182; October 1 is day 274.
$274 - 182 = 92$ days
Use Table 13.2.
Compound amount

$2965.72(1.008860519) = \$2992$

9. Look in Table 13.3 for 6% and 2 years.

Compound amount
$= \$5000(1.12748573)$
$= \$5637.43$

11. Look in Table 13.3 for 7% and 3 years.

Compound amount
$= \$14,000(1.23365322)$
$= \$17,271.15$

13. Look in Table 13.3 for 7% and 3 years.

Compound amount
$= \$20,000(1.23365322)$
$= \$24,673.06$

Interest

$24,673.06 - \$20,000 = \$4673.06$

15. Look in Table 13.3 for 8% and 4 years.

Compound interest
$= \$3800(1.37707948)$
$= \$5232.90$

Interest

$5232.90 - \$3800 = \$1432.90$

17. Answers will vary.

19. October 1 is day 274;
July 1 is day 182.
$274 - 182 = 92$ days
$4300(1.008860519) = \$4338.10$

July 30 is day 211
$274 - 211 = 63$ days
$1000(1.006059089) = \$1006.06$

September 5 is day 248
$274 - 248 = 26$ days
$500(1.002496141) = \$501.25$

Total in account

$4338.10 + \$1006.06 + \$501.25$
$\quad = \$5845.41$

21. $17,500 - \$5000 - \$980 = \$11,520$
earned interest from April 1 (day 91)
to July 1 (day 182).
$182 - 91 = 91$ days

$11,520(1.008763788) = \$11,620.96$
$11,620.96 - \$11,520 = \$100.96$

$5000 earned interest for 21 days.
$5000(1.002015631) = \$5010.08$
$5010.08 - \$5000 = \$10.08$

$980 earned interest for
$91 - 12 = 79$ days
$980(1.007603742) = \$987.45$
$987.45 - \$980 = \$7.45$

(a) Interest earned

$100.96 + \$10.08 + \$7.45$
$\quad = \$118.49$

(b) Balance on July 1

$17,500 + \$118.49 - (\$5000 + \$980)$
$\quad = \$11,638.49$

23. No interest was earned because the money was withdrawn within 3 months.

25. 12 $(15 - 3)$ months interest will be paid $3\frac{1}{2}$% passbook rate.
Use Table 13.2.

$5000(1.035121585) = \$5175.61$

Interest earned

$5175.61 - \$5000 = \$175.61$

27. $P = \$8000$; $r = 0.06$; $y = 2$

$$M = Pe^{yr}$$
$$= \$8000e^{2(0.06)}$$
$$= \$8000e^{0.12}$$
$$= \$8000(1.12749685)$$
$$= \$9019.97$$

Interest

$$\$9019.97 - \$8000 = \$1019.97$$

29. $P = \$4100.70$; $r = 0.08$; $y = \dfrac{9}{12}$

$$M = Pe^{yr}$$
$$= \$4100.70e^{0.75(0.08)}$$
$$= \$4100.70e^{0.06}$$
$$= \$4100.70(1.06183655)$$
$$= \$4354.27$$

Interest

$$\$4354.27 - 4100.70 = \$253.57$$

31. Use Table 13.3.

(a) Compound amount

$$\$800,000(1.12748573) = \$901,988.58$$

(b) Interest earned

$$\$901,988.58 - \$800,000 = \$101,988.58$$

33. 10% compounded semiannually

$$P = \$17,000; \ i = \frac{10\%}{2} = 5\%; \ n = 1\tfrac{1}{2} \times 2 = 3$$

Use row 3 of the 5% interest table in Appendix D.

$$M = \$17,000(1.15762500) = \$19,679.63$$

$9\tfrac{1}{2}\%$ compounded continuously

$$P = \$17,000; \ r = 0.095; \ y = 1.5$$

$$M = Pe^{yr}$$
$$= \$17,000e^{1.5(0.095)}$$
$$= \$17,000e^{0.1425}$$
$$= \$17,000(1.15315308)$$
$$= \$19,603.60$$

Choosing 10% semiannual interest will earn $76.03 ($19,679.63 − $19,603.60) extra interest.

35. Answers will vary.

13.3 Finding Time and Rate

1. $P = \$6200$; $M = \$7384.30$; $i = 6\%$

$$M = P(1 + i)^n$$
$$\$7384.30 = \$6200(1 + 0.06)^n$$
$$\$7384.30 = \$6200(1.06)^n$$
$$\frac{\$7384.30}{\$6200} = (1.06)^n$$
$$1.191016129 = (1.06)^n$$

Use column A of the 6% interest table in Appendix D to find the value closest to 1.19101600; so $n = 3$ years.

3. $P = \$3600$; $M = \$4824.34$; $i = \dfrac{10\%}{2} = 5\%$

$$M = P(1 + i)^n$$
$$\$4824.34 = \$3600(1 + 0.05)^n$$
$$\$4824.34 = \$3600(1.05)^n$$
$$\frac{\$4824.34}{\$3600} = (1.05)^n$$
$$1.340094444 = (1.05)^n$$

Use column A of the 5% interest table to find the value closest to 1.340094444. It is 1.34009564; so $n = 6$ periods or 3 years.

5. $M = \$11,082.73$; $i = \dfrac{8\%}{2} = 4\%$; $n = 7 \times 2 = 14$

$$P = \frac{M}{(1 + i)^n}$$
$$= \frac{\$11,082.73}{(1 + 0.04)^{14}}$$
$$= \frac{\$11,082.73}{(1.04)^{14}}$$
$$= \frac{\$11,082.73}{1.73167645}$$
$$= \$6400.00$$

7. $M = \$15,149.72$; $P = \$12,000$; $n = 4$

$$M = P(1 + i)^4$$
$$\$15,149.72 = \$12,000(1 + i)^4$$
$$\frac{\$15,149.72}{\$12,000} = (1 + i)^4$$
$$1.262476667 = (1 + i)^4$$

Look only in row 4 across the pages for a value close to 1.262476667. It is 1.26247696 on the 6% page.

The interest rate is 6%.

9. $M = \$13,403.64$; $P = \$8500$; $n = 5\frac{3}{4} \times 4 = 23$

$$M = P(1+i)^n$$
$$\$13,403.64 = \$8500(1+i)^{23}$$
$$\frac{\$13,403.64}{\$8500} = (1+i)^{23}$$
$$1.576898824 = (1+i)^{23}$$

Look in row 23 across the pages for a value close to 1.576898824. It is 1.57689926 on the 2% page. The interest rate is $4 \times 2\% = 8\%$.

11. Answers will vary.

13. (a) $P = \$12,000$; $i = \dfrac{6\%}{4} = 1\frac{1}{2}\%$; $n = 3 \times 4 = 12$

$$\begin{aligned} M &= P(1+i)^n \\ &= \$12,000(1+0.015)^{12} \\ &= \$12,000(1.015)^{12} \\ &= \$12,000(1.19561817) \\ &= \$14,347.42 \end{aligned}$$

(b) $P = \$12,000$; $i = \dfrac{6\%}{4} = 1\frac{1}{2}\%$; $n = 5 \times 4 = 20$

$$\begin{aligned} M &= \$12,000(1+0.015)^{20} \\ &= \$12,000(1.015)^{20} \\ &= \$12,000(1.34685501) \\ &= \$16,162.26 \end{aligned}$$

15. $M = \$58,708.95$; $P = \$46,000$; $n = 2\frac{1}{2} \times 2 = 5$

$$M = P(1+i)^n$$
$$\$58,708.95 = \$46,000(1+i)^5$$
$$\frac{\$58,708.95}{\$46,000} = (1+i)^5$$
$$1.276281522 = (1+i)^5$$

Look only in row 5 across the pages for a value close to 1.276281522. It is 1.27628156 on the 5% page.

The interest is $2 \times 5\% = 10\%$.

17. $M = \$5934.06$; $P = \$5200$; $i = \dfrac{9\%}{2} = 4\frac{1}{2}\%$

$$M = P(1+i)^n$$
$$\$5934.06 = \$5200(1+0.045)^n$$
$$\$5934.06 = \$5200(1.045)^n$$
$$\frac{\$5934.06}{\$5200} = (1.045)^n$$
$$1.141165385 = (1.045)^n$$

Use the $4\frac{1}{2}\%$ page. Look down column A and find a value close to 1.141165385.

It is 1.14116613, which corresponds to 3 periods or $1\frac{1}{2}$ years.

19. $M = \$2$; $P = \$1$; $i = 2\frac{1}{2}\%$

$$M = P(1+i)^n$$
$$2 = 1(1+0.025)^n$$
$$2 = (1.025)^n$$

Use the $2\frac{1}{2}\%$ page. Look down column A to find the number closest to 2. It is 1.99649502, which corresponds to 28 periods or 28 years.

21. $M = \$2$; $P = \$1$; $i = 3\frac{1}{2}\%$

$$M = P(1+i)^n$$
$$2 = 1(1+0.035)^n$$
$$2 = (1.035)^n$$

Use the $3\frac{1}{2}\%$ page. Look down column A to find the number closest to 2. It is 1.98978886, which corresponds to 20 periods or 20 years.

23. $M = 2$; $P = 1$; $i = 2\%$

$$M = P(1+i)^n$$
$$2 = 1(1+0.02)^n$$
$$2 = (1.02)^n$$

Use the 2% page. Look down column A to find the value closest to 2. It is 1.99988955, which corresponds to 35 periods or 35 years.

25. The initial \$10,000 deposit earned interest for the entire 5-period.

$$P = \$10,000; \quad i = \frac{8\%}{4} = 2\%; \quad n = 5 \times 4 = 20$$
$$\begin{aligned} M &= P(1+i)^n \\ &= \$10,000(1+0.02)^{20} \\ &= \$10,000(1.02)^{20} \\ &= \$10,000(1.48594740) \\ &= \$14,859.47 \end{aligned}$$

The deposit of \$20,000 earned interest for three years.

$P = \$20,000; \; i = \dfrac{8\%}{4} = 2\%, \; n = 3 \times 4 = 12$

$P = P(1 + i)^n$
$= \$20,000(1 + 0.02)^{12}$
$= \$20,000(1.02)^{12}$
$= \$20,000(1.26824179)$
$= \$25,364.84$

Total in account

$\$14,859.47 + \$25,364.84$
$= \$40,224.31$

27. Do this problem in two parts.

(1) Find the value of $\$4200$ at the end of $3\frac{3}{4}$ years.

$P = \$4200; \; i = \dfrac{8\%}{4} = 2\%; \; n = 3\frac{3}{4} \times 4 = 15$

Use the 2% page, column A, row 15.

$M = \$4200(1.34586834)$
$= \$5652.65$

(2) Find the value of the certificate of deposit at the end of 3 years.
Look at Table 13.3 for 8% and 3 years.

$P = \$5652.65; \; i = 8\%; \; n = 3$

$M = \$5652.65(1.27121572)$
$= \$7185.73$

29. Do the problem in four parts.

(1) Find the value of $\$11,000$ at the end of 4 years.

$P = \$11,000; \; i = \dfrac{10\%}{2} = 5\%; \; n = 4 \times 2 = 8$

Use the 5% page, column A, row 8 of the interest table.

$M = \$11,000(1.47745544)$
$= \$16,252.00984$

(2) Find the value of a $\$5000$ certificate of deposit after 3 years.
Look at Table 13.3 for 7% and 3 years.

$M = \$5000(1.23365322)$
$= \$6168.2661$

(3) Find the value of the balance remaining in the 10% account.

$P = \$16,252.00984 - \$5000$
$= \$11,252.00984$

$i = \dfrac{10\%}{2} = 5\%; \; n = 3 \times 2 = 6$

Use the 5% page, column A, row 6 of the interest table.

$M = \$11,252.00984(1.34009564)$
$= \$15,078.76933$

(4) The total in both accounts is

$\$6168.27 + \$15,078.77 = \$21,247.04.$

13.4   Present Value at Compound Interest

1. $M = \$4800; \; i = 8\%; \; n = 3$

$P = M \cdot \dfrac{1}{(1 + i)^n}$

$P = \$4800 \cdot \dfrac{1}{(1 + 0.08)^3}$

$P = \$4800 \cdot \dfrac{1}{(1.08)^3}$

Use the 8% page, column B, row 3 of the interest table.

$P = \$4800(0.79383224)$
$= \$3810.39$

$I = \$4800 - \$3810.39$
$= \$989.61$

3. $M = \$12,200; \; i = \dfrac{12\%}{4} = 3\%; \; n = 2\frac{1}{2} \times 4 = 10$

$P = M \cdot \dfrac{1}{(1 + i)^n}$

$P = \$12,200 \cdot \dfrac{1}{(1 + 0.03)^{10}}$

$P = \$12,200 \cdot \dfrac{1}{(1.03)^{10}}$

Use the 3% page, column B, row 10 of the interest table.

$P = \$12,200(0.74409391)$
$= \$9077.95$

$I = \$12,200 - \$9077.95$
$= \$3122.05$

5. $M = \$8500$; $i = \dfrac{12\%}{12} = 1\%$; $1 \times 12 = 12$

$$P = M \cdot \frac{1}{(1+i)^n}$$

$$= \$8500 \cdot \frac{1}{(1+0.01)^{12}}$$

$$= \$8500 \cdot \frac{1}{(1.01)^{12}}$$

Use the 1% page, column B, row 12 of the interest table.

$P = \$8500(0.88744923)$
  $= \$7543.32$

$I = \$8500 - \$7543.32$
  $= \$956.68$

7. Answers will vary.

9. $M = \$5000$; $i = 7\%$; $n = 5$

$$P = M \cdot \frac{1}{(1+i)^n}$$

$$= \$5000 \cdot \frac{1}{(1+0.07)^5}$$

$$= \$5000 \cdot \frac{1}{(1.07)^5}$$

Use the 7% table, column B, row 5 of the interest table.

$P = \$5000(0.71298618)$
  $= \$3564.93$

$I = \$5000 - \$3564.93$
  $= \$1435.07$

11. $M = \$37,500$; $i = \dfrac{12\%}{4} = 3\%$; $n = 4 \times 4 = 16$

$$P = M \cdot \frac{1}{(1+i)^n}$$

$$= \$37,500 \cdot \frac{1}{(1.03)^{16}}$$

Use the 3% page, column B, row 16 of the interest table.

$P = \$37,500(0.62316694)$
  $= \$23,368.76$

13. $M = \$9000$; $i = \dfrac{8\%}{4} = 2\%$; $n = 3 \times 4 = 12$

$$P = M \cdot \frac{1}{(1+i)^n}$$

$$= \$9000 \cdot \frac{1}{(1.02)^{12}}$$

Use the 2% page, column B, row 12 of the interest table.

$P = \$9000(0.78849318)$
  $= \$7096.44$

15. Find the present value of \$3800.

$$M = \$3800; \; i = \frac{8\%}{4} = 2\%; \; n = 5 \times 4 = 20$$

$$P = M \cdot \frac{1}{(1+i)^n}$$

$$= \$3800 \times \frac{1}{(1.02)^{20}}$$

Use the 2% page, column B, row 20 of the interest table.

$P = \$3800(0.67297133)$
  $= \$2557.29$

Since \$2557.29 is larger than \$2500, \$3800 in 5 years is larger.

17. Future value

$$P = \$30,000; \; i = \frac{10\%}{2} = 5\%; \; n = 2\tfrac{1}{2} \times 2 = 5$$

Use the 5% page, column A, row 5 of the interest table.

$M = \$30,000(1.27628156)$
  $= \$38,288.45$

Present value

$$M = \$38,288.45; \; i = \frac{8\%}{4} = 2\%; \; n = 2\tfrac{1}{2} \times 4 = 10$$

$$P = M \cdot \frac{1}{(1+i)^n}$$

$$= \$38,288.45 \cdot \frac{1}{(1.02)^{10}}$$

Use the 2% table, column B, row 10 of the interest table.

$P = \$38,288.45(0.82034830)$
  $= \$31,409.86$

19. First, find the interest.

$$I = PRT$$
$$= \$16{,}800 \times 0.10 \times 4$$
$$= \$6720$$

Then, find the maturity value of the note.

$$M = P + I$$
$$= \$16{,}800 + \$6720$$
$$= \$23{,}520$$

Find the present value of this amount.

$$M = \$23{,}520; \; i = \frac{6\%}{4} = 1\tfrac{1}{2}\%; \; n = 4 \times 4 = 16$$

$$M = M \cdot \frac{1}{(1+i)^n}$$

$$= \$23{,}520 \cdot \frac{1}{(1.015)^{16}}$$

Use the $1\tfrac{1}{2}\%$ page, column B, row 16 of the interest table.

$$P = \$23{,}520(0.78803104)$$
$$= \$18{,}534.49$$

21. (a) Future value

$$P = \$20{,}000; \; i = 10\%; \; n = 3$$

Use the 10% page, column A, row 3 of the interest table.

$$M = P(1+i)^n$$
$$= \$20{,}000(1.10)^3$$
$$= \$20{,}000(1.33100000)$$
$$= \$26{,}620$$

(b) Find the present value of this amount.

$$M = \$26{,}620; \; i = \frac{8\%}{4} = 2\%; \; n = 3 \times 4 = 12$$

Use the 2% page, column B, row 12 of the interest table.

$$P = M \cdot \frac{1}{(1+i)^n}$$

$$= \$26{,}620 \cdot \frac{1}{(1.02)^{12}}$$

$$= \$26{,}620(0.78849318)$$
$$= \$20{,}989.69$$

Chapter 13 Review Exercises

1. $P = \$12{,}400; \; i = 6\tfrac{1}{2}\%; \; n = 6$

Use the $6\tfrac{1}{2}\%$ page, column A, row 6 of the interest table.

$$M = P(1+i)^n$$
$$= \$12{,}400(1 + 0.065)^6$$
$$= \$12{,}400(1.065)^6$$
$$= \$12{,}400(1.45914230)$$
$$= \$18{,}093.36$$

$$I = M - P$$
$$= \$18{,}093.36 - \$12{,}400$$
$$= \$5693.36$$

2. $P = \$7000; \; i = \frac{6\%}{2} = 3\%; \; n = 4 \times 2 = 8$

Use the 3% page, column A, row 8 of the interest table.

$$M = P(1+i)^n$$
$$= \$7000(1 + 0.03)^8$$
$$= \$7000(1.03)^8$$
$$= \$7000(1.26677008)$$
$$= \$8867.39$$

$$I = M - P$$
$$= \$8867.39 - \$7000$$
$$= \$1867.39$$

3. $P = \$4800; \; i = \frac{10\%}{4} = 2\tfrac{1}{2}\%; \; n = 3 \times 4 = 12$

Use the $2\tfrac{1}{2}\%$ page, column A, row 12 of the interest table.

$$M = P(1+i)^n$$
$$= \$4800(1 + 0.025)^{12}$$
$$= \$4800(1.025)^{12}$$
$$= \$4800(1.34488882)$$
$$= \$6455.47$$

$$I = M - P$$
$$= \$6455.47 - \$4800$$
$$= \$1655.47$$

4. $P = \$18{,}000;\ i = \dfrac{7\%}{4} = 1\tfrac{3}{4}\%;\ n = 4 \times 4 = 16$

Use the $1\tfrac{3}{4}\%$ page, column A, row 16 of the interest table.

$$\begin{aligned} M &= P(1+i)^n \\ &= \$18{,}000(1+0.0175)^{16} \\ &= \$18{,}000(1.0175)^{16} \\ &= \$18{,}000(1.31992935) \\ &= \$23{,}758.73 \end{aligned}$$

$$\begin{aligned} I &= M - P \\ &= \$23{,}758.73 - \$18{,}000 \\ &= \$5758.73 \end{aligned}$$

5. $P = \$9000;\ i = \dfrac{9\%}{12} = \dfrac{3}{4}\%;\ n = 2\tfrac{1}{2} \times 12 = 30$

Use the $\tfrac{3}{4}\%$ page, column A, row 30 of the interest table.

$$\begin{aligned} M &= P(1+i)^n \\ &= \$9000(1+0.0075)^{30} \\ &= \$9000(1.0075)^{30} \\ &= \$9000(1.25127176) \\ &= \$11{,}261.45 \end{aligned}$$

$$\begin{aligned} I &= M - P \\ &= \$11{,}261.45 - \$9000 \\ &= \$2261.45 \end{aligned}$$

6. $P = \$12{,}000;\ i = \dfrac{6\%}{12} = \dfrac{1}{2}\%;\ n = 3\tfrac{1}{4} \times 12 = 39$

Use the $\tfrac{1}{2}\%$ page, column A row 39 of the interest table.

$$\begin{aligned} M &= P(1+i)^n \\ &= \$12{,}000(1+0.005)^{39} \\ &= \$12{,}000(1.005)^{39} \\ &= \$12{,}000(1.21472063) \\ &= \$14{,}576.65 \end{aligned}$$

$$\begin{aligned} I &= M - P \\ &= \$14{,}576.65 - \$12{,}000 \\ &= \$2576.65 \end{aligned}$$

7. 7% compounded quarterly

$$\dfrac{7\%}{4} = 1\tfrac{3}{4}\% \text{ for 4 periods}$$

Use the $1\tfrac{3}{4}\%$ page, column A, row 4 of the interest table. Subtract 1.

$$1.07185903 - 1 = 0.07185903$$

The effective rate of interest is 7.19%.

8. 8% compounded quarterly

$$\dfrac{8\%}{4} = 2\% \text{ for 4 periods}$$

Use the 2% page, column A, row 4 of the interest table. Subtract 1.

$$1.08243216 - 1 = 0.08243216$$

The effective rate of interest is 8.24%.

9. 7% compounded semiannually

$$\dfrac{7\%}{2} = 3\tfrac{1}{2}\% \text{ for 2 periods}$$

Use the $3\tfrac{1}{2}\%$ page, column A, row 2 of the interest table. Subtract 1.

$$1.07122500 - 1 = 0.07122500$$

The effective rate of interest is 7.12%.

10. 9% compounded monthly

$$\dfrac{9\%}{12} = \dfrac{3}{4}\% \text{ for 12 periods}$$

Use the $\tfrac{3}{4}\%$ page, column A, row 12 of the interest table. Subtract 1.

$$1.09380690 - 1 = 0.09380690$$

The effective interest rate is 9.38%.

11. March 4 is day 63;
March 30 is day 150.
$150 - 63 = 87$ days

Use Table 13.2.

$$\$2900(1.008376958) = \$2924.29$$

$$\begin{aligned} I &= \$2924.29 - \$2900 \\ &= \$24.29 \end{aligned}$$

12. May 20 is day 140; July 1 is day 182.
$182 - 140 = 42$ days

Use Table 13.2.

$$\$6000(1.004035324) = \$6024.21$$

$$\begin{aligned} I &= \$6024.21 - \$6000 \\ &= \$24.21 \end{aligned}$$

13. July 15 is day 196;
October 1 is day 274.
$274 - 196 = 78$ days

Use Table 13.2.

$$\$3020.80(1.007507132) = \$3043.48$$

$$\begin{aligned} I &= \$3043.48 - \$3020.80 \\ &= \$22.68 \end{aligned}$$

14. January 22 is day 22;
    April 1 is day 91.
    $91 - 22 = 69$ days

    Use Table 13.2.

    $\$3500(1.006638056) = \$3523.23$

15. April 22 is day 112;
    July 1 is day 182.
    $182 - 112 = 70$ days

    Use Table 13.2.

    $\$7200.35(1.006734583) = \$7248.84$

16. August 10 is day 222;
    October 1 is day 274.
    $274 - 222 = 52$ days

    Use Table 13.2.

    $\$9600.40(1.004998513) = \$9648.39$

17. Look in Table 13.3 for 7% and 3 years.

    Compound amount
    $= \$4000(1.23365322)$
    $= \$4934.61$

18. Look in Table 13.3 for 6% and 2 years.

    Compound amount
    $= \$6500(1.12748573)$
    $= \$7328.66$

19. Look in Table 13.3 for 8% and 4 years.

    Compound amount
    $= \$8800(1.37707948)$
    $= \$12,118.30$

20. $P = \$12,600$; $r = 0.08$; $y = 7$

    Use Appendix B.

    $M = Pe^{yr}$
    $= \$12,600e^{7(0.08)}$
    $= \$12,600e^{0.56}$
    $= \$12,600(1.75067250)$
    $= \$22,058.47$

    Interest

    $\$22,058.47 - \$12,600 = \$9458.47$

21. $P = \$5000$; $r = 0.07$; $y = 5$

    Use Appendix B.

    $M = Pe^{yr}$
    $= \$5000e^{5(0.07)}$
    $= \$5000e^{(0.35)}$
    $= \$5000(1.41906755)$
    $= \$7095.34$

    Interest

    $\$7095.34 - \$5000 = \$2095.34$

22. $P = \$4300$; $M = \$5754.37$; $n = 5$

    $$M = P(1 + i)^n$$
    $$\$5754.37 = \$4300(1 + i)^5$$
    $$\frac{\$5754.37}{\$4300} = (1 + i)^5$$
    $$1.338225581 = (1 + i)^5$$

    Look only in row 5 and read across the pages for a value close to 1.338225581. It is 1.33822558 on the 6% page.
    The interest rate is 6%.

23. $P = \$8600$; $M = \$11,566.04$; $n = 3 \times 4 = 12$

    $$M = P(1 + i)^n$$
    $$\$11,566.04 = \$8600(1 + i)^{12}$$
    $$\frac{\$11,566.04}{\$8600} = (1 + i)^{12}$$
    $$1.344888372 = (1 + i)^{12}$$

    Look only in row 12 and read across the pages for a value close to 1.344888372. It is 1.34488882 on the $2\frac{1}{2}$% page.
    The interest rate is $4 \times 2\frac{1}{2}\% = 10\%$.

24. $P = \$7500$; $M = \$9914.25$; $n = 3\frac{1}{2} \times 4 = 14$

    $$M = P(1 + i)^n$$
    $$\$9914.25 = \$7500(1 + i)^{14}$$
    $$\frac{\$9914.25}{\$7500} = (1 + i)^{14}$$
    $$1.3219 = (1 + i)^{14}$$

    Look only in row 14 and read across the pages for a value close to 1.3219. It is 1.31947876 on the 2% page.
    The interest rate is $4 \times 2\% = 8\%$.

25. $P = \$6000$; $M = \$7986$; $i = 10\%$

$$M = P(1+i)^n$$
$$\$7986 = \$6000(1.10)^n$$
$$\frac{\$7986}{\$6000} = (1.10)^n$$
$$1.331 = (1.10)^n$$

Use column A of the 10% interest table to find the value closest to 1.331. It is 1.33100000, so $n = 3$ years.

26. $P = \$8400$; $M = \$10{,}357.20$; $i = \dfrac{6\%}{12} = \dfrac{1}{2}\%$

$$M = P(1+i)^n$$
$$\$10{,}357.20 = \$8400(1.005)^n$$
$$\frac{\$10{,}357.20}{\$8400} = (1.005)^n$$
$$1.233 = (1.005)^n$$

Use column A of $\frac{1}{2}\%$ interest table to find the value closest to 1.233. It is 1.23303270 so $n = 42$ periods or $3\frac{1}{2}$ years.

27. $M = \$14{,}300$; $i = 5\frac{1}{2}\%$; $n = 3$

$$P = M \cdot \frac{1}{(1+i)^n}$$
$$P = \$14{,}300 \cdot \frac{1}{(1.055)^3}$$

Use the $5\frac{1}{2}$ page, column B, row 3 of the interest table.

$$P = \$14{,}300(0.85161366)$$
$$= \$12{,}178.08$$

$$I = M - P$$
$$= \$14{,}300 - \$12{,}178.08$$
$$= \$2121.92$$

28. $M = \$4000$; $i = \dfrac{9\%}{2} = 4\frac{1}{2}\%$; $n = 3 \times 2 = 6$

$$P = M \cdot \frac{1}{(1+i)^n}$$
$$= \$4000 \cdot \frac{1}{(1.045)^6}$$

Use the $4\frac{1}{2}\%$ page, column B, row 6 of the interest table.

$$P = \$4000(0.76789574)$$
$$= \$3071.58$$

$$I = M - P$$
$$= \$4000 - \$3071.58$$
$$= \$928.42$$

29. $M = \$6000$; $i = \dfrac{10\%}{4} = 2\frac{1}{2}\%$; $n = 5 \times 4 = 20$

$$P = M \cdot \frac{1}{(1+i)^n}$$
$$= \$6000 \cdot \frac{1}{(1.025)^{20}}$$

Use the $2\frac{1}{2}\%$ page, column B, row 20 of the interest table.

$$P = \$6000(0.61027094)$$
$$= \$3661.63$$

$$I = M - P$$
$$= \$6000 - \$3661.63$$
$$= \$2338.37$$

30. $M = \$3000$; $i = \dfrac{6\%}{12} = \dfrac{1}{2}\%$; $n = 4 \times 12 = 48$

$$P = M \cdot \frac{1}{(1+i)^n}$$
$$= \$3000 \cdot \frac{1}{(1.005)^{48}}$$

Use the $\frac{1}{2}\%$ page, column B, row 48 of the interest table.

$$P = \$3000(0.78709841)$$
$$= \$2361.30$$

$$I = M - P$$
$$= \$3000 - \$2361.30$$
$$= \$638.70$$

31. April 1 is day 91;
January 1 is day 1.
$91 - 1 = 90$ days

$$\$1800(1.008667067) = \$1815.60$$

March 12 is day 71.
$91 - 71 = 20$ days

$$\$2300(1.001919556) = \$2304.41$$

Total account

$$\$1815.60 + \$2304.41 = \$4120.01$$

32. September 1 is day 244;
June 10 is day 161.
$244 - 161 = 83$ days

$\$4000(1.007990276) = \$4031.96$

July 6 is day 187.
$244 - 187 = 57$ days

$\$1200(1.005480454) = \$1206.58$

Total account

$\$4031.96 + \$1206.58 = \$5238.54$

33. Look in Table 13.3 for 8% and 10 years.

Compound amount
$= \$18,000(2.22534585)$
$= \$40,056.23$

Interest

$\$40,056.23 - \$18,000 = \$22,056.23$

34. $12(15 - 3)$ months or interest will be paid at $3\frac{1}{2}\%$ passbook rate.

Use Table 13.2.

$\$7350(1.035121585) = \$7608.14$

Interest

$\$7608.14 - \$7350 = \$258.14$

35. $M = \$47,500;\ i = \dfrac{12\%}{12} = 1\%;\ n = 4 \times 12 = 48$

$P = M \cdot \dfrac{1}{(1+i)^n}$

$\quad = \$47,500 \cdot \dfrac{1}{(1.01)^{48}}$

Use the 1% page, column B, row 48 of the interest table.

$P = \$47,500(0.62026041)$
$\quad = \$29,462.37$

36. (a) Use the 10% page, column A, row 2 of the interest table.

$M = 1.21000(\$12,540)$
$\quad = \$15,173.40$

(b) $M = \$15,173.40;\ i = \dfrac{6\%}{2} = 3\%;\ n = 2 \times 2 = 4$

$P = M \cdot \dfrac{1}{(1+i)^n}$

$\quad = \$15,173.40 \cdot \dfrac{1}{(1.03)^4}$

Use the 3% page, column B, row 4 of the interest table.

$P = \$15,173.40(0.88848705)$
$\quad = \$13,481.37$

37. $P = \$15,000;\ M = \$18,937.15;\ i = 6\%$

$$M = P(1+i)^n$$
$$\$18,937.15 = \$15,000(1.06)^n$$
$$\dfrac{\$18,937.15}{\$15,000} = (1.06)^n$$
$$1.262476667 = (1.06)^n$$

Use column A of the 6% interest table to find the value closest to 1.262476667. It is 1.26247696; so $n = 4$ periods or 4 years.

38. $M = 2;\ P = 1;\ i = 6\%$

$$M = P(1+i)^n$$
$$2 = 1(1 + 0.06)^n$$
$$2 = (1.06)^n$$

Use the 6% page. Look down column A to find the value closest to 2. It is 2.01219647, which corresponds to 12 years.

Chapter 13 Summary Exercise

(a) Future value

$P = \$2,300,000; \; i = 12\%; \; n = 5$

Use the 12% page, column A, row 5 of the interest table.

$M = P(1+i)^n$
$\quad = \$2,300,000(1.12)^5$
$\quad = \$2,300,000(1.76234168)$
$\quad = \$4,053,386$

Present value

$M = \$4,053,386; \; i = \dfrac{6\%}{2} = 3\%; \; n = 5 \times 2 = 10$

Use the 3% page, column B, row 10 of the interest table.

$P = M \cdot \dfrac{1}{(1+i)^n}$

$\quad = \$4,053,386 \cdot \dfrac{1}{(1.03)^{10}}$

$\quad = \$4,053,386(0.74409391)$
$\quad = \$3,016,100$

(b) Future value with 2% growth

$P = \$2,300,000; \; i = 2\%; \; n = 5$

Use the 2% page, column A, row 5 of the interest table.

$M = P(1+i)^n$
$\quad = \$2,300,000(1.02)^5$
$\quad = \$2,300,000(1.10408080)$
$\quad = \$2,539,386$

Present value

$M = \$2,539,386; \; i = \dfrac{6\%}{2} = 3\%; \; n = 5 \times 2 = 10$

Use the 3% page, column B, row 10 of the interest table.

$P = M \cdot \dfrac{1}{(1+i)^n}$

$\quad = \$2,539,386 \cdot \dfrac{1}{(1.03)^{10}}$

$\quad = \$2,539,386(0.74409391)$
$\quad = \$1,889,542$

(c) Future value

$P = \$2,300,000; \; i = 4\%; \; n = 5$

Use the 4% page, column A, row 5 of the interest table.

$M = P(1+i)^n$
$\quad = \$2,300,000(1.04)^5$
$\quad = \$2,300,000(1.21665290)$
$\quad = \$2,798,302$

Present value

$M = \$2,798,302; \; i = \dfrac{6\%}{2} = 3\%; \; n = 5 \times 2 = 10$

Use the 3% page, column B, row 10 of the interest table.

$P = M \cdot \dfrac{1}{(1+i)^n}$

$\quad = \$2,798,302 \cdot \dfrac{1}{(1.03)^{10}}$

$\quad = \$2,798,302(0.74409391)$
$\quad = \$2,082,199$

ANNUITIES AND SINKING FUNDS

14.1 Amount of an Annuity

For this section, use column C of the interest table in Appendix D.

1. $s_{\overline{15}|0.03}$

Use row 15 of the 3% table.

18.59891389

3. $s_{\overline{10}|0.09}$

Use row 10 of the 9% table.

15.19292972

5. $S = \$850 \cdot s_{\overline{28}|0.06}$

Use row 28 of the 6% table.

$S = \$850(68.52811162)$
$ = \$58,248.89$

Interest

$\$58,248.89 - (28 \times \$850) = \$34,448.89$

7. $S = \$1000 \cdot s_{\overline{25}|0.08}$

Use row 25 of the 8% table.

$S = \$1000(73.10593995)$
$ = \$73,105.94$

Interest

$\$73,105.94 - (25 \times \$1000) = \$48,105.94$

9. $i = \dfrac{10\%}{4} = 2\frac{1}{2}\%$; $n = 8 \times 4 = 32$ periods

$S = \$1400 \cdot s_{\overline{32}|0.025}$

Use row 32 of the $2\frac{1}{2}\%$ table.

$S = \$1400(48.15027751)$
$ = \$67,410.39$

Interest

$\$67,410.39 - (32 \times \$1400) = \$22,610.39$

11. $i = \dfrac{9\%}{12} = \dfrac{3}{4}\%$; $n = 4 \times 12 = 48$ periods

$S = \$800 \cdot s_{\overline{48}|0.0075}$

Use row 48 of the $\frac{3}{4}\%$ table.

$S = \$800(57.52071111)$
$ = \$46,016.57$

Interest

$\$46,016.57 - (48 \times \$800) = \$7616.57$

13. For an annuity due, use one additional period. Here, there are 8 periods, so look up $8 + 1 = 9$ in the table. Then, subtract one payment of \$1200.

$i = 7.5\% = 0.075$
$S = \$1200 \cdot s_{\overline{9}|0.075} - \$1200$

Use row 9 of the $7\frac{1}{2}\%$ table.

$S = \$1200(12.22984883) - \$1200$
$ = \$13,475.82$

Interest

$\$13,475.82 - (8 \times \$1200) = \$3875.82$

15. For an annuity due, use one additional period. There are 6 periods, so look up $6 + 1 = 7$ in the table. Then, subtract one payment of \$17,544.

$i = 8\% = 0.08$
$S = \$17,544 \cdot s_{\overline{7}|0.08} - \$17,544$

Use row 7 of the 8% table.

$S = \$17,544(8.92280336) - \$17,544$
$ = \$138,997.66$

Interest

$\$138,997.66 - (6 \times \$17,544) = \$33,733.66$

17. $\$900 \cdot s_{\overline{10}|0.05}$
$ = \$900(12.57789254)$
$ = \$11,320.10$

19. $\$900 \cdot s_{\overline{10}|0.09}$
$ = \$900(15.19292972)$
$ = \$13,673.64$

21. $\$900 \cdot s_{\overline{10}|0.13}$
$ = \$900(18.41974915)$
$ = \$16,577.77$

23. $900 \cdot s_{\overline{20}|0.07}$
  $= \$900(40.99549232)$
  $= \$36,895.94$

25. $900 \cdot s_{\overline{20}|0.11}$
  $= \$900(64.20283215)$
  $= \$57,782.55$

27. $900 \cdot s_{\overline{30}|0.05}$
  $= \$900(66.43884750)$
  $= \$59,794.96$

29. $900 \cdot s_{\overline{30}|0.09}$
  $= \$900(136.30753855)$
  $= \$122,676.78$

31. $900 \cdot s_{\overline{30}|0.13}$
  $= \$900(293.19921506)$
  $= \$263,879.29$

33. Answers will vary.

35. For an annuity due, use one additional period. There are $4 \times 10 = 40$ periods, so look up $40 + 1 = 41$ in the table. Then, subtract one payment of $500.

$$i = \frac{7\%}{4} = 1\tfrac{3}{4}\% = 0.0175$$

$$S = \$500 \cdot s_{\overline{41}|0.0175} - \$500$$
$$= \$500(59.23573124) - \$500$$
$$= \$29,117.87$$

Interest

$$\$29,117.87 - (40 \times \$500) = \$9117.87$$

37. For an annuity due, use one additional period. There are $4 \times 9 = 36$ periods, so look up $36 + 1 = 37$ in the table. Then, subtract one payment of $100.

$$i = \frac{8\%}{4} = 2\% = 0.02$$

$$S = \$100 \cdot s_{\overline{37}|0.02} - \$100$$
$$= \$100(54.03425453) - \$100$$
$$= \$5303.43$$

Interest

$$\$5303.43 - (36 \times \$100) = \$1703.43$$

39. $S = \$20,000;\ i = 7\tfrac{1}{2}\% = 0.075;\ n = 10$

$$R = \frac{\$20,000}{s_{\overline{10}|0.075}} = \frac{\$20,000}{14.14708750}$$

$$= \$1413.72$$

41. $S = \$50,000;\ i = \dfrac{12\%}{4} = 3\%;\ n = 8 \times 4 = 32$

$$R = \frac{\$50,000}{s_{\overline{32}|0.03}}$$

$$= \frac{\$50,000}{52.50275852}$$

$$= \$952.33$$

43. $S = \$15,000;\ R = \$450;\ i = \dfrac{6\%}{4} = 1\tfrac{1}{2}$

$$s_{\overline{n}|0.015} = \frac{\$15,000}{\$450} = 33.\overline{3}$$

Look at the $1\tfrac{1}{2}\%$ page. Go down column C for the first number that is at least equal to $33.\overline{3}$. Row 28 gives 34.48147867, so 28 payments will be needed. These 28 payments will produce a total of $450(34.48147867) = \$15,516.67$.

45. $S = \$40,000;\ R = \$750;\ i = \dfrac{10\%}{12} = \tfrac{5}{6}\%$

$$s_{\overline{n}|0.008\overline{3}} = \frac{\$40,000}{\$750} = 53.\overline{3}$$

Look at the $\tfrac{5}{6}\%$ page. Go down column C for the first number that is at least equal to $53.\overline{3}$. Row 45 gives 54.32787575, so 45 payments will be needed. These 45 payments will produce a total of $750(54.32787575) = \$40,745.91$.

47. $i = \dfrac{8\%}{4} = 2\%;\ n = 3\tfrac{1}{2} \times 4 = 14$

$$S = \$450 \cdot s_{\overline{14}|0.02}$$
$$= \$450(15.97393815)$$
$$= \$7188.27$$

49. $i = \dfrac{12\%}{12} = 1\%;\ n = 4 \times 12 = 48$

(a) $S = \$300 \cdot s_{\overline{48}|0.01}$
  $= \$300(61.22260777)$
  $= \$18,366.78$

(b) Interest

  $\$18,366.78 - (48 \times \$300) = \$3966.78$

51. This is an annuity due, so use one additional period, and subtract the amount of one payment.

$$S = \$2435 \cdot s_{\overline{9}|0.06} - \$2435$$
$$= \$2435(11.49131598) - \$2435$$
$$= \$25,546.35$$

Now find the compound amount for an additional 5 years. Use row 5 of column A on the 6% page.

$25,546.35(1.33822558) = $34,186.78

Interest

$34,186.78 − (8 × $2435) = $14,706.78

53. $i = \dfrac{4\%}{12} = \dfrac{1}{3}\%$

$s_{\overline{n}|0.00\overline{3}} = \dfrac{\$4000}{\$125} = 32$

$n = 31$ months (Use column C.)

The total amount is

$125(32.60113110) = $4075.14.

55. **(a)** $i = \dfrac{8\%}{4} = 2\%$; $n = 10 \times 4 = 40$

$S = \$250 \cdot s_{\overline{40}|0.02}$
  $= \$250(60.40198318)$
  $= \$15,100.50$

(b) $i = \dfrac{6\%}{4} = 1\frac{1}{2}\%$; $n = 40$

$S = \$250 \cdot s_{\overline{40}|0.015}$
  $= \$250(54.26789391)$
  $= \$13,566.97$

14.2   Present Value of an Annuity

For this section, use column D of the interest table in Appendix D.

1. $a_{\overline{15}|0.075}$

Use row 15, column D of the $7\frac{1}{2}\%$ table.

8.82711975

3. $a_{\overline{15}|0.12}$

Use row 15, column D of the 12% table.

6.81086449

5. $R = \$2400$; $n = 9$; $i = 8\% = 0.08$

$A = \$2400 \cdot a_{\overline{9}|0.08}$
  $= \$2400(6.24688791)$
  $= \$14,992.53$

7. $R = \$800$; $n = 10 \times 2 = 20$;

$i = \dfrac{6\%}{2} = 3\% = 0.03$

$A = \$800 \cdot a_{\overline{20}|0.03}$
  $= \$800(14.87747486)$
  $= \$11,901.98$

9. $R = \$400$; $n = 5 \times 4 = 20$;

$i = \dfrac{8\%}{4} = 2\% = 0.02$

$A = \$400 \cdot a_{\overline{20}|0.02}$
  $= \$400(16.35143334)$
  $= \$6540.57$

11. Answers will vary.

13. $R = \$85,480$; $n = 20$; $i = 8\% = 0.08$

$A = \$85,480 \cdot a_{\overline{20}|0.08}$
  $= \$85,480(9.81814741)$
  $= \$839,255.24$

15. $R = \$8000$; $n = 10 \times 4 = 40$;

$i = \dfrac{6\%}{4} = 1\frac{1}{2}\% = 0.015$

$A = \$8000 \cdot a_{\overline{40}|0.015}$
  $= \$8000(29.91584520)$
  $= \$239,326.76$

17. $R = \$1200$; $n = 7 \times 2 = 14$;

$i = \dfrac{8\%}{2} = 4\% = 0.04$

$A = \$1200 \cdot a_{\overline{14}|0.04}$
  $= \$1200(10.56312293)$
  $= \$12,675.75$

Interest

$(14 \times \$1200) − \$12,675.75 = \$4124.25$

19. $R = \$15,000$; $n = 25$;

(a) $i = 8\% = 0.08$

$A = \$15,000 \cdot a_{\overline{25}|0.08}$
  $= \$15,000(10.67477619)$
  $= \$160,121.64$

(b) $i = 12\% = 0.12$

$A = \$15,000 \cdot a_{\overline{25}|0.12}$
  $= \$15,000(7.84313911)$
  $= \$117,647.09$

21. (a) $R = \$3600$; $n = 4 \times 4 = 16$;

$$i = \frac{8\%}{4} = 2\% = 0.02$$

$$A = 3600 \cdot a_{\overline{16}|0.02}$$
$$= \$3600(13.57770931)$$
$$= \$48,879.75$$

(b) $R = \$700$; $n = 8 \times 4 = 32$;

$$i = \frac{8\%}{4} = 2\% = 0.02$$

$$S = \$700 \cdot s_{\overline{32}|0.02}$$
$$= \$700(44.22702961)$$
$$= \$30,958.92$$

No, he will not have enough money available.

23. Use the present value formula.
$R = \$4000$; $n = 20 \times 2 = 40$;

$$i = \frac{10\%}{2} = 5\% = 0.05$$

$$A = \$4000 \cdot a_{\overline{40}|0.05}$$
$$= \$4000(17.15908635)$$
$$= \$68,636.35$$

Equivalent cash price

$$\$11,000 + \$68,636.35 = \$79,636.35$$

25. $R = \$8000$; $n = 12$; $i = 10\% = 0.10$

$$A = \$8000a \cdot {}_{\overline{12}|0.10}$$
$$= \$8000(6.81369182)$$
$$= \$54,509.53$$

Equivalent cash price

$$\$51,000 + \$54,509.53 = \$105,509.53$$

Accept the second offer because it is more than the first offer of \$100,000.

27. $R = \$21,000$; $n = 5 \times 4 = 20$;
$i = \frac{10}{4}\% = 2\frac{1}{2}\% = 0.025$

$$A = \$21,000 \cdot a_{\overline{20}|0.025}$$
$$= \$21,000(15.58916229)$$
$$= \$327,372.41$$

Equivalent cash price

$$\$80,000 + \$327,372.41 = \$407,372.41$$

Accept the cash offer of \$420,000 today.

29. Find the amount of annuity. (Use column C.)

$$R = \$12,000; n = 5 \times 2 = 10;$$

$$i = \frac{8\%}{2} = 4\% = 0.04$$

$$S = R \cdot s_{\overline{10}|0.04}$$
$$= \$12,000(12.00610712)$$
$$= \$144,073.29$$

Now, find the present value of the annuity. (Use column D)

$$R = \$10,000; n = 15 \times 2 = 30;$$

$$i = \frac{8\%}{2} = 4\% = 0.04$$

$$A = \$10,000 \cdot a_{\overline{30}|0.04}$$
$$= \$10,000(17.29203330)$$
$$= \$172,920.33$$

The shortage is

$$\$28,847.04 \ (\$172,920.33 - \$144,073.29).$$

31. First, find the lump sum deposit (present value) needed. (Use column D.)

$$R = \$12,000; n = 4; i = 8\% = 0.08$$

$$A = \$12,000 \cdot a_{\overline{4}|0.08}$$
$$= \$12,000(3.31212684)$$
$$= \$39,745.52$$

Now, find the present value of this lump sum deposit.

$$A = \$39,745.52; n = 5; i = 8\% = 0.08 \text{ (Use column B.)}$$

$$P = A \cdot \frac{1}{(1+i)^n}$$

$$= \$39,745.52(0.68058320)$$
$$= \$27,050.13$$

14.3   Sinking Funds

For this section, use column E of the interest table in Appendix D.

1. $\dfrac{1}{s_{\overline{12}|0.075}}$

Use row 12 of the $7\frac{1}{2}\%$ table.

0.05427783

3. $\dfrac{1}{s_{\overline{40}|0.09}}$

Use row 40 of the 9% table.

0.00295961

5. $S = \$8500$; $n = 4$; $i = 5\frac{1}{2}\% = 0.055$

$$R = \$8500\left(\dfrac{1}{s_{\overline{4}|0.055}}\right)$$
$$= \$8500(0.23029449)$$
$$= \$1957.50$$

7. $S = \$14,000$; $n = 20$; $i = \dfrac{8\%}{4} = 2\% = 0.02$

$$R = \$14,000\left(\dfrac{1}{s_{\overline{20}|0.02}}\right)$$
$$= \$14,000(0.04115672)$$
$$= \$576.19$$

9. Answers will vary.

11. (a) $R = \$10,000$; $n = 25$; $i = 8\% = 0.08$

$$A = \$10,000 \cdot a_{\overline{25}|0.08}$$
$$= \$10,000(10.67477619)$$
$$= \$106,747.76$$

(b) $S = \$106,747.76$; $n = 25 \times 2 = 50$;

$$i = \dfrac{8\%}{2} = 4\% = 0.04$$

$$R = \$106,747.76\left(\dfrac{1}{s_{\overline{50}|0.04}}\right)$$
$$= \$106,747.76(0.00655020)$$
$$= \$699.22$$

13. (a) $R = \$12,000$; $n = 9 \times 4 = 36$;

$$i = \dfrac{8\%}{4} = 2\% = 0.02$$

$$A = \$12,000 \cdot a_{\overline{36}|0.02}$$
$$= \$12,000(25.48884248)$$
$$= \$305,866.11$$

(b) $S = \$305,866.11$; $n = 25 \times 2 = 50$;

$$i = \dfrac{8\%}{2} = 4\% = 0.04$$

$$R = \$305,866.11\left(\dfrac{1}{s_{\overline{50}|0.04}}\right)$$
$$= \$305,866.11(0.00655020)$$
$$= \$2003.48$$

15. $S = \$28,000$; $n = 3 \times 4 = 12$;

$$i = \dfrac{8\%}{4} = 2\% = 0.02$$

(a) $R = \$28,000\left(\dfrac{1}{s_{\overline{12}|0.02}}\right)$

$$= \$28,000(0.07455960)$$
$$= \$2087.67$$

(b)   Interest

$$\$28,000 - (12 \times \$2087.67) = \$2947.96$$

17. $S = \$110,000$; $n = 9$; $i = 6\% = 0.06$;

$$R = \$110,000\left(\dfrac{1}{s_{\overline{9}|0.06}}\right)$$
$$= \$110,000(0.08702224)$$
$$= \$9572.45$$

19. $S = \$4,000,000$; $n = 8$; $i = 6\% = 0.06$

$$R = \$4,000,000\left(\dfrac{1}{s_{\overline{8}|0.06}}\right)$$
$$= \$4,000,000(0.10103594)$$
$$= \$404,143.76$$

Interest

$$\$4,000,000 - (8 \times \$404,143.76) = \$766,849.92$$

21.

	Begining of Period		End of Period	
Period	Accumulated Amount	Periodic Deposit	Interest Earned	Accumulated Amount
1	$0	$213,090.95	$0	$213,090.95
2	$213,090.95	$213,090.95	$17,047.28	$443,229.18
3	$443,229.18	$213,090.95	$35,458.33	$691,778.46
4	$691,778.46	$213,090.98	$55,342.28	$960,211.72

23. (a) $S = \$2,300,000$; $n = 3 \times 12 = 36$; $i = \dfrac{12\%}{12} = 1\% = 0.01$

$$R = \$2,300,000 \left(\frac{1}{s_{\overline{36}|0.01}} \right)$$

$$= \$2,300,000(0.02321431)$$
$$= \$53,392.91$$

(b) $S = \$2,300,000$; $n = 4 \times 12 = 48$; $i = \dfrac{12\%}{12} = 1\% = 0.01$

$$R = \$2,300,000 \left(\frac{1}{s_{\overline{48}|0.01}} \right)$$

$$= \$2,300,000(0.01633384)$$
$$= \$37,567.83$$

Chapter 14 Review Exercises

1. $R = \$1500$; $n = 22$, $i = 6\frac{1}{2}\%$

$S = R \cdot s_{\overline{n}|i}$
$= \$1500 \cdot s_{\overline{22}|0.065}$
$= \$1500(46.10163573)$
$= \$69,152.45$

2. $R = \$1000$; $n = 12 \times 2 = 24$; $i = \dfrac{6\%}{2} = 3\%$

$S = R \cdot s_{\overline{n}|i}$
$= \$1000 \cdot s_{\overline{24}|0.03}$
$= \$1000(34.42647022)$
$= \$34,426.47$

3. $R = \$3000$; $n = 6\frac{3}{4} \times 4 = 27$;

$i = \dfrac{10\%}{4} = 2\frac{1}{2}\%$

$S = R \cdot s_{\overline{n}|i}$
$= \$3000 \cdot s_{\overline{27}|0.025}$
$= \$3000(37.91200073)$
$= \$113,736.00$

4. $R = \$1000$; $n = 3\frac{1}{2} \times 12 = 42$;

$i = \dfrac{9\%}{12} = \dfrac{3}{4}\%$

$S = R \cdot s_{\overline{n}|i}$
$= \$1000 \cdot s_{\overline{42}|0.0075}$
$= \$1000(49.15329148)$
$= \$49,153.29$

5. $R = \$18,000;\ n = 7 + 1 = 8;\ i = 6\frac{1}{2}\%$

$$\begin{aligned} S &= R \cdot s_{\overline{n}|i} - R \\ &= \$18,000 \cdot s_{\overline{8}|0.065} - \$18,000 \\ &= \$18,000(10.07685648) - \$18,000 \\ &= \$181,383.42 - \$18,000 \\ &= \$163,383.42 \end{aligned}$$

6. $R = \$3500;\ n = 5\frac{1}{2} \times 4 + 1 = 23;$

$$i = \frac{8\%}{4} = 2\%$$

$$\begin{aligned} S &= R \cdot s_{\overline{n}|i} - R \\ &= \$3500 \cdot s_{\overline{23}|0.02} - \$3500 \\ &= \$3500(28.84496321) - \$3500 \\ &= \$100,957.37 - \$3500 \\ &= \$97,457.37 \end{aligned}$$

7. $R = \$803.47;\ n = 3\frac{3}{4} \times 4 = 15;$

$$i = \frac{12\%}{4} = 3\%$$

$$\begin{aligned} S &= R \cdot s_{\overline{n}|i} \\ &= \$803.47 s_{\overline{15}|0.03} \\ &= \$803.47(18.59891389) \\ &= \$14,943.67 \end{aligned}$$

Interest

$$\begin{aligned} I &= \$14,943.67 - (15 \times \$803.47) \\ &= \$2891.62 \end{aligned}$$

8. $R = \$7500;\ n = 7\frac{1}{2} \times 2 = 15;$

$$i = \frac{5\%}{2} = 2\frac{1}{2}\%$$

$$\begin{aligned} S &= R \cdot s_{\overline{n}|i} \\ &= \$7500 s_{\overline{15}|0.025} \\ &= \$7500(17.93192666) \\ &= \$134,489.45 \end{aligned}$$

Interest

$$\begin{aligned} I &= \$134,489.45 - (15 \times \$7500) \\ &= \$21,989.45 \end{aligned}$$

9. $R = \$4200;\ n = 15;\ i = 7\%$

$$\begin{aligned} A &= R \cdot a_{\overline{n}|i} \\ &= \$4200 \cdot a_{\overline{15}|0.07} \\ &= \$4200(9.10791401) \\ &= \$38,253.24 \end{aligned}$$

10. $R = \$800;\ n = 4\frac{1}{2} \times 2 = 9;\ i = \frac{8\%}{2} = 4\%$

$$\begin{aligned} A &= R \cdot a_{\overline{n}|i} \\ &= \$800 \cdot a_{\overline{9}|0.04} \\ &= \$800(7.43533161) \\ &= \$5948.27 \end{aligned}$$

11. $R = \$450;\ n = 5\frac{1}{4} \times 4 = 21;\ i = \frac{6\%}{4} = 1\frac{1}{2}\%$

$$\begin{aligned} A &= R \cdot a_{\overline{n}|i} \\ &= \$450 \cdot a_{\overline{21}|0.015} \\ &= \$450(17.90013673) \\ &= \$8055.06 \end{aligned}$$

12. $R = \$125;\ n = 4\frac{1}{6} \times 12 = 50;\ i = \frac{10\%}{12} = \frac{5}{6}\%$

$$\begin{aligned} A &= R \cdot a_{\overline{n}|i} \\ &= \$125 \cdot a_{\overline{50}|0.008\overline{3}} \\ &= \$125(40.75442288) \\ &= \$5094.30 \end{aligned}$$

13. $S = \$85,000;\ n = 6;\ i = 9\%$

$$\begin{aligned} R &= S \cdot \left(\frac{1}{s_{\overline{n}|i}}\right) \\ &= \$85,000\left(\frac{1}{s_{\overline{6}|0.09}}\right) \\ &= \$85,000(0.13291978) \\ &= \$11,298.18 \end{aligned}$$

14. $S = \$42,000;\ n = 26;\ i = \frac{6\%}{4} = 1\frac{1}{2}\%$

$$\begin{aligned} R &= S \cdot \left(\frac{1}{s_{\overline{n}|i}}\right) \\ &= \$42,000\left(\frac{1}{s_{\overline{26}|0.015}}\right) \\ &= \$42,000(0.03173196) \\ &= \$1332.74 \end{aligned}$$

15. $S = \$100,000;\ n = 9;\ i = \frac{12\%}{2} = 6\%$

$$\begin{aligned} R &= S \cdot \left(\frac{1}{s_{\overline{n}|i}}\right) \\ &= \$100,000\left(\frac{1}{s_{\overline{9}|0.06}}\right) \\ &= \$100,000(0.08702224) \\ &= \$8702.22 \end{aligned}$$

16. $S = \$35,000$; $n = 47$; $i = \dfrac{9\%}{12} = \dfrac{3}{4}\%$

$$R = S \cdot \left(\dfrac{1}{s_{\overline{n}|i}}\right)$$
$$= \$35,000 \left(\dfrac{1}{s_{\overline{47}|0.0075}}\right)$$
$$= \$35,000(0.01782532)$$
$$= \$623.89$$

17. $R = \$800$; $n = 10 \times 4 = 40$; $i = \dfrac{10\%}{4} = 2\frac{1}{2}\%$

$$S = R \cdot s_{\overline{n}|i}$$
$$= \$800 \cdot s_{\overline{40}|0.025}$$
$$= \$800(67.40255354)$$
$$= \$53,922.04$$

18. $S = \$100,000$; $R = \$600$; $i = 10\%$

$$s_{\overline{n}|i} = \dfrac{S}{R}$$
$$s_{\overline{n}|0.10} = \dfrac{\$100,000}{\$600}$$
$$= 166.\overline{6}$$

Look at the 10% page. Go down column C for the first number that is at least equal to $166.\overline{6}$. Row 31 gives 181.94342496, so 31 payments will be needed. 31 years are required.

19. $S = \$45,000$; $n = 4 \times 12 = 48$; $i = \dfrac{12\%}{12} = 1\%$

$$R = \left(\dfrac{1}{s_{\overline{n}|i}}\right)$$
$$= \$45,000 \left(\dfrac{1}{s_{\overline{48}|0.01}}\right)$$
$$= \$45,000(0.01633384)$$
$$= \$735.02$$

20. $M = \$7500$; $n = 3 \times 2 = 6$; $i = \dfrac{10\%}{2} = 5\%$

$$P = \dfrac{A}{(1+i)^n} \quad \text{(Use column B.)}$$
$$= \dfrac{\$7500}{(1+0.05)^6}$$
$$= \$7500(0.74621540)$$
$$= \$5596.62$$

21. $A = \$28,000$; $n = 17$; $i = \dfrac{6\%}{12} = \dfrac{1}{2}\%$

$$P = \dfrac{A}{(1+i)^n}$$
$$= \dfrac{\$28,000}{(1+0.005)^{17}}$$
$$= \$28,000(0.91870684)$$
$$= \$25,723.79$$

22. $A = \$48,000$; $n = 4 \times 4 = 16$; $i = \dfrac{6\%}{4} = 1\frac{1}{2}\%$

$$R = S \cdot \left(\dfrac{1}{s_{\overline{n}|i}}\right)$$
$$= S \cdot \left(\dfrac{1}{s_{\overline{16}|0.015}}\right)$$
$$= \$48,000(0.05576508)$$
$$= \$2676.72$$

23. $\$680,000 - \$240,000 = \$440,000$

$S = \$440,000$; $n = 5\frac{1}{2} \times 2 = 11$; $i = \dfrac{8\%}{2} = 4\%$

$$R = S \cdot \left(\dfrac{1}{s_{\overline{n}|i}}\right)$$
$$= \$440,000 \cdot \left(\dfrac{1}{s_{\overline{11}|0.04}}\right)$$
$$= \$440,000(0.07414904)$$
$$= \$32,625.58$$

24. $R = \$500$; $n = (24 \times 2) + 1 = 49$; $i = \dfrac{10\%}{2} = 5\%$

$$S = R \cdot s_{\overline{n}|i} - R$$
$$= R \cdot s_{\overline{49}|0.05} - R$$
$$= \$500(198.42666259) - \$500$$
$$= \$99,213.33 - \$500$$
$$= \$98,713.33$$

25. $R = \$11,546.48$; $n = 5 \times 4 = 20$; $i = \dfrac{10\%}{4} = 2\frac{1}{2}\%$

$$A = R \cdot a_{\overline{n}|i}$$
$$= \$11,546.48 \cdot a_{\overline{20}|0.025}$$
$$= \$11,546.48(15.58916229)$$
$$= \$179,999.95$$

$$A = \$179{,}999.95; \quad n = 3 \times 12 = 36; \quad i = \frac{12\%}{12} = 1\%$$

$$P = \frac{A}{(1+i)^n}$$

$$= \frac{\$179{,}999.95}{(1+0.01)^{36}}$$

$$= \$179{,}999.95(0.69892495)$$

$$= \$125{,}806.46$$

26.

Period	Begining of Period Accumulated Amount	Periodic Deposit	End of Period Interest Earned	Accumulated Amount
1	\$0	\$21,866.87	\$0	\$21,866.87
2	\$21,866.87	\$21,866.87	\$1968.02	\$45,701.76
3	\$45,701.76	\$21,866.87	\$4113.16	\$71,681.79
4	\$71,681.79	\$21,866.85	\$6451.36	\$100,000.00

Chapter 14 Summary Exercise

(a) $R = \dfrac{1}{2} \times \$1800 = \$900; \quad n = 33$

    (1) $i = 12\%$

$$S = \$900 \cdot s_{\overline{33}|0.12}$$
$$= \$900(342.42944555)$$
$$= \$308{,}186.50$$

    (2) $i = 8\%$

$$S = \$900 \cdot s_{\overline{33}|0.08}$$
$$= \$900(145.95062044)$$
$$= \$131{,}355.56$$

Total amount

$$\$308{,}186.50 + \$131{,}355.56 = \$439{,}542.06$$

(b) $P = \$20{,}000; \quad n = 33; \quad i = 3\%$

$$A = P(1+i)^n$$
$$= \$20{,}000(1 + 0.03)^{33}$$
$$= \$20{,}000(1.03)^{33}$$
$$= \$20{,}000(2.65233524)$$
$$= \$53{,}046.70$$

(c) $R = \$53{,}046.70; \quad n = 20; \quad i = 8\%$

$$A = R \cdot a_{\overline{n}|i}$$
$$= \$53{,}046.70\, a_{\overline{20}|0.08}$$
$$= \$53{,}046.70(9.81814741)$$
$$= \$520{,}820.32$$

(d)   No, she will be short

$81,278.26($520,820.32 − $439,542.06)

BUSINESS AND CONSUMER LOANS

15.1 Open-End Credit

1. $836.15 \times 0.012 = \$10.03$

3. $389.95 \times 0.0125 = \$4.87$

5. Previous balance: $139.56

$139.56 - \$45.00 = \$94.56$
$94.56 + \$37.25 = \$131.81$

Unpaid Balance		Number of Days	Total
$139.56	×	8	$1116.48
$94.56	×	1	$ 94.56
$131.81	×	21	$2768.01
		30	$3979.05

Average daily balance

$$\frac{\$3979.05}{30} = \$132.64$$

Finance charge

$132.64 \times 0.015 = \$1.99$

Balance at the end of billing cycle

$131.81 + \$1.99 = \$133.80$

7. Previous balance: $684.32

$684.32 - \$50.00 = \$634.32$
$634.32 + \$75.75 = \$710.07$

Unpaid Balance		Number of Days	Total
$684.32	×	11	$ 7527.52
$634.32	×	4	$ 2537.28
$710.07	×	16	$11,361.12
		31	$21,425.92

Average daily balance
$$\frac{\$21,425.92}{31} = \$691.16$$
Finance charge

$691.16 \times 0.015 = \$10.37$

Balance at the end of billing cycle

$710.07 + \$10.37 = \$720.44$

9. Previous balance: $312.78

$312.78 - \$106.45 = \$206.33$
$206.33 + \$115.73 = \$322.06$
$322.06 + \$74.19 = \$396.25$
$396.25 - \$115.00 = \$281.25$

Unpaid Balance		Number of Days	Total
$312.78	×	4	$1251.12
$206.33	×	5	$1031.65
$322.06	×	4	$1288.24
$396.25	×	9	$3566.25
$281.25	×	8	$2250.00
		30	$9387.26

Average daily balance

$$\frac{\$9387.26}{30} = \$312.91$$

Finance charge

$312.91 \times 0.015 = \$4.69$

Balance at the end of billing cycle

$281.25 + \$4.69 = \$285.94$

11. Previous balance: $714.58

$714.58 + \$ 26.94 = \$741.52$
$741.52 - \$ 25.41 = \$716.11$
$716.11 + \$ 31.82 = \$747.93$
$747.93 - \$128.00 = \$619.93$
$619.93 - \$ 71.14 = \$548.79$
$548.79 + \$110.00 = \$658.79$
$658.79 + \$100.00 = \$758.79$

Unpaid Balance		Number of Days	Total
$714.58	×	4	$ 2858.32
$741.52	×	2	$ 1483.04
$716.11	×	4	$ 2864.44
$747.93	×	4	$ 2991.72
$619.93	×	9	$ 5579.37
$548.79	×	2	$ 1097.58
$658.79	×	3	$ 1976.37
$758.79	×	3	$ 2276.37
		31	$21,127.21

Average daily balance

$$\frac{\$21{,}127.21}{31} = \$681.52$$

Finance charge

$$\$681.52 \times 0.015 = \$10.22$$

Balance at the end of billing cycle

$$\$758.79 + \$10.22 = \$769.01$$

13. (a) $\$2800.35 \times 0.015 = \$42.01$

   (b) $\$2800.35 \times 0.01 = \$28.00$

   (c) $\$42.01 - \$28.00 = \$14.01$

15. (a) $\$4850.39 \times 0.015 = \$72.76$

   (b) $\$4850.39 \times 0.008 = \$38.80$

   (c) $\$72.76 - \$38.80 = \$33.96$

17. Answers will vary.

19. Answers will vary.

15.2   Installment Loans

1. Total installment cost

$$\$5000 + (60 \times \$264.94) = \$20{,}896.40$$

Finance charge

$$\$20{,}896.40 - \$17{,}400 = \$3496.40$$

3. Total installment cost

$$12 \times \$15 = \$180$$

Finance charge

$$\$180 - \$150 = \$30$$

5. Total installment cost

$$\$25 + (10 \times \$7.50) = \$100$$

Finance charge

$$\$100 - \$90 = \$10$$

7. $\dfrac{\text{Finance charge}}{\text{Amount financed}} \times 100 = \dfrac{\$157.30}{\$1850} \times 100$

$$= 8.50$$

Look at row 15 (15 payments) in Table 15.2 and find 8.50. Since 8.50 is closest to 8.53, APR = 12.5%.

9. $\dfrac{\text{Finance charge}}{\text{Amount financed}} \times 100 = \dfrac{\$28.68}{\$442} \times 100$

$$= 6.49$$

Look at row 14 (14 payments) in Table 15.2 and find 6.49. Since 6.49 is closest to 6.52, APR = 10.25%.

11. $\dfrac{\text{Finance charge}}{\text{Amount financed}} \times 100 = \dfrac{\$13.25}{\$145} \times 100$

$$= 9.14$$

Look at row 18 (18 payments) in Table 15.2 and find 9.14.
APR = 11.25%.

13. Answers will vary.

15. Cost = \$375; Payment = \$33.16;
Number of payments = 12

Finance charge

$$(12 \times \$33.16) - \$375 = \$22.92$$

$\dfrac{\text{Finance charge}}{\text{Amount financed}} \times 100 = \dfrac{\$22.92}{\$375} \times 100$

$$= 6.11$$

Look at row 12 in Table 15.2 and find 6.11. Since 6.11 is closest to 6.06, APR = 11%.

17. $\$400 \times 0.25 = \$100$ *down payment*
     $\$400 - \$100 = \$300$ *amount financed*
$\$100 + (8 \times \$39.32) = \$414.56$ *total paid*
     $\$414.56 - \$400 = \$14.56$ *finance charge*

$\dfrac{\text{Finance charge}}{\text{Amount financed}} \times 100 = \dfrac{\$14.56}{\$300} \times 100$

$$= 4.85$$

Look at row 8 (8 payments) in Table 15.2 and find 4.85.
APR = 12.75%

19. Cost = 650,000 pesos; Down payment = 100,000 pesos; Payment = 26,342.18 pesos; 24 payments

(a) Total cost

$$100{,}000 + (24 \times 26{,}342.18)$$
$$= 732{,}212.32 \text{ pesos}$$

(b) $\dfrac{\text{Finance charge}}{\text{Amount financed}} \times 100$

$$= \frac{24(26{,}342.18) - 550{,}000}{550{,}000} \times 100$$

$$= 14.95$$

Look at row 24 (24 payments) in Table 15.2.
APR = 13.75%

21. Cost = \$180,000; Payment = \$6974.66; 30 payments

(a) 30 × \$6974.66 = \$209,239.80

(b) $\dfrac{\text{Finance charge}}{\text{Amount financed}} \times 100$

$= \dfrac{(30 \times \$6974.66) - \$180,000}{\$180,000} \times 100$

$= 16.24$

Look at row 30 (30 payments) in Table 15.2.
APR = 12%

23. Answers will vary.

15.3   Early Payoffs of Loans

1. Interest to day 56

$I = PRT$

$= \$6500 \times 0.08 \times \dfrac{56}{360}$

$= \$80.89$

Payment − interest due = reduced principal

\$1500 − \$80.89 = \$1419.11
\$6500 − \$1419.11 = \$5080.89   *balance owed*

$I = PRT$

$= \$5080.89 \times 0.08 \times \dfrac{44}{360}$

$= \$49.68$

Balance due

\$5080.89 + \$49.68 = \$5130.57

Total interest

\$80.89 + \$49.68 = \$130.57

3. Interest to day 45

$I = PRT$

$= \$8500 \times 0.12 \times \dfrac{45}{360}$

$= \$127.50$

Payment − interest due = reduced principal

\$5000 − \$127.50 = \$4872.50
\$8500 − \$4872.50 = \$3627.50 *balance owed*

$I = PRT$

$= \$3627.50 \times 0.12 \times \dfrac{105}{360}$

$= \$126.96$

Balance due

\$3627.50 + \$126.96 = \$3754.46

Total interest

\$127.50 + \$126.96 = \$254.46

5. Interest to day 120:

$I = PRT$

$= \$10,000 \times 0.0825 \times \dfrac{120}{360}$

$= \$275$

Payment − interest due = reduced principal

\$6000 − \$275 = \$5725
\$10,000 − \$5725 = \$4275   *balanced owed*

$I = PRT$

$= \$4275 \times 0.0825 \times \dfrac{60}{360}$

$= \$58.78$

Balance due

\$4275 + \$58.78 = \$4333.78

Total interest

\$275 + \$58.78 = \$333.78

For Exercises 7-11, use the formula

$$U = F\left(\dfrac{N}{P}\right)\left(\dfrac{1+N}{1+P}\right)$$

where U = unearned interest, F = finance charge, N = number of payments remaining, and P = total number of payments.

7. Unearned interest

$\$975 \times \left(\dfrac{30}{48}\right) \times \left(\dfrac{1+30}{1+48}\right)$

$= \$975 \times \left(\dfrac{30 \times 31}{48 \times 49}\right)$

$= \$385.52$

9. Unearned interest

$$\$460 \times \left(\frac{4}{20}\right) \times \left(\frac{1+4}{1+20}\right)$$

$$= \$460 \times \left(\frac{4 \times 5}{20 \times 21}\right)$$

$$= \$21.90$$

11. Unearned interest

$$\$3653.82 \times \left(\frac{9}{48}\right) \times \left(\frac{1+9}{1+48}\right)$$

$$= \$3653.82 \times \left(\frac{9 \times 10}{48 \times 49}\right)$$

$$= \$139.81$$

13. Answers will vary.

15. Answers will vary.

17. $P = \$125,000$; $i = 11\%$; time $= 250$ days

(a) $I = PRT$

$$= \$125,000 \times 0.11 \times \frac{172}{360}$$

$$= \$6569.44$$

(b) Balance due

$$\$125,000 + \$6569.44 = \$131,569.44$$

19. Finance charge: $240

Total number of payments: 36

Remaining number of payments: 21

Unearned interest

$$\$240 \times \left(\frac{21}{36}\right) \times \left(\frac{1+21}{1+36}\right)$$

$$= \$240 \times \left(\frac{21 \times 22}{36 \times 37}\right)$$

$$= \$83.24$$

21. Total amount scheduled to be paid

$$= \$100 + (18 \times \$45.20) = \$913.60$$

Finance charge

$$\$913.60 - \$800 = \$113.60$$

Total number of payments: 18

Remaining number of payments: 12

(a) Unearned interest

$$\$113.60 \times \left(\frac{12}{18}\right) \times \left(\frac{1+12}{1+18}\right)$$

$$= \$113.60 \times \left(\frac{12 \times 13}{18 \times 19}\right)$$

$$= \$51.82$$

(b) Amount necessary to pay loan

$$= (12 \times \$45.20) - \$51.82$$

$$= \$490.58$$

23. $P = \$104,500$; $i = 11\%$;

Feb. 18-May 15 = 86 days

Feb. 18-Mar. 20: 30 days

$I = PRT$

$$= \$104,500 \times 0.11 \times \frac{30}{360}$$

$$= \$957.92$$

Payment − interest due = reduced principal

$\$38,000 - \$957.92 = \$37,042.08$

$\$104,500 - \$37,042.08 = \$67,457.92$ *balance owed*

Mar. 20-April 16: 27 days

$I = PRT$

$$= \$67,457.92 \times 0.11 \times \frac{27}{360}$$

$$= \$556.53$$

Payment − interest due = reduced principal

$\$27,200 - \$556.53 = \$26,643.47$

$\$67,457.92 - \$26,643.47 = \$40,814.45$ *balance owed*

$86 - 30 - 27 = 29$ *days remaining*

$I = PRT$

$$= \$40,814.45 \times 0.11 \times \frac{29}{360}$$

$$= \$361.66$$

Balance due

$$\$40,814.45 + \$361.66 = \$41,176.11$$

Total interest

$$\$957.92 + \$556.53 + \$361.66 = \$1876.11$$

15.4   Personal Property Loans

For this section, use column F of the interest table in Appendix D.

1. $\dfrac{1}{a\,\overline{15}|0.075}$

Use row 15 of the $7\frac{1}{2}\%$ table.

0.11328724

3. $\dfrac{1}{a_{\overline{36}|0.10}}$

Use row 36 of the 10% table.

0.10334306

5. $A = \$1850$; $n = 2$; $i = 7\frac{1}{2}\%$

$$R = A\left(\frac{1}{a_{\overline{2}|0.075}}\right)$$
$$= \$1850(0.55692771)$$
$$= \$1030.32$$

7. $A = \$4500$; $n = 2 \times 7\frac{1}{2} = 15$; $i = \dfrac{8\%}{2} = 4\%$

$$R = A\left(\frac{1}{a_{\overline{15}|0.04}}\right)$$
$$= \$4500(0.08994110)$$
$$= \$404.73$$

9. $A = \$96,000$; $n = 4 \times 7\frac{3}{4} = 31$; $i = \dfrac{8\%}{4} = 2\%$

$$R = A\left(\frac{1}{a_{\overline{31}|0.02}}\right)$$
$$= \$96,000(0.04359635)$$
$$= \$4185.25$$

11. $A = \$4876$; $n = 12 \times 3 = 36$; $i = \dfrac{12\%}{12} = 1\%$

$$R = A\left(\frac{1}{a_{\overline{36}|0.01}}\right)$$
$$= \$4876(0.03321431)$$
$$= \$161.95$$

13. Using Table 15.3, read down 9%
and across 36 (36 payments): 0.0318.
Payment: $\$4800(0.0318) = \$152.64$
Total paid: $36(\$152.64) = \$5495.04$
Finance charge: $\$5495.04 - \$4800 = \$695.04$

15. Using Table 15.3, read across 13% and down
48 (48 payments): 0.026827.
Payment: $\$12,000(0.026827) = \$321.92$
Total paid: $48(\$321.92) = \$15,452.16$
Finance charge: $\$15,452.16 - \$12,000 = \$3452.16$

17. Amount financed

$$\$340,000 - \$40,000 = \$300,000$$

$$n = 4 \times 7 = 28; \quad i = \frac{12\%}{4} = 3\%$$

Quarterly payment

$$R = A\left(\frac{1}{a_{\overline{28}|0.03}}\right)$$
$$= \$300,000(0.05329323)$$
$$= \$15,987.97$$

Total interest

$$(28 \times \$15,987.97) - \$300,000 = \$147,663.16$$

19. $R = A\left(\dfrac{1}{a_{\overline{4}|0.08}}\right)$

$$= \$4000(0.30192080)$$
$$= \$1207.68$$

Payment Number	Amount of Payment	Interest for Period	Portion to Principal	Principal at End of Period
0	----	----	----	$4000.00
1	$1207.68	$320.00	$ 887.68	$3112.32
2	$1207.68	$248.99	$ 958.69	$2153.63
3	$1207.68	$172.29	$1035.39	$1118.24
4	$1207.70	$ 89.46	$1118.24	$0

21. $(\$3500 \times 7) - \$10,000 = \$14,500$

Using Table 15.3, read across 11% and down
48 (48 payments): 0.025846.
$\$14,500(0.025846) = \$374.77$

Payment Number	Amount of Payment	Interest for Period	Portion to Principal	Principal at End of Period
0	----	----	----	$14,500.00
1	$374.77	$132.92	$241.85	$14,258.15
2	$374.77	$130.70	$244.07	$14,014.08
3	$374.77	$128.46	$246.31	$13,767.77
4	$374.77	$126.20	$248.57	$13,519.20
5	$374.77	$123.93	$250.84	$13,268.36

23. Answers will vary.

25. $8000; $n = 15$; $i = \dfrac{12\%}{12} = 1\%$

$$R = A\left(\dfrac{1}{a_{\overline{15}|0.01}}\right)$$
$$= \$8000(0.07212378)$$
$$= \$576.99$$

Payment Number	Amount of Payment	Interest for Period	Portion to Principal	Principal at End of Period
0	----	----	----	$8000.00
1	$576.99	$80.00	$496.99	$7503.01
2	$576.99	$75.03	$501.96	$7001.05
3	$576.99	$70.01	$506.98	$6494.07
⋮	⋮	⋮	⋮	⋮

15.5   Real Estate Loans

1. Using Table 15.4, read down $7\frac{1}{4}\%$ and across 30 (years): 6.83
($78,000 \div 1000) \times 6.83 = \$532.74$

3. Using Table 15.4, read down $8\frac{1}{2}\%$ and across 15 (years): 9.85.
($112,800 \div 1000) \times 9.85 = \$1111.08$

5. Using Table 15.4, read down $9\frac{3}{4}\%$ and across 15 (years): 10.60.
($96,500 \div 1000) \times 10.60 = \$1022.90$

7. Answers will vary.

9. Using Table 15.4, read down 7% and across 30 (years): 6.66.
($69,000 \div 1000) \times 6.66 = \$459.54$

Payment + taxes + insurance

$\$459.54 + \dfrac{\$1850 + \$450}{12} = \$651.21$

11. Using Table 15.4, read down 8% and across 30 (years): 7.34.
($58,600 \div 1000) \times 7.34 = \$430.12$

Payment + taxes + insurance

$\$430.12 + \dfrac{\$745 + \$380}{12} = \$523.87$

13. Using Table 15.4, read down $8\frac{1}{4}\%$ and across 25 (years): 7.89.
($91,580 \div 1000) \times 7.89 = \$722.57$
Payment + taxes + insurance

$\$722.57 + \dfrac{\$1326 + \$489}{12} = \$873.82$

15. Amount of loan

$145,000 - \$20,000 = \$125,000$

Using Table 15.4, read down $7\frac{1}{2}\%$ and across 15 (years): 9.27.
($125,000 \div 1000) \times 9.27 = \$1158.75$

Payment + taxes + insurance

$\$1158.75 + \dfrac{\$780 + \$2950}{12} = \$1469.58$

Yes, they are qualified for the loan.

17. Loan = $122,500; $n = 15$; $i = 7\frac{1}{2}\%$
Use Table 15.4, $7\frac{1}{2}\%$, 15 (years): 9.27

Payment number 1

($122,500 \div 1000) \times 9.27 = \$1135.58$

Interest payment

$\$122,500 \times 0.075 \times \dfrac{1}{12} = \$765.63$

Principal payment

$\$1135.58 - \$765.63 = \$369.95$

Balance of principal

$\$122,500 - \$369.95 = \$122,130.05$

Payment number 2

Interest payment

$\$122,130.05(0.075)\left(\dfrac{1}{12}\right) = \$763.31$

Principal payment

$\$1135.58 - \$763.31 = \$372.27$

Balance of principal

$\$122,130.05 - \$372.27 = \$121,757.78$

Payment Number	Total Payment	Interest Payment	Principal Payment	Balance of Principal
0	----	----	----	$122,500.00
1	$1135.58	$765.63	$369.95	$122,130.05
2	$1135.58	$763.31	$372.27	$121,757.78

19. Since the down payment is 20%, the amount financed is 80% of $110,000.

Loan $= 0.80(\$110,000) = \$88,000$

$n = 20;\ i = 8\frac{1}{2}\%$

Use Table 15.4, $8\frac{1}{2}\%$, 20 (years): 8.68

Monthly payment

$(\$88,000 \div 1000) \times 8.68 = \$763.84$

20 years $\times$ 12 payments per year $= 240$ payments

Total cost

$\$763.84 \times 240 = \$183,321.60$

Interest charges

$\$183,321.60 - \$88,000 = \$95,321.60$

21. 9%, 30 year: 8.05

Number of thousands of debt	$\times$	Factor from Table 15.4	$=$	Monthly payment
Number of thousands of debt	$\times$	8.05	$=$	$650

Number of thousands of debt $= \dfrac{\$650}{8.05} = 80.745 \approx 81$

$81 \times \$1000 = \$81,000$ *maximum mortgage*

23. Answers will vary.

Chapter 15 Review Exercises

1. $\$243 \times 0.015 = \$3.65$

2. $\$3240.60 \times 0.0125 = \$40.51$

3. $\$875.12 \times 0.0162 = \$14.18$

4. Previous balance: $634.25

$\$634.25 - \$125 = \$509.25$
$\$509.25 + \$34.26 = \$543.51$

Unpaid Balance		Number of Days	Total
$634.25	$\times$	8	$ 5074.00
$509.25	$\times$	13	$ 6620.25
$543.51	$\times$	10	$ 5435.10
		31	$17,129.35

Average daily balance

$\dfrac{\$17,129.35}{31} = \$552.56$

Finance charge

$\$552.56 \times 0.015 = \$8.29$

Balance at the end of billing cycle

$\$543.51 + \$8.29 = \$551.80$

5. Previous balance: $236.26

$\$236.26 + \$28.25 = \$264.51$
$\$264.51 - \$75.00 = \$189.51$
$\$189.51 + \$35.00 = \$224.51$
$\$224.51 - \$24.36 = \$200.15$

Unpaid Balance		Number of Days	Total
$236.26	$\times$	5	$1181.30
$264.51	$\times$	5	$1322.55
$189.51	$\times$	11	$2084.61
$224.51	$\times$	5	$1122.55
$200.15	$\times$	5	$1000.75
		31	$6711.76

Average daily balance

$\dfrac{\$6711.76}{31} = \$216.51$

Finance charge

$\$216.51 \times 0.015 = \$3.25$

Balance at the end of billing cycle.

$\$200.15 + \$3.25 = \$203.40$

6. Cash price $=$ Down payment $+$ Amount financed
$= \$400 + \$2300$
$= \$2700$

Total installment price
$=$ Down payment $+$
(Number of payments $\times$ Amount of payment)
$= \$400 + (18 \times \$139.73)$
$= \$2915.14$

Finance charge
$=$ Total installment price $-$ Cash price
$= \$2915.14 - \$2700$
$= \$215.14$

7. Cash price = Down payment + Amount financed
$$= \$800 + \$3800$$
$$= \$4600$$

Total installment price
= Down payment
  + (Number of payments × Amount of payment)
$$= \$800 + (20 \times \$212)$$
$$= \$5040$$

Finance charge
= Total installment price − Cash price
$$= \$5040 - \$4600$$
$$= \$440$$

8. Cash price = Down payment + Amount financed
$$= \$1500 + \$6500$$
$$= \$8000$$

Total installment price
= Down payment
  + (Number of payments × Amount of payment)
$$= \$1500 + (36 \times \$225)$$
$$= \$9600$$

Finance charge
= Total installment price − Cash price
$$= \$9600 - \$8000$$
$$= \$1600$$

9. $\dfrac{\text{Finance charge}}{\text{Amount financed}} \times 100 = \dfrac{\$435}{\$4100} \times 100$
$$= 10.61$$

Low at row 18 (18 payments) in Table 15.2 for 10.61.
APR = 13%

10. $\dfrac{\text{Finance charge}}{\text{Amount financed}} \times 100 = \dfrac{\$698}{\$5600} \times 100$
$$= 12.46$$

Low at row 20 (20 payments) in Table 15.2 for 12.46.
APR = 13.75%

11. $\dfrac{\text{Finance charge}}{\text{Amount financed}} \times 100 = \dfrac{\$766.48}{\$8800} \times 100$
$$= 8.71$$

Look at row 15 (15 payments) in Table 15.2 for 8.71.
APR = 12.75%

12. $\dfrac{\text{Finance charge}}{\text{Amount financed}} \times 100 = \dfrac{\$1065}{\$10,270} \times 100$
$$= 10.37$$

Look at row 20 (20 payments) in Table 15.2 for 10.37.
APR = 11.5%

13. $I = PRT$
$$= \$12,400 \times 0.085 \times \frac{62}{360}$$
$$= \$181.52$$

Payment:	$2000.00
Interest:	− 181.52
Principal reduction:	$1818.48

Amount owed:	$12,400.00
Principal reduction:	− 1818.48
Balance owed:	$10,581.52

$I = PRT$
$$= \$10,581.52 \times 0.085 \times \frac{118}{360}$$
$$= \$294.81$$

Balance owed:	$10,581.52
Interest:	+ 294.81
Balance at maturity:	$10,876.33

Since the note continues for an additional 118 days after the last payment, the interest is be added.

Interest for 62 days:	$181.52
Interest for 118 days:	+ 294.81
Total interest:	$476.33

14. $I = PRT$
$$= \$9000 \times 0.12 \times \frac{40}{360}$$
$$= \$120$$

Payment:	$2000
Interest:	− 120
Principal reduction:	$1880

Amount owed:	$9000
Principal reduction:	− 1880
Balance owed:	$7120

$I = PRT$
$$= \$7120 \times 0.12 \times \frac{80}{360}$$
$$= \$189.87$$

Balance owed:	=	$7120.00
Interest:	= +	189.87
Balance at maturity:	=	$7309.87

Since the note continues for an additional 80 days after the last payment, the interest is added.

Interest for 40 days:	$120.00
Interest for 80 days:	+ 189.87
Total interest:	$309.87

15. $I = PRT$

$$= \$6000 \times 0.11 \times \frac{30}{360}$$

$$= \$55$$

Payment:	$3200
Interest:	− 55
Principal reduction:	$3145

Amount owed:	$6000
Principal reduction:	− 3145
Balance owed:	$2855

$I = PRT$

$$= \$2855 \times 0.11 \times \frac{60}{360}$$

$$= \$52.34$$

Payment:	$2000
Interest:	− 52.34
Principal reduction:	$1947.66

Amount owed:	$2855.00
Principal reduction:	− 1947.66
Balance owed:	$ 907.34

$I = PRT$

$$= \$907.34 \times 0.11 \times \frac{30}{360}$$

$$= \$8.32$$

Balance owed:	$907.34
Interest:	+ 8.32
Balance at maturity:	$915.66

Since the note continues for additional 30 days after the last payment, the interest is added.

Interest for 30 days:	$55.00
Interest for 60 days:	52.34
Interest for 30 days:	+ 8.32
Total interest:	$115.66

16. $I = PRT$

$$= \$9000 \times 0.09 \times \frac{30}{360}$$

$$= \$67.50$$

Payment:	$3000.00
Interest:	− 67.50
Principal reduction:	$2932.50

Amount owed:	$9000.00
Principal reduction:	− 2932.50
Balance owed:	$6067.50

$I = PRT$

$$= \$6067.50 \times 0.09 \times \frac{15}{360}$$

$$= \$22.75$$

Payment:	$1500.00
Interest:	− 22.75
Principal reduction:	$1477.25

Amount owed:	$6067.50
Principal reduction:	− 1477.25
Balance owed:	$4590.25

$I = PRT$

$$= \$4590.25 \times 0.09 \times \frac{45}{360}$$

$$= \$51.64$$

Balance owed:	$4590.25
Interest:	+ 51.64
Balance at maturity:	$4641.89

Since the note continues for an additional 45 days after the last payment, the interest is added.

Interest for 30 days:	$67.50
Interest for 15 days:	22.75
Interest for 45 days:	+ 51.64
Total interest	$141.89

17. $24,000; n = 8; i = \dfrac{9\%}{2} = 4\frac{1}{2}\%$

$$R = A \left(\frac{1}{a_{\overline{8}|0.045}} \right)$$

$$= \$24,000(0.15160965)$$

$$= \$3638.63$$

18. $18,500; $n = 10$; $i = \dfrac{10\%}{2} = 5\%$

$$R = A\left(\frac{1}{a\,\overline{10}|0.05}\right)$$

$$= \$18,500(0.12950457)$$

$$= \$2395.83$$

19. $12,400; $n = 20$; $i = \dfrac{10\%}{4} = 2\frac{1}{2}\%$

$$R = A\left(\frac{1}{a\,\overline{20}|0.025}\right)$$

$$= \$12,400(0.06414713)$$

$$= \$795.42$$

20. $8600; $n = 24$; $i = \dfrac{9\%}{12} = \dfrac{3}{4}\%$

$$R = A\left(\frac{1}{a\,\overline{24}|0.0075}\right)$$

$$= \$8600(0.04568474)$$

$$= \$392.89$$

21. Use the number from Table 15.3 for 12% and 24 months: 0.047075.

Monthly payment

$9400 \times 0.047075 = \$442.51$

Finance charge

$(\$442.51 \times 24) - \$9400 = \$1220.24$

22. Use the number from Table 15.3 for 16% and 42 months: 0.03125.

Monthly payment

$7500 \times 0.03125 = \$234.38$

Finance charge

$(\$234.38 \times 42) - \$7500 = \$2343.96$

23. Use the number from Table 15.3 for 14% and 48 months: 0.027327.

Monthly payment

$9000 \times 0.027327 = \$245.94$

Finance charge

$(\$245.94 \times 48) - \$9000 = \$2805.12$

24. Use the number from Table 15.3 for 11% and 30 months: 0.038277.

Monthly payment

$15,000 \times 0.038277 = \$574.16$

Finance charge

$(\$574.16 \times 30) - \$15,000 = \$2224.80$

25. Use the numbers from Table 15.4 for $8\frac{3}{4}\%$ and 30 years: 7.87.

$(\$74,000 \div 1000) \times 7.87 = \$582.38$

26. Use the numbers from Table 15.4 for $7\frac{1}{2}\%$ and 20 years: 8.06.

$(\$65,000 \div 1000) \times 8.06 = \$523.90$

27. Use the numbers from Table 15.4 for 8% and 15 years: 9.56.

$(\$100,000 \div 1000) \times 9.56 = \$956$

28. Use the numbers from Table 15.4 for $7\frac{1}{4}\%$ and 30 years: 6.83.

$(\$120,000 \div 1000) \times 6.83 = \$819.60$

29. (a) $I = PRT$

$$= \$5600 \times 0.11 \times \frac{30}{360}$$

$$= \$51.33$$

Payment:	$1330.00
Interest:	− 51.33
Principal reduction:	$1278.67

Amount owed:	$5600.00
Principal reduction:	− 1278.67
Balance owed:	$4321.33

$I = PRT$

$$= \$4321.33 \times 0.11 \times \frac{45}{360}$$

$$= \$59.42$$

Payment:	$1655.00
Interest:	− 59.42
Principal reduction:	$1595.58

Amount owed:	$4321.33
Principal reduction:	− 1595.58
Balance owed:	$2725.75

$I = PRT$

$$= \$2725.75 \times 0.11 \times \frac{75}{360}$$

$$= \$62.47$$

Balance owed:　　　　$2725.75
Interest:　　　　　　+　62.47
Balance at maturity:　$2788.22

(b) Since the note continues for an additional 75 days after the last payment, the interest is added.

Interest for 30 days:　　$51.33
Interest for 45 days:　　59.42
Interest for 75 days:　+　62.47
Total interest:　　　　$173.22

30. Previous balance: $52.45

$52.45 − $15.00 = $37.45
$37.45 + $17.40 = $54.85
$54.85 + $23.00 = $77.85

Unpaid Balance		Number of Days	Total
$52.45	×	10	$524.50
$37.45	×	5	$187.25
$54.85	×	7	$383.95
$77.85	×	8	+ $622.80
		30	$1718.50

Average daily balance

$$\frac{\$1718.50}{30} = \$57.28$$

Finance charge

$$\$57.28 \times 0.015 = \$0.86$$

31. Total Installment cost
= Down payment
　+ (Number of payments × Amount of payment)
= $2780 + (48 × $332.84) = $18,756.32

Finance charge

= Total installment price-Cash price
= $18,756.32 − $15,780
= $2976.32

Amount financed

$15,780 − $2780 = $13,000

$$\frac{\text{Finance charge}}{\text{Amount financed}} \times 100 = \frac{\$2976.32}{\$13,000} \times 100$$

$$= 22.89$$

Look at row 48 (48 payments) in Table 15.2 for 22.89.
APR = 10.5%

32. $A = \$28,100$; $n = 4 \times 3 = 12$; $i = \dfrac{12\%}{4} = 3\%$

$$R = A\left(\frac{1}{a\,\overline{12}|0.03}\right)$$
$$= \$28,100(0.10046209)$$
$$= \$2822.98$$

33. Since the down payment is 20%, the amount financed is 80% of $90,000.
Loan = 0.80($90,000) = $72,000
Use the numbers from Table 15.4 for 30 years and 8%: 7.34

$$(\$72,000 \div 1000) \times 7.34 = \$528.48$$

Payment + taxes + insurance

$$\$528.48 + \frac{\$960 + \$252}{12} = \$629.48$$

34. (a) Use the numbers from Table 15.4 for 20 years and 8%: 8.37.

$$(\$122,500 \div 1000) \times 8.37 = \$1025.33$$

Total monthly payment

$$\$1025.33 + \frac{\$3200 + \$1275}{12} = \$1398.25$$

(b) Total payments

20 years × 12 payments per year
　= 240 payments

Total cost

$$(240 \times \$1398.25) + \$25,000 = \$360,580$$

Chapter 15 Summary Exercise

(a) Honda Accord

$$A = \$18,800; \quad n = 4 \times 12 = 48; \quad i = \frac{12\%}{12} = 1\%$$

$$R = A\left(\frac{1}{a\,\overline{48}|0.01}\right)$$

$$= \$18,800(0.02633384)$$
$$= \$495.08$$

Ford Truck

$$A = \$14,300; \quad n = 4 \times 12 = 48; \quad i = \frac{18\%}{12} = 1\tfrac{1}{2}\%$$

$$R = A\left(\frac{1}{a_{\overline{48}|0.015}}\right)$$

$$= \$14,300(0.02937500)$$

$$= \$420.06$$

Home

Use the numbers from Table 15.4 for $8\tfrac{1}{2}\%$ and 15 years: 9.85

$$(\$96,500 \div 1000) \times 9.85 = \$950.53$$

2nd mortgage on home

$$A = \$4500; \quad n = 3 \times 12 = 36; \quad i = \frac{12\%}{12} = 1\%$$

$$R = \left(\frac{1}{a_{\overline{36}|0.01}}\right)$$

$$= \$4500(0.03321431)$$

$$= \$149.46$$

Total monthly payment

$$\begin{array}{r} \$ \ 495.08 \\ 420.06 \\ 950.53 \\ + \ 149.46 \\ \hline \$2015.13 \end{array}$$

(b) Monthly expense for taxes on home

$$\frac{\$2530}{12} = \$210.83$$

Total monthly outlay

Payments on debt from (a)	$2015.13
Car insurance	215.00
Health insurance	120.00
Taxes on home	+ 210.83
	$2560.96

(c) Honda Accord

$$A = \$14,900; \quad n = 4 \times 12 = 48; \quad i = \frac{12\%}{12} = 1\%$$

$$R = A\left(\frac{1}{a_{\overline{48}|0.01}}\right)$$

$$= \$14,900(0.02633384)$$

$$= \$392.37$$

Ford Truck

$$A = \$8600; \quad n = 3 \times 12 = 36; \quad i = \frac{12\%}{12} = 1\%$$

$$R = A\left(\frac{1}{a_{\overline{36}|0.01}}\right)$$

$$= \$8600(0.03321431)$$

$$= \$285.64$$

Home

Use the numbers from Table 15.4 for 8% and 30 years: 7.34

$$(\$94,800 \div 1000) \times 7.34 = \$695.83$$

2nd mortgage on home

remains at $149.46

Car insurance

$$\$215 - \$28 = \$187$$

Health insurance

remains at $120

Taxes on home

remains at $210.83

Total monthly outlay

$$\begin{array}{r} \$ \ 392.37 \\ 285.64 \\ 695.83 \\ 149.46 \\ 187.00 \\ 120.00 \\ + \ 210.83 \\ \hline \$2041.13 \end{array}$$

(d) Reduction in monthly payments

$$\$2560.96 - \$2041.13 = \$519.83$$

Cumulative Review Exercises (Chapters 11-15)

1. $I = PRT$

   $I = \$4500 \times 0.08 \times \dfrac{5}{12}$

   $I = \$150$

2. $I = PRT$

   $I = \$6200 \times 0.097 \times \dfrac{250}{360}$

   $I = \$417.64$

3. $P = \dfrac{I}{RT}$

   $P = \dfrac{\$46.67}{0.07 \times \frac{100}{360}}$

   $P = \$2400.17$

4. $P = \dfrac{I}{RT}$

   $P = \dfrac{\$302.60}{0.125 \times \frac{70}{360}}$

   $P = \$12,449.83$

5. $R = \dfrac{I}{PT}$

   $R = \dfrac{\$50.93}{\$2100 \times \frac{90}{360}}$

   $R = 0.0970 = 9.7\%$

6. $R = \dfrac{I}{PT}$

   $R = \dfrac{\$306}{\$6800 \times \frac{120}{360}}$

   $R = 0.135 = 13.5\%$

7. $T = \dfrac{I}{PR}$

   $T = \dfrac{\$202.22}{\$9100 \times 0.10} \times 360$

   $T = 80$ days

8. $T = \dfrac{I}{PR}$

   $T = \dfrac{\$915}{\$18,300 \times 0.12} \times 360$

   $T = 150$ days

9. $B = MDT$

   $B = \$9000 \times 0.12 \times \dfrac{90}{360}$

   $B = \$270$

   $P = M - B$
   $P = \$9000 - \$270$
   $P = \$8730$

10. $B = MDT$

    $B = \$875 \times 0.065 \times \dfrac{210}{360}$

    $B = \$33.18$

    $P = M - B$
    $P = \$875 - \$33.18$
    $P = \$841.82$

11. $R = \dfrac{D}{1 - DT}$

    $R = \dfrac{0.10}{1 - (0.10 \times \frac{90}{360})}$

    $R = 0.1025 = 10.3\%$

12. $D = \dfrac{R}{1 + RT}$

    $D = \dfrac{0.12}{1 + (0.12 \times \frac{180}{360})}$

    $D = 0.1132 = 11.3\%$

For Exercises 13 and 14, use column A of the interest tables in Appendix D.

13. $P = \$1000$; $i = 4\%$; $n = 17$
    Use row 17 of the 4% interest table.

    $M = \$1000(1.94790050)$
    $\quad = \$1947.90$

14. $P = \$3520$; $i = 8\%$; $n = 10$
    Use row 10 of the 8% interest table.

    $M = \$3520(2.15892500)$
    $\quad = \$7599.42$

15. March 24 is day 83;
    June 3 is day 154.
    $154 - 83 = 71$ days
    Use Table 13.2.

    $\$12,600(1.006831119)$
    $\quad = \$12,686.07$

    $I = \$12,686.07 - \$12,600$
    $\quad = \$86.07$

16. November 20 is day 324;
February 14 is day 45.
$(365 - 324) + 45 = 86$ days
Use Table 13.2.

$$\$7500(1.008280273) = \$7562.10$$

$$I = \$7562.10 - \$7500$$
$$= \$62.10$$

17. $M = \$1000$; $i = 8\%$; $n = 7$

$$P = M \cdot \frac{1}{(1+i)^n}$$

$$= \$1000 \cdot \frac{1}{(1.08)^7}$$

Use the 8% page, column B, row 7 of the interest table.

$$P = \$1000(0.58349040)$$
$$= \$583.49$$

$$I = \$1000 - \$583.49$$
$$= \$416.51$$

18. $M = \$19,000$; $i = \dfrac{5\%}{2} = 2\frac{1}{2}\%$; $n = 9 \times 2 = 18$

$$P = M \cdot \frac{1}{(1+i)^n}$$

$$= \$19,000 \cdot \frac{1}{(1.025)^{18}}$$

Use the $2\frac{1}{2}\%$ page, column B, row 18 of the interest table.

$$P = \$19,000(0.64116591)$$
$$= \$12,182.15$$

$$I = \$19,000 - \$12,182.15$$
$$= \$6817.85$$

For Exercises 19 and 20, use column C of the interest table in Appendix D.

19. $S = \$1000 \cdot s_{\overline{8}|0.04}$

Use row 8 of the 4% table.

$$S = \$1000(9.21422626)$$
$$= \$9214.23$$

20. $S = \$2000 \cdot s_{\overline{6}|0.06}$

Use row 6 of the 6% table.

$$S = \$2000(6.97531854)$$
$$= \$13,950.64$$

For Exercises 21 and 22, use column D of the interest table in Appendix D.

21. $R = \$925$; $n = 2 \times 11 = 22$; $i = \dfrac{8\%}{2} = 4\%$

$$A = \$925 \cdot a_{\overline{22}|0.04}$$

Use row 22 of the 4% table.

$$A = \$925(14.45111533)$$
$$= \$13,367.28$$

22. $R = \$27,235$; $n = 4 \times 8 = 32$; $i = \dfrac{8\%}{4} = 2\%$

$$A = \$27,235 \cdot a_{\overline{32}|0.02}$$

Use row 32 of the 2% table.

$$A = \$27,235(23.46833482)$$
$$= \$639,160.10$$

For Exercises 23 and 24, use column E of the interest table in Appendix D.

23. $S = \$3600$; $n = 7$; $i = 8\%$

$$R = \$3600\left(\frac{1}{s_{\overline{7}|0.08}}\right)$$

Use row 7 of the 8% table.

$$R = \$3600(0.11207240)$$
$$= \$403.46$$

24. $S = \$4500$; $n = 4 \times 7 = 28$; $i = \dfrac{10\%}{4} = 2\frac{1}{2}\%$

$$R = \$4500\left(\frac{1}{s_{\overline{28}|0.025}}\right)$$

Use row 28 of the $2\frac{1}{2}\%$ table.

$$R = \$4500(0.02508793)$$
$$= \$112.90$$

25. $P = \$5000$; $i = \dfrac{8\%}{2} = 4\%$; $n = 2 \times 9 = 18$

Using column A, look at row 18 of the 4% interest table.

$$M = \$5000(2.02581652)$$
$$= \$10,129.08$$

26. $M = \$2800; \ i = \dfrac{18\%}{12} = 1\frac{1}{2}\%; \ n = 17$

$P = M \cdot \dfrac{1}{(1+i)^n}$

$\quad = \$2800 \cdot \dfrac{1}{(1.015)^{17}}$

Using column B, look at row 17 of the $1\frac{1}{2}\%$ interest table.

$P = \$2800(0.77638526)$
$\quad = \$2173.88$

27. (a) Maturity value

$M = P(1 + RT)$

$M = \$12{,}000(1 + 0.08 \times \dfrac{40}{360})$

$M = \$12{,}106.67$

(b) Interest

$I = M - P$
$I = \$12{,}106.67 - \$12{,}000$
$I = \$106.67$

28. $B = MDT$

$B = \$15{,}000 \times 0.10 \times \dfrac{100}{360}$

$B = \$416.67$

$P = M - B$
$P = \$15{,}000 - \$416.67$
$P = \$14{,}583.33$

29. $M = P(1 + RT)$

$M = \$7850(1 + 0.06 \times \dfrac{5}{12})$

$M = \$8046.25$

$B = MDT$

$B = \$8046.25 \times 0.0792 \times \dfrac{1}{12})$

$B = \$53.11$

$P = M - B$
$P = \$8046.25 - \$53.11$
$P = \$7993.14$

30. Increase in wages of 4.1%.

$0.041 \times \$23{,}500 = \$963.50$

Inflation of 3.7%.

$0.037 \times \$23{,}500 = \$869.50$

Net gain in purchasing power

$\$963.50 - \$869.50 = \$94$

31. (a) Finance charge at 1.5%

$\$6327.12 \times 0.015 = \$94.91$

(b) Finance charge at 0.8%

$\$6327.12 \times 0.008 = \$50.62$

(c) Savings

$\$94.91 - \$50.62 = \$44.29$

32. (a) Total installment cost

$\$1000 + (36 \times \$315) = \$12{,}340$

(b) Finance charge

$\$12{,}340 - \$10{,}500 = \$1840$

(c) Amount financed

$\$10{,}500 - \$1000 = \$9500$

(d) Annual percentage rate

$\dfrac{\text{Finance charge}}{\text{Amount financed}} \times 100 = \dfrac{\$1840}{\$9500} \times 100$

$\qquad\qquad\qquad\qquad = 19.37$

Look at row 36 (36 payments) in Table 15.2 and find 19.37.
APR = 12%

33. July 7 is day 188;
September 30 is day 273.

$273 - 188 = 85$ days

Interest to day 85

$I = PRT$

$\quad = \$5000 \times 0.09 \times \dfrac{85}{360}$

$\quad = \$106.25$

Payment-interest due = reduced principal

$\$1200 - \$106.25 = \$1093.75$
$\$5000 - \$1093.75 = \$3906.25$

34. $S = \$60{,}000{,}000; \ n = 5; \ i = 10\%$

$R = \$60{,}000{,}000(\dfrac{1}{s_{\overline{5}|0.10}})$

Using column E, look at row 5 of the 10% table.

$R = \$60{,}000{,}000(0.16379748)$
$\quad = \$9{,}827{,}848.8$
$\quad = \$9.828$ million

35. $S = \$300;\ i = \dfrac{10\%}{4} = 2.5\%$

    (a) At age 65 $(n = 4 \times 7 = 28)$

$$S = \$300 \cdot s_{\overline{28}|0.025}$$

Using column C, look at row 28 of the $2\frac{1}{2}\%$ interest table.

$S = \$300(39.85980075)$
$\ \ \ = \$11{,}957.94$

    (b) At age 70 $(n = 4 \times 12 = 48)$

$$S = \$300 \cdot s_{\overline{48}|0.025}$$

Using column C, look at row 48 of the $2\frac{1}{2}\%$ interest table.

$S = \$300(90.85958243)$
$\ \ \ = \$27{,}257.87$

36. $A = \$18{,}500;\ n = 48;\ i = 10\%$

Use Table 15.3.

$\$18{,}500(0.025363) = \$469.22$

37. Use Table 13.2

$\$3200(1.006252041) = \$3220.01$

$I = \$3220.01 - \$3200$
$\ \ \ = \$20.01$

38. Since the down payment is 20%, the amount financed is 80% of $140,000.

Loan $= 0.80(\$140{,}000) = \$112{,}000$
$n = 30;\ i = 8\frac{3}{4}\%$

Use Table 15.4, $8\frac{3}{4}\%$, 30 (years): **7.87**

Monthly payment

$(\$112{,}000 \div 1000) \times 7.87 = \$881.44$

DEPRECIATION

16.1 Straight-Line Method

For Exercises 1-11, use the formula

$$\text{rate} = \frac{1}{\text{years of life}} \quad \text{or} \quad \text{rate} = \frac{1}{n}.$$

1. $\text{rate} = \frac{1}{n} = \frac{1}{5} = 20\%$

3. $\text{rate} = \frac{1}{n} = \frac{1}{8} = 12.5\%$

5. $\text{rate} = \frac{1}{20} = 5\%$

7. $\text{rate} = \frac{1}{15} = 6\frac{2}{3}\%$

9. $\text{rate} = \frac{1}{80} = 1.25\%$

11. $\text{rate} = \frac{1}{50} = 2\%$

Throughout this chapter, round depreciation amounts to the nearest dollar.

13. $d = \frac{c-s}{n} = \frac{\$9000-0}{20} = \$450$

15. $d = \frac{c-s}{n} = \frac{\$2700-\$300}{3} = \$800$

17. $d = \frac{c-s}{n} = \frac{\$4200-0}{5} = \$840$

For Exercises 19-21, find the book value after one year by subtracting one year's of depreciation from the original cost. Use the formula

Book value = Cost − Accumulated depreciation.

Find d using the formula

$$d = \frac{c-s}{n}.$$

19. $d = \frac{c-s}{n} = \frac{\$3200-\$400}{8} = \$350$

$b = c - d = \$3200 - \$350 = \$2850$

21. $d = \frac{c-s}{n} = \frac{\$5400-\$600}{12} = \$400$

$b = c - d = \$5400 - \$400 = \$5000$

For Exercises 23-25, find the book value after 5 years by subtracting 5 times one year's of depreciation from the cost, or $b = c - 5d$.

23. $d = \frac{c-s}{n} = \frac{\$4800-\$750}{10} = \$405$

$b = c - 5d = \$4800 - (5 \times \$405) = \$2775$

25. $d = \frac{c-s}{n} = \frac{\$80,000-\$10,000}{50} = \$1400$

$b = c - 5d = \$80,000 - (5 \times \$1400) = \$73,000$

27. Answers will vary.

29. $c = \$12,000; \; n = 3; \; s = \$3000$

$d = \frac{c-s}{n} = \frac{\$12,000-\$3000}{3} = \$3000$

$\text{rate} = \frac{1}{n} = \frac{1}{3} = 33\frac{1}{3}\%$

Year	Computation	Amt. of Deprec.	Accum. Deprec.	Book Value
0	----	----	----	$12,000
1	$(33\frac{1}{3}\% \times \$9000)$	$3000	$3000	$9000
2	$(33\frac{1}{3}\% \times \$9000)$	$3000	$6000	$6000
3	$(33\frac{1}{3}\% \times \$9000)$	$3000	$9000	$3000

31. $c = \$9400; \; n = 6; \; s = \$2200$

$d = \frac{c-s}{n} = \frac{\$9400-\$2200}{6} = \$1200$

$\text{rate} = \frac{1}{n} = \frac{1}{6} = 16\frac{2}{3}\%$

Year	Computation	Amt. of Deprec.	Accum. Deprec.	Book Value
0	----	----	----	$9400
1	$(16\frac{2}{3}\% \times \$7200)$	$1200	$1200	$8200
2	$(16\frac{2}{3}\% \times \$7200)$	$1200	$2400	$7000
3	$(16\frac{2}{3}\% \times \$7200)$	$1200	$3600	$5800
4	$(16\frac{2}{3}\% \times \$7200)$	$1200	$4800	$4600
5	$(16\frac{2}{3}\% \times \$7200)$	$1200	$6000	$3400
6	$(16\frac{2}{3}\% \times \$7200)$	$1200	$7200	$2200

33. $c = \$1,300,000;\ n = 20;\ s = \$200,000$

    (a) $d = \dfrac{c-s}{n} = \dfrac{\$1,300,000 - \$200,000}{20} = \$55,000$

    (b) $b = c - 5d = 1,300,000 - (5 \times \$55,000)$
        $= \$1,025,000$

35. $c = \$37,900;\ n = 15;\ s = 0$

    (a) $d = \dfrac{c-s}{n} = \dfrac{\$37,900 - 0}{15} = \$2527$

    (b) $b = c - 7d = \$37,900 - (7 \times \$2527)$
        $= \$20,211$

37. $c = \$880;\ n = 8;\ s = \$160$

    (a) rate $= \dfrac{1}{n} = \dfrac{1}{8} = 12.5\%$

    (b) $d = \dfrac{c-s}{n} = \dfrac{\$880 - \$160}{8} = \$90$

    (c) $b = c - d = \$880 - 90 = \$790$

16.2   Declining-Balance Method

For Exercises 1–11, find the double declining balance rate by doubling the straight line rate. Use the formula

$$\text{rate} = \frac{2}{n}.$$

1. rate $= \dfrac{2}{n} = \dfrac{2}{5} = 40\%$

3. rate $= \dfrac{2}{n} = \dfrac{2}{8} = 25\%$

5. rate $= \dfrac{2}{15} = 13\frac{1}{3}\%$

7. rate $= \dfrac{2}{10} = 20\%$

9. rate $= \dfrac{2}{6} = \dfrac{1}{3} = 33\frac{1}{3}\%$

11. rate $= \dfrac{2}{50} = 4\%$

For Exercises 13–17, find the first year;s of depreciation using the formula, $d = r \times b$. (Notice that the scrap value is not used.)

13. $c = \$18,000;\ n = 10;\ s = \$3000$

    rate $= \dfrac{2}{n} = \dfrac{2}{10} = 20\%$

    $d = r \times b = 0.20 \times \$18,000 = \$3600$

15. $c = \$10,500;\ n = 5;\ s = \$500$

    rate $= \dfrac{2}{n} = \dfrac{2}{5} = 40\%$

    $d = r \times b = 0.40 \times \$10,500 = \$4200$

17. $c = \$3800;\ n = 4;\ s = 0$

    rate $= \dfrac{2}{n} = \dfrac{2}{4} = 50\%$

    $d = r \times b = 0.50 \times \$3800 = \$1900$

For Exercises 19–21, find the book value after one year by subtracting the first year's of depreciation from the cost. Use $b = c - d$.

19. $c = \$4200;\ n = 10;\ s = \$1000$

    rate $= \dfrac{2}{n} = \dfrac{2}{10} = 20\%$

    $d = r \times b = 0.20 \times \$4200 = \$840$
    $b = c - d = \$4200 - \$840 = \$3360$

21. $c = \$1620;\ n = 8;\ s = 0$

    rate $= \dfrac{2}{n} = \dfrac{2}{8} = 25\%$

    $d = r \times b = 0.25 \times \$1620 = \$405$
    $b = c - d = \$1620 - \$405 = \$1215$

23. $c = \$16,200;\ n = 8;\ s = \$1500$

    rate $= \dfrac{2}{n} = \dfrac{2}{8} = 25\%$

Year	Computation	Amt. of Deprec.	Accum. Deprec.	Book Value
0	-------	----	----	$16,200
1	(25% × $16,200)	$4050	$4050	$12,150
2	(25% × $12,150)	$3038	$7088	$9112
3	(25% × $9112)	$2278	$9366	$6834

25. $c = \$6000;\ n = 3;\ s = \$750$

    rate $= \dfrac{2}{n} = \dfrac{2}{3} = 66\frac{2}{3}\%$

Year	Computation	Amt. of Deprec.	Accum. Deprec.	Book Value
0	-------	----	----	$6000
1	(66⅔% × $6000)	$4000	$4000	$2000
2	(66⅔% × $6000)	$1250*	$5250	$750
3	-------	----	$5250	$750

Only $1250 is used so that the book value does not fall below the salvage value ($750).

27. Answers will vary.

29. $c = \$14{,}400$; $n = 4$; $s = 0$

$$\text{rate} = \frac{2}{n} = \frac{2}{4} = 50\%$$

Year	Computation	Amt. of Deprec.	Accum. Deprec.	Book Value
0	-------	----	----	$14,400
1	(50% × $14,400)	$7200	$7200	$7200
2	(50% × $7200)	$3600	$10,800	$3600
3	(50% × $3600)	$1800	$12,600	$1800
4	(50% × $1800)	$1800*	$14,400	$0

*Fully depreciate in last year.

31. $c = \$25{,}500$; $n = 8$; $s = \$3500$

$$\text{rate} = \frac{2}{n} = \frac{2}{8} = 25\%$$

Year	Computation	Amt. of Deprec.	Accum. Deprec.	Book Value
0	-------	----	----	$25,500
1	(25% × $25,500)	$6375	$6375	$19,125
2	(25% × 19,125)	$4781	$11,156	$14,344
3	(25% × 14,344)	$3586	$14,742	$10,758
4	(25% × 10,758)	$2690	$17,432	$8068
5	(25% × 8068)	$2017	$19,449	$6051
6	(25% × 6051)	$1513	$20,962	$4538
7	(25% × 4538)	$1038*	$22,000	$3500
8	-------	----	$22,000	$3500

*Depreciate to scrap amount.

33. $c = \$8200$; $n = 8$; $s = \$1250$

$$\text{rate} = \frac{2}{n} = \frac{2}{8} = 25\%$$

Year	Computation	Amt. of Deprec.	Accum. Deprec.	Book Value
0	----	----	----	$8200
1	(25% × $8200)	$2050	$2050	$6150
2	(25% × $6150)	$1538	$3588	$4612
3	(25% × $4612)	$1153	$4741	$3459

The amount of depreciation in the third year was $1153.

35. $c = \$1090$; $n = 5$; $s = 0$

$$\text{rate} = \frac{2}{n} = \frac{2}{5} = 40\%$$

Year	Computation	Amt. of Deprec.	Accum. Deprec.	Book Value
0	----	----	----	$1090
1	(40% × $1090)	$436	$436	$654
2	(40% × $ 654)	$262	$698	$392
3	(40% × $ 392)	$157	$855	$235

The book value at the end of the third year is $235.

37. $c = \$5800$; $n = 8$; $s = \$1000$

(a) $\text{rate} = \dfrac{2}{n} = \dfrac{2}{8} = 25\%$

Year	Computation	Amt. of Deprec.	Accum. Deprec.	Book Value
0	----	----	----	$5800
1	(25% × $5800) **(b)**	$1450	$1450	$4350
2	(25% × $4350)	$1088	$2538	$3262
3	(25% × $3262)	$816	$3354	$2446
4	(25% × $2446)	$612	$3966	$1834
5	(25% × $1834)	$459 **(c)**	$4425	**(d)** $1375

16.3 Sum-of-the-Years'-Digits Method

For Exercises 1-7, find the denominator of the depreciation fraction with the formula,

$$\frac{n(n+1)}{2}.$$

The numerator for the first year is the number of years of life of the asset.

1. $\dfrac{n(n+1)}{2} = \dfrac{4 \times 5}{2} = 10$; $r = \dfrac{4}{10}$

3. $\dfrac{n(n+1)}{2} = \dfrac{6 \times 7}{2} = 21$; $r = \dfrac{6}{21}$

5. $\dfrac{n(n+1)}{2} = \dfrac{7 \times 8}{2} = 28$; $r = \dfrac{7}{28}$

7. $\dfrac{n(n+1)}{2} = \dfrac{10 \times 11}{2} = 55$; $r = \dfrac{10}{55}$

For the rest of the exercises, use the formula

$$d = r \times (c - s).$$

9. $c = \$4800$; $n = 4$; $s = \$700$

$$\frac{n(n+1)}{2} = \frac{4 \times 5}{2} = 10;\ r = \frac{4}{10}$$

$$d = r \times (c - s) = \frac{4}{10} \times (\$4800 - \$700)$$

$$= \$1640$$

11. $c = \$60,000;\; n = 10;\; s = \$5000$

$$\frac{n(n+1)}{2} = \frac{10 \times 11}{2} = 55;\; r = \frac{10}{55}$$

$$d = r \times (c - s) = \frac{10}{55} \times (\$60,000 - \$5000)$$

$$= \$10,000$$

13. $c = \$1350;\; n = 3;\; s = \$150$

$$\frac{n(n+1)}{2} = \frac{3 \times 4}{2} = 6;\; r = \frac{3}{6}$$

$$d = r \times (c - s) = \frac{3}{6} \times (\$1350 - \$150)$$

$$= \$600$$

15. $c = \$9500;\; n = 8;\; s = \$1400$

$$\frac{n(n+1)}{2} = \frac{8 \times 9}{2} = 36;\; r = \frac{8}{36}$$

$$d = r \times (c - s) = \frac{8}{36} \times (\$9500 - \$1400)$$

$$= \$1800$$

$$b = c - d = \$9500 - \$1800$$
$$= \$7700$$

17. $c = \$3800;\; n = 5;\; s = \$500$

$$\frac{n(n+1)}{2} = \frac{5 \times 6}{2} = 15;\; r = \frac{5}{15}$$

$$d = r \times (c - s) = \frac{5}{15} \times (\$3800 - \$500)$$

$$= \$1100$$

$$b = c - d = \$3800 - \$1100$$
$$= \$2700$$

19. $c = \$2240;\; n = 6;\; s = \$350$

$$\frac{n(n+1)}{2} = \frac{6 \times 7}{2} = 21$$

Year	Computation	Amt. of Deprec.	Accum. Deprec.	Book Value
0	----	----	----	$2240
1	$\left(\frac{6}{21} \times \$1890\right)$	$540	$540	$1700
2	$\left(\frac{5}{21} \times \$1890\right)$	$450	$990	$1250
3	$\left(\frac{4}{21} \times \$1890\right)$	$360	$1350	$890

The book value at the end of 3 years is $890.

21. $c = \$4500;\; n = 8;\; s = \$900$

$$\frac{n(n+1)}{2} = \frac{8 \times 9}{2} = 36$$

Year	Computation	Amt. of Deprec.	Accum. Deprec.	Book Value
0	----	----	----	$4500
1	$\left(\frac{8}{36} \times \$3600\right)$	$800	$800	$3700
2	$\left(\frac{7}{36} \times \$3600\right)$	$700	$1500	$3000
3	$\left(\frac{6}{36} \times \$3600\right)$	$600	$2100	$2400

The book value after 3 years is $2400.

23. Answers will vary.

25. $c = \$10,800;\; n = 6;\; s = \$2400$

$$\frac{n(n+1)}{2} = \frac{6 \times 7}{2} = 21$$

Year	Computation	Amt. of Deprec.	Accum. Deprec.	Book Value
0	----	----	----	$10,800
1	$\left(\frac{6}{21} \times \$8400\right)$	$2400	$2400	$8400
2	$\left(\frac{5}{21} \times \$8400\right)$	$2000	$4400	$6400
3	$\left(\frac{4}{21} \times \$8400\right)$	$1600	$6000	$4800
4	$\left(\frac{3}{21} \times \$8400\right)$	$1200	$7200	$3600
5	$\left(\frac{2}{21} \times \$8400\right)$	$800	$8000	$2800
6	$\left(\frac{1}{21} \times \$8400\right)$	$400	$8400	$2400

27. $c = \$2700$; $n = 6$; $s = \$600$

$$\frac{n(n+1)}{2} = \frac{6 \times 7}{2} = 21$$

Year	Computation	Amt. of Deprec.	Accum. Deprec.	Book Value
0	----	----	----	$2700
1	$(\frac{6}{21} \times \$2100)$	$600	$600	$2100
2	$(\frac{5}{21} \times \$2100)$	$500	$1100	$1600
3	$(\frac{4}{21} \times \$2100)$	$400	$1500	$1200
4	$(\frac{3}{21} \times \$2100)$	$300	$1800	$900
5	$(\frac{2}{21} \times \$2100)$	$200	$2000	$700
6	$(\frac{1}{21} \times \$2100)$	$100	$2100	$600

29. $c = \$32,000$; $n = 8$; $s = \$5000$

$$\frac{n(n+1)}{2} = \frac{8 \times 9}{2} = 36$$

Year	Computation	Amt. of Deprec.	Accum. Deprec.	Book Value
0	----	----	----	$32,000
1	$(\frac{8}{36} \times \$27,000)$	$6000	$6000	$26,000

The depreciation the first year is $6000.

31. $c = \$31,880$; $n = 20$; $s = \$5000$

$$\frac{n(n+1)}{2} = \frac{20 \times 21}{2} = 210$$

Year	Computation	Amt. of Deprec.	Accum. Deprec.	Book Value
0	----	----	----	$31,880
1	$(\frac{20}{210} \times \$26,880)$	$2560	$2560	$29,320
2	$(\frac{19}{210} \times \$26,880)$	$2432	$4992	$26,888
3	$(\frac{18}{210} \times \$26,880)$	$2304	$7296	$24,584

The book value after 3 years is $24,584.

33. $c = \$6360$; $n = 5$; $s = 0$

$$\frac{n(n+1)}{2} = \frac{5 \times 6}{2} = 15$$

Year	Computation	Amt. of Deprec.	Accum. Deprec.	Book Value
0	----	----	----	$6360
1	$(\frac{5}{15} \times \$6360)$	**(a)** $2120	$2120	$4240
2	$(\frac{4}{15} \times \$6360)$	**(b)** $1696	$3816	$2544

35. $c = \$12,420$; $n = 8$; $s = \$1800$

$$\frac{n(n+1)}{2} = \frac{8 \times 9}{2} = 36; \quad r = \frac{8}{36}$$

Year	Computation	Amt. of Deprec.	Accum. Deprec.	Book Value
0	----	----	----	$12,420
1	$(\frac{8}{36} \times \$10,620)$	**(b)** $2360	$2360	$10,060
2	$(\frac{7}{36} \times \$10,620)$	$2065	$4425	$7995
3	$(\frac{6}{36} \times \$10,620)$	$1770	$6195	$6225
4	$(\frac{5}{36} \times \$10,620)$	$1475	$7670	**(d)** $4750
5	$(\frac{4}{36} \times \$10,620)$	$1180	$8850	$3570
6	$(\frac{3}{36} \times \$10,620)$	$885	$9735	$2685
7	$(\frac{2}{36} \times \$10,620)$	$590	$10,325	$2095
8	$(\frac{1}{36} \times \$10,620)$	$295	**(c)** $10,620	$1800

16.4 Units-of-Production Method and Partial-Year Depreciation

Round the depreciation per unit to the nearest thousandth of a dollar (or tenth of a cent) in this section.

1. $c = \$22,500$; $s = \$1500$; $n = 60,000$

Depreciation amount

$c - s = \$22,500 - \$1500 = \$21,000$

Depreciation per unit

$$\frac{c - s}{n} = \frac{\$22,500 - \$1500}{60,000} = \$0.35$$

3. $c = \$3750$; $s = \$250$; $n = 120{,}000$

Depreciation amount

$c - s = \$3750 - \$250 = \$3500$

Depreciation per unit

$\dfrac{c - s}{n} = \dfrac{\$3750 - \$250}{120{,}000} = \$0.029$

5. $c = \$300{,}000$; $s = \$25{,}000$; $n = 4000$

Depreciation amount

$c - s = \$300{,}000 - \$25{,}000 = \$275{,}000$

Depreciation per unit

$\dfrac{c - s}{n} = \dfrac{\$300{,}000 - \$25{,}000}{4000} = \$68.75$

7. $c = \$175{,}000$; $s = \$25{,}000$; $n = 5000$

Depreciation amount

$c - s = \$175{,}000 - \$25{,}000 = \$150{,}000$

Depreciation per unit

$\dfrac{c - s}{n} = \dfrac{\$175{,}000 - \$25{,}000}{5000} = \$30$

9. Amount of depreciation

$78{,}000 \times \$0.23 = \$17{,}940$

11. Amount of depreciation

$32{,}000 \times \$0.54 = \$17{,}280$

13. Amount of depreciation

$15{,}000 \times 0.185 = \$2775$

15. Amount of depreciation

$17{,}400 \times 0.40 = \$6960$

17. Answers will vary.

19. Straight line method

$c = \$9700$; $s = \$700$; $n = 4$

Depreciation amount

$c - s = \$9700 - \$700 = \$9000$

Depreciation per year

$\dfrac{c - s}{n} = \dfrac{\$9700 - \$700}{4} = \$2250$

June 1 to Dec. 31 is 7 months.

$\$2250 \times \dfrac{7}{12} = \$1313$ *partial year 1*

$\$2250$ *year 2*

21. Double-declining method

$c = \$20{,}000$; $s = \$1000$; $n = 20$

rate $= \dfrac{2}{n} = \dfrac{2}{20} = 10\%$

$d = r \times b = 0.10 \times \$20{,}000 = \$2000$

March 1 to Dec. 31 is 10 months.

$\$2000 \times \dfrac{10}{12} \approx \$1667$ *partial year 1*

$0.10 \times (\$20{,}000 - \$1667) = \$1833$ *year 2*

23. Sum-of-years'-digits method

$c = \$3150$; $s = \$0$; $n = 6$

$\dfrac{n(n + 1)}{2} = 21$; rate $= \dfrac{6}{21}$

$d = r \times (c - s) = \dfrac{6}{21} \times (\$3150 - 0)$

$= \$900$

July 12 to Dec. 31 is 6 months.

$\$900 \times \dfrac{6}{12} = \$450$ *partial year 1*

$\$900 \times \dfrac{6}{12} = \$450$ *1st half of year 2*

$\dfrac{5}{21} \times \$3150 = \$750$

$\$750 \times \dfrac{6}{12} = \$375$ *2nd half of year 2*

$\$375 + \$450 = \$825$ *year 2*

25. Sum-of-years'-digits method

$c = \$6300$; $s = \$900$; $n = 8$

$\dfrac{n(n + 1)}{2} = 36$; rate $= \dfrac{8}{36}$

$d = r \times (c - s) = \dfrac{8}{36} \times (\$6300 - \$900)$

$= \$1200$

Mar. 28 to Dec. 31 is 9 months

$\$1200 \times \dfrac{9}{12} = \$900$ *partial year 1*

$\$1200 \times \dfrac{3}{12} = \$300$ $\frac{1}{4}$ *of year 2*

$\dfrac{7}{36} \times (\$6300 - \$900) = \$1050$

$\$1050 \times \dfrac{9}{12} = \$788$ $\frac{3}{4}$ *of year 2*

$\$300 + \$788 = \$1088$ *year 2*

27. $c = \$6800$; $n = 5000$; $s = \$500$

Depreciation per unit

$$\frac{c - s}{n} = \frac{\$6800 - \$500}{5000} = \$1.26$$

Year	Computation	Amt. of Deprec.	Accum. Deprec.	Book Value
0	---	---	---	$6800
1	(1350 × $1.26)	$1701	$1701	$5099
2	(1820 × $1.26)	$2293	$3994	$2806
3	(730 × $1.26)	$920	$4914	$1886
4	(1100 × $1.26)	$1386	$6300	$500

29. $c = \$185,000$; $s = \$30,000$; $n = 20,000$

$$\frac{c - s}{n} = \frac{\$185,000 - \$30,000}{20,000} = \$7.75 \text{ per hour}$$

$$3400 \times \$7.75 = \$26,350$$

31. $c = \$156,000$; $n = 10$; $s = \$10,000$

$$d = \frac{c - s}{n} = \frac{\$156,000 - \$10,000}{10} = \$14,600$$

Oct. 1 to Dec. 31 is 3 months.

$$\$14,600 \times \frac{3}{12} = \quad \$3650 \quad \text{partial year 1}$$
$$\$14,600 \quad \text{year 2}$$

33. $c = \$4500$; $n = 5$; $s = \$0$

$$\text{rate} = \frac{2}{n} = \frac{2}{5} = 40\%$$

$$d = r \times b = 0.40 \times \$4500 = \$1800$$

Oct. 8 to Dec. 31 is 3 months.

$$\$1800 \times \frac{3}{12} = \$450 \quad \text{partial year 1}$$

$$0.40 \times (\$4500 - \$450) = \$1620 \quad \text{year 2}$$

35. $c = \$44,400$; $n = 15$; $s = \$0$

$$\frac{n(n+1)}{2} = 120; \text{rate} = \frac{15}{120}$$

$$d = r \times (c - s) = \frac{15}{120} \times (\$44,400 - 0)$$
$$= \$5550$$

June 27 to Dec. 31 is 6 months

$$\$5550 \times \frac{6}{12} = \$2775 \quad \text{partial year 1}$$

$$\$5550 \times \frac{6}{12} = \$2775 \quad \text{first half of year 2}$$

$$\frac{14}{120} \times \$44,400 = \$5180$$

$$\$5180 \times \frac{6}{12} = \$2590 \quad \text{second half of year 2}$$

$$\$2775 + \$2590 = \$5365 \quad \text{year 2}$$

16.5   Modified Accelerated Cost Recovery System

Use Table 16.1 to find the rates of depreciation.

1. 19.2%   **3.** 11.52%   **5.** 20%

7. 3.636%   **9.** 5.76%   **11.** 2.564%

13. $c = \$12,250$
$d = r \times c = 14.29\% \times \$12,250 = \$1751$

15. $c = \$9680$
$d = r \times c = 33.33\% \times \$9680 = \$3226$

17. $c = \$48,000$
$d = r \times c = 10\% \times \$48,000 = \$4800$

19. $c = \$9380$
$d = r \times c = 33.33\% \times \$9380 = \$3126$
$b = c - d = \$9380 - \$3126 = \$6254$

21. $c = \$18,800$
$d = r \times c = 10\% \times \$18,800 = \$1880$
$b = c - d = \$18,800 - \$1880 = \$16,920$

23. $c = \$9570$

Year 1 rate: 20.00%
Year 2 rate: 32.00%
Year 3 rate: <u>19.20%</u>
                71.20%

$d = r \times c = 71.2\% \times \$9570 = \$6814$
$b = c - d = \$9570 - \$6814 = \$2756$

25. $c = \$87,300$

Year 1 rate: 1.818%
Year 2 rate: 3.636%
Year 3 rate: <u>3.636%</u>
                9.090%

$d = r \times c = 9.09\% \times \$87,300 = \$7936$
$b = c - d = \$87,300 - \$7936 = \$79,364$

27. Answers will vary.

29.

Year	Computation	Amt. of Deprec.	Accum. Deprec.	Book Value
0	----	----	----	$10,980
1	(33.33% × $10,980)	$3660	$3660	$7320
2	(44.45% × $10,980)	$4881	$8541	$2439
3	(14.81% × $10,980)	$1626	$10,167	$813
4	(7.41% × $10,980)	$813*	$10,980	$0

*Due to rounding in prior years

31.

Year	Recovery Percent Rate	Deprec. Amt.	Accum. Deprec.	Book Value
0	----	----	----	$122,700
1	(10.00% × $122,700)	$12,270	$12,270	$110,430
2	(18.00% × $122,700)	$22,086	$34,356	$88,344
3	(14.40% × $122,700)	$17,669	$52,025	$70,675
4	(11.52% × $122,700)	$14,135	$66,160	$56,540
5	(9.22% × $122,700)	$11,313	$77,473	$45,227
6	(7.37% × $122,700)	$9043	$86,516	$36,184
7	(6.55% × $122,700)	$8037	$94,553	$28,147
8	(6.55% × $122,700)	$8037	$102,590	$20,110
9	(6.56% × $122,700)	$8049	$110,639	$12,061
10	(6.55% × $122,700)	$8037	$118,676	$4024
11	(3.28% × $122,700)	$4024*	$122,700	$0

*Due to rounding in prior years

33. $c = \$3700$

Year 1 rate: 20.00%
Year 2 rate: 32.00%
Year 3 rate: 19.20%
            71.20%

$d = r \times c = 71.2\% \times \$3700 = \$2634$
$b = c - d = \$3700 - \$2634 = \$1066$

35. $c = \$480,000$; 39-year property

Year	Rate	$r \times c = d$
1	2.568%	$0.02568 \times \$480,000 = \$12,326$
2	2.564%	$0.02564 \times \$480,000 = \$12,307$
3	2.564%	$0.02564 \times \$480,000 = \$12,307$
4	2.564%	$0.02564 \times \$480,000 = \$12,307$
5	2.564%	$0.02564 \times \$480,000 = \$12,307$

Chapter 16 Review Exercises

1. Straight-line rate: $\frac{1}{n} = \frac{1}{5} = 20\%$

   Double-declining rate: $\frac{2}{n} = \frac{2}{5} = 40\%$

   Sum-of-years'-digits: $\frac{n(n+1)}{2}$

   $= \frac{5 \times 6}{2} = 15$

   1st year depreciation fraction: $\frac{5}{15}$

2. Straight-line rate: $\frac{1}{n} = \frac{1}{6} = 16\frac{2}{3}\%$

   Double-declining rate: $\frac{2}{n} = \frac{2}{6} = 33\frac{1}{3}\%$

   Sum-of-years'-digits: $\frac{n(n+1)}{2}$

   $= \frac{6 \times 7}{2} = 21$

   1st year depreciation fraction: $\frac{6}{21}$

3. Straight-line rate: $\frac{1}{n} = \frac{1}{4} = 25\%$

   Double-declining rate: $\frac{2}{n} = \frac{2}{4} = 50\%$

   Sum-of-years'-digits: $\frac{n(n+1)}{2}$

   $= \frac{4 \times 5}{2} = 10$

   1st year depreciation fraction: $\frac{4}{10}$

4. Straight-line rate: $\frac{1}{n} = \frac{1}{10} = 10\%$

   Double-declining rate: $\frac{2}{n} = \frac{2}{10} = 20\%$

   Sum-of-years'-digits: $\frac{n(n+1)}{2}$

   $= \frac{10 \times 11}{2} = 55$

   1st year depreciation fraction: $\frac{10}{55}$

5. Straight-line rate: $\frac{1}{n} = \frac{1}{20} = 5\%$

   Double-declining rate: $\frac{2}{20} = 10\%$

   Sum-of-years'-digits: $\frac{n(n+1)}{2}$

   $= \frac{20 \times 21}{2} = 210$

   1st year depreciation fraction: $\frac{20}{210}$

6. Straight-line rate: $\frac{1}{n} = \frac{1}{8} = 12.5\%$

   Double-declining rate: $\frac{2}{n} = \frac{2}{8} = 25\%$

   Sum-of-years'-digits: $\frac{n(n+1)}{2}$

   $= \frac{8 \times 9}{2} = 36$

   1st year depreciation fraction: $\frac{8}{36}$

7. 11.52%   8. 33.33%

9. 4.522%   10. 17.49%

11. 2.564%   12. 3.174%

13. $c = \$12,400$; $n = 10$; $s = \$3000$

   $d = \frac{c-s}{n} = \frac{\$12,400 - \$3000}{10} = \$940$

14. $c = \$74,000$; $n = 20$; $s = \$12,000$

   $d = \frac{c-s}{n} = \frac{\$74,000 - \$12,000}{20} = \$3100$

   $b = c - 10d = \$74,000 - (10 \times \$3100)$
   $= \$43,000$

15. $c = \$38,000$; $n = 8$

   rate $= \frac{2}{n} = \frac{2}{8} = 25\%$

Year	Computation	Amt. of Deprec.	Accum. Deprec.	Book V.
0	----	----	----	$38,000
1	(0.25 × $38,000)	$9500	$9500	$28,500
2	(0.25 × $28,500)	$7125	$16,625	$21,375

16. $c = \$18,500$; $n = 5$; $s = 0$

   rate $= \frac{2}{n} = \frac{2}{5} = 40\%$

   $d = r \times b = 0.40 \times \$18,500 = \$7400$

17. $c = \$8250$; $n = 4$; $s = \$1500$

$$\frac{n(n+1)}{2} = \frac{4 \times 5}{2} = 10$$

Year	Computation	Amt. of Deprec.	Accum. Deprec.	Book Value
0	----	----	----	$8250
1	($\frac{4}{10} \times \$6750$)	$2700	$2700	$5550
2	($\frac{3}{10} \times \$6750$)	$2025	$4725	$3525
3	($\frac{2}{10} \times \$6750$)	$1350	$6075	$2175
4	($\frac{1}{10} \times \$6750$)	$675	$6750	$1500

18. $c = \$7375$; $n = 10$; $s = \$500$

$$\frac{n(n+1)}{2} = \frac{10 \times 11}{2} = 55$$

Year	Computation	Amt. of Deprec.	Accum. Deprec.	Book Value
0	----	----	----	$7375
1	($\frac{10}{55} \times \$6875$)	$1250	$1250	$6125
2	($\frac{9}{55} \times \$6875$)	$1125	$2375	$5000
3	($\frac{8}{55} \times \$6875$)	$1000	$3375	$4000

19. $c = \$11,000$; $s = \$2500$; $n = 5000$

Depreciation per hour

$$\frac{c - s}{n} = \frac{\$11,000 - \$2500}{5000} = \$1.70$$

Total depreciation

$$\$1.70 \times 900 = \$1530$$

20. c = $20,100; $n = 30,000$; $s = \$1500$

Depreciation per hour

$$\frac{c - s}{n} = \frac{\$20,100 - \$1500}{30,000} = \$0.62$$

Year	Computation	Amt. of Deprec.	Accum. Deprec.	Book Value
0	----	----	----	$20,100
1	(7800 × 0.62)	$4836	$4836	$15,264
2	(4300 × 0.62)	$2666	$7502	$12,598
3	(4850 × 0.62)	$3007	$10,509	$9591
4	(7600 × 0.62)	$4712	$15,221	$4879

21. $c = \$22,400$; $n = 5$; $s = \$3500$

$$d = \frac{c - s}{n} = \frac{\$22,400 - \$3500}{5} = \$3780$$

Accumulated depreciation at the end of the fourth year

$$4d = 4 \times \$3780 = \$15,120$$

22. $c = \$85,000$; $n = 5$; $s = \$13,000$

$$\frac{n(n+1)}{2} = \frac{5 \times 6}{2} = 15$$

Year	Computation	Amt. of Deprec.	Accum. Deprec.	Book Value
0	----	----	----	$85,000
1	($\frac{5}{15} \times \$72,000$)	$24,000	$24,000	$61,000
2	($\frac{4}{15} \times \$72,000$)	$19,200	$43,200	$41,800
3	($\frac{3}{15} \times \$72,000$)	$14,400	$57,600	$27,400
4	($\frac{2}{15} \times \$72,000$)	$9600	$67,200	$17,800
5	($\frac{1}{15} \times \$72,000$)	$4800	$72,000	$13,000

23. $c = \$48,000$; $n = 5$; $s = \$4500$

Depreciation per year

$$\frac{c - s}{n} = \frac{\$48,000 - \$4500}{5} = \$8700$$

Sept. 10 to Dec. 31 is 4 months.

$$\$8700 \times \frac{4}{12} = \$2900$$

24. $c = \$9720$; $n = 8$; $s = \$0$

$$\frac{n(n+1)}{2} = 36; \text{ rate} = \frac{8}{36}$$

$$d = r \times (c - s) = \frac{8}{36} \times (\$9720 - \$0)$$

$$= \$2160$$

June 1 to Dec. 31 is 7 months

$$\$2160 \times \frac{7}{12} = \$1260 \text{ } partial \text{ } year \text{ } 1$$

$$\$2160 \times \frac{5}{12} = \$900 \text{ } \tfrac{5}{12} \text{ } of \text{ } year \text{ } 2$$

$$\frac{7}{36} \times (\$9720 - \$0) = \$1890$$

$$\$1890 \times \frac{7}{12} = \$1103 \text{ } \tfrac{7}{12} \text{ } of \text{ } year \text{ } 2$$

$$\$900 + \$1103 = \$2003 \text{ } year \text{ } 2$$

25. $c = \$28,400$; $n = 10$; $s = \$6000$

rate $= \dfrac{2}{n} = \dfrac{2}{10} = 20\%$

$d = r \times b = 0.20 \times \$28,400 = \$5680$

June 19 to December 31 is 6 months.

$\$5680 \times \dfrac{6}{12} = \$2840$ *partial year 1*

$0.20 \times (\$28,400 - \$2840) = \$5112$ *year 2*

26. $c = \$2800 \times 4 = \$11,200$; $n = 10$; $s = \$0$

$\dfrac{n(n+1)}{2} = \dfrac{10 \times 11}{2} = 55$

Year 1 Depreciation fraction: $\dfrac{10}{55}$

Year 2 Depreciation fraction: $\dfrac{9}{55}$

Year 3 Depreciation fraction: $\dfrac{8}{55}$

Total: $\dfrac{27}{55}$

$d = r \times (c - s) = \dfrac{27}{55} \times (\$11,200 - 0)$

$= \$5498$

$d = c - d = \$11,200 - \$5498 = \$5702$

27. $c = \$15,000$; $n = 8$; $s = \$0$

rate $= \dfrac{2}{n} = \dfrac{2}{8} = 25\%$

Year	Computation	Amt. of Deprec.	Accum. Deprec.	Book Value
0	– – – –	– – – –	– – – –	$15,000
1	(25% × $15,000)	$3750	$3750	$11,250
2	(25% × $11,250)	$2813	$6563	$8437
3	(25% × $8437)	$2109	$8672	$6328
4	(25% × $6328)	$1582	$10,254	$4746

28. $c = \$74,125$; $n = 10$; $s = \$15,000$

$\dfrac{n(n+1)}{2} = \dfrac{10 \times 11}{2} = 55$

Year	Computation	Amt. of Deprec.	Accum. Deprec.	Book Value
0	– – – –	– – – –	– – – –	$74,125
1	($\frac{10}{55}$ × $59,125)	$10,750	$10,750	$63,375
2	($\frac{9}{55}$ × $59,125)	$9675	$20,425	$53,700
3	($\frac{8}{55}$ × $59,125)	$8600	$29,025	$45,100
4	($\frac{7}{55}$ × $59,125)	$7525	$36,550	$37,575

29. $c = \$14,750$; $n = 8$; $s = \$0$

rate $= \dfrac{2}{n} = \dfrac{2}{8} = 25\%$

Year	Computation	Amt. of Deprec.	Accum. Deprec.	Book Value
0	– – – –	– – – –	– – – –	$14,750
1	(25% × $14,750)	$3688	$3688	$11,062
2	(25% × $11,062)	$2766	$6454	$8296
3	(25% × $8296)	$2074	$8528	$6222

30. The third year of a 15-year recovery period has a depreciation rate of 8.55%.

$c = \$56,000$

$d = r \times c = 8.55\% \times \$56,000 = \$4788$

31. Cost = $8100   7-year property

Year	Rate	r	×	c	=	d
1	(14.29%)	0.1429	×	$8100	=	$1157
2	(24.49%)	0.2449	×	$8100	=	$1984
3	(17.49%)	0.1749	×	$8100	=	$1417
4	(12.49%)	0.1249	×	$8100	=	$1012
5	(8.93%)	0.0893	×	$8100		$723

32. $c = \$2,800,000$ 39-year property

Year 1 rate: 2.568%
Year 2 rate: 2.564%
Year 3 rate: 2.564%
Year 4 rate: 2.564%
Year 5 rate: 2.564%
Total: 12.824%

$d = r \times c = 12.824\% \times \$2,800,000$

$= \$359,072$

$b = c - d = \$2,800,000 - \$359,072$

$= \$2,440,928$

Chapter 16 Summary Exercise

$c = \$285,000$; $n = 5$; $s = \$0$

(a) $d = \dfrac{c - s}{n} = \dfrac{\$285,000 - \$0}{5} = \$57,000$

$b = c - 3d = \$285,000 - (3 \times \$57,000)$

$= \$114,000$

(b) rate $= \dfrac{2}{n} = \dfrac{2}{5} = 40\%$

Year	Computation	Amt. of Deprec.	Accum. Deprec.	Book Value
0	----	----	----	$285,000
1	(40% × $285,000)	$114,000	$114,000	$171,000
2	(40% × $171,000)	$68,400	$182,400	$102,600
3	(40% × $102,600)	$41,040	$223,440	$61,560

(c) $\dfrac{n(n+1)}{2} = \dfrac{5 \times 6}{2} = 15$

Year 1 Depreciation fraction: $\dfrac{5}{15}$

Year 2 Depreciation fraction: $\dfrac{4}{15}$

Year 3 Depreciation fraction: $\dfrac{3}{15}$

Total: $\dfrac{12}{15}$

$d = r \times (c - s) = \dfrac{12}{15} \times (\$285{,}000 - \$0)$

$= \$228{,}000$

(d) Straight-line depreciation

$57,000

Double-declining-balance depreciation

$40\% \times \$61{,}560 = \$24{,}624$

Sum-of-the-year's-digits depreciation

$\dfrac{2}{15} \times \$285{,}000 = \$38{,}000$

FINANCIAL STATEMENTS AND RATIOS

17.1   The Income Statement

1. Net sales = gross sales − returns = $685,900 − $2350 = $683,550

   (a) Gross profit = net sales − cost of goods sold = $683,550 − $367,200 = $316,350

   (b) Net income before taxes = gross profit − operating expenses = $316,350 − $228,300 = $88,050

   (c) Net income after taxes = income before taxes − income taxes = $88,050 − $22,700 = $65,350

3. $263,400 + $343,500 + $4800 = $611,700
   $611,700 − $287,500 = $324,200

For Exercises 5 and 7, follow the steps on pages 612 and 613 in the textbook.

5.

FUTURE TECH COMPUTING INCOME STATEMENT YEAR ENDING DECEMBER 31		
Gross Sales	$284,000	
Returns	− $6000	
Net Sales		$278,000
Inventory, January 1	$58,000	
Cost of Goods		
Purchased	$232,000	
Freight	+ $3000	
Total Cost of Goods Purchased	+ $235,000	
Total of Goods Available for Sale	$293,000	
Inventory, December 31	− $69,000	
Cost of Goods Sold		− $224,000
Gross Profit		$54,000
Expenses		
Salaries and Wages	$15,000	
Rent	$6000	
Advertising	$2000	
Utilities	$1000	
Taxes on Inventory, Payroll	$3000	
Miscellaneous Expenses	+ $4000	
Total Expenses		− $31,000
NET INCOME BEFORE TAXES		$23,000
Income Taxes		− $2400
NET INCOME		$20,600

7.

```
BETTY THOMAS, CONSULTANT
INCOME STATEMENT
YEAR ENDING DECEMBER 31
```

Gross Sales	$170,500	
Returns	− 0	
Net Sales		$170,500
		↓
Inventory, January 1	$ 0	
Cost of Goods		
Purchased	$ 0	
Freight	$ 0	
	↘	
Total Cost of Goods Purchased	$ 0	
Total of Goods Available for Sale	$ 0	
Inventory, December 31	$ 0	↘ $ 0
Cost of Goods Sold		
Gross Profit		$170,500
Expenses		
Salaries and Wages	$63,000	
Rent	$28,000	
Advertising	$12,000	
Utilities	$4000	
Taxes on Inventory, Payroll	$3800	
Miscellaneous Expenses	+ $9400	
	↘	
Total Expenses		− $120,200
NET INCOME BEFORE TAXES		$50,300
Income Taxes		− $6800
NET INCOME		$43,500

9. Answers will vary.   **11.** Answers will vary.   **13.** Answers will vary.

17.2   Analyzing the Income Statement

1. Cost of goods sold

$$\frac{\$243,570}{\$480,300} = 0.5071 = 50.7\%$$

Operating expenses

$$\frac{\$140,450}{\$480,300} = 0.2924 = 29.2\%$$

3. Capital Appliance Center

	Amount	Percent	Percent from Table 17.1
Net Sales	$900,000	100.0%	100.0%
Cost of Goods Sold	$617,000	68.6%	66.9%
Gross Profit	$283,000	31.4%	33.1%
Wages	$108,900	12.1%	11.9%
Rent	$20,700	2.3%	2.4%
Advertising	$27,000	3.0%	2.5%
Total Expenses	$216,000	24.0%	26.0%
Net Income before taxes	$67,000	7.4%	7.2%

5. Best Tires, Inc.
Comparative Income Statement

	This Year		Last Year	
	Amount	Percent	Amount	Percent
Gross Sales	$1,856,000	100.3%	$1,692,000	100.7%
Returns	$6000	0.3%	$12,000	0.7%
Net Sales	$1,850,000	100.0%	$1,680,000	100.0%
Cost of Goods Sold	$1,202,000	65.0%	$1,050,000	62.5%
Gross Profit	$648,000	35.0%	$630,000	37.5%
Wages	$152,000	8.2%	$148,000	8.8%
Rent	$82,000	4.4%	$78,000	4.6%
Advertising	$111,000	6.0%	$122,000	7.3%
Utilities	$32,000	1.7%	$17,000	1.0%
Taxes on Inv., Payroll	$17,000	0.9%	$18,000	1.1%
Miscellaneous Expenses	$62,000	3.4%	$58,000	3.5%
Total Expenses	$456,000	24.6%	$441,000	26.3%
Net Income	$192,000	10.4%	$189,000	11.3%

	Type of Store	Cost of Goods	Gross Profit	Total Operating Expenses	Net Income	Wages	Rent	Advertising
7.	Supermarkets	84.5%	15.5%	14.4%	1.1%	6.4%	2.1%	0.9%
	Net income too low Rent is high.	82.7%	17.3%	13.9%	3.4%	6.5%	0.8%	1.0%
9.	Drugstore	71.2%	28.8%	26.5%	2.3%	12.9%	5.3%	2.0%
	Net income too low Cost of goods, operating expenses and rent are high.	67.9%	32.1%	23.5%	8.6%	12.3%	2.4%	1.4%

11. Answers will vary.

13. Answers will vary.

17.3   The Balance Sheet

1.

APPLE CONSTRUCTION BALANCE SHEET FOR DECEMBER 31		
Assets		
Currents Assets		
Cash	$273	
Notes Receivable	$312	
Accounts Receivable	$264	
Inventory	$180	
Total Current Assets		$1029
Plant Assets		
Land	$466	
Buildings	$290	
Fixtures	$28	
Total Plant Assets		$784
TOTAL ASSETS		$1813
Liablities		
Current Liabilities		
Notes Payable	$312	
Accounts Payable	$63	
Total Current Liabilities		$375
Long-term Liabilities		
Mortgages Payable	$212	
Long-term Notes Payable	$55	
Total Long-term Liabilities		$267
Total Liabilities		$642
Owner's Equity		
Owner's Equity	$1171	
TOTAL LIABILITIES AND OWNER'S EQUITY		$1813

3. Answers will vary.

5. Answers will vary.

17.4   Analyzing the Balance Sheet: Financial Ratios

1.

INTERSTATE RUBBER SUPPLY COMPARATIVE BALANCE SHEET	Amount This Year	Percent This Year	Amount Last Year	Percent Last Year
Assets				
Current Assets				
Cash	$52,000	13.0%	$42,000	13.1%
Notes Receivable	$8000	2.0%	$6000	1.9%
Accounts Receivable	$148,000	37.0%	$120,000	37.5%
Inventory	$153,000	38.3%	$120,000	37.5%
Total Current Assets	$361,000	90.3%	$288,000	90.0%
Plant Assets				
Land	$10,000	2.5%	$8000	2.5%
Buildings	$14,000	3.5%	$11,000	3.4%
Fixtures	$15,000	3.8%	$13,000	4.1%
Total Plant Assets	$39,000	9.8%	$32,000	10.0%
TOTAL ASSETS	$400,000	100.0%	$320,000	100.0%
Liabilities				
Current Liabilities	$3000	0.8%	$4000	1.3%
Accounts Payable	$201,000	50.3%	$152,000	47.5%
Notes Payable	$204,000	51.0%	$156,000	48.8%
Total Current Liabilities				
Long-term Liabilities				
Mortgages Payable	$20,000	5.0%	$16,000	5.0%
Long-term Notes Payable	$58,000	14.5%	$42,000	13.1%
Total Long-term Liabilities	$78,000	19.5%	$58,000	18.1%
Total Liabilities	$282,000	70.5%	$214,000	66.9%
Owner's Equity	$118,000	29.5%	$106,000	33.1%
TOTAL LIABILITIES AND OWNER'S EQUITY	$400,000	100.0%	$320,000	100.0%

For Exercises 3-5, use the formula for the current ratio and the acid-test ratio.

3. (a) Current ratio $= \dfrac{\$361,000}{\$204,000} = 1.77$

Liquid assets

$\$52,000 + \$8000 + \$148,000 = \$208,000$

(b) Acid-test ratio $= \dfrac{\$208,000}{\$204,000} = 1.02$

(c) No, current ratio is low.

5. (a) Current ratio $= \dfrac{\$2,210,350}{\$1,232,500} = 1.79$

Liquid assets

$\$480,500 + \$279,050 = \$759,550$

(b) Acid-test ratio $= \dfrac{\$759,550}{\$1,232,500} = 0.62$

(c) No, both ratios are low.

Use the following information for Exercise 7.

	Amount This Year	Percent This Year	Amount Last Year	Percent Last Year
Current Assets				
Cash	$12,000	15.0%	$15,000	20.0%
Notes Rec.	$ 4000	5.0%	$ 6000	8.0%
Accts. Rec.	$22,000	27.5%	$18,000	24.0%
Inventory	$26,000	32.5%	$24,000	32.0%
Total Cur. Assets	$64,000	80.0%	$63,000	84.0%
Total Plant and Equip.	$16,000	20.0%	$12,000	16.0%
Total Assets	$80,000	100.0%	$75,000	100.0%
Total Current Liabilities	$30,000	37.5%	$25,000	33.3%

7. Current ratio $= \dfrac{\$64,000}{\$30,000} = 2.13$

Liquid assets

$\$12,000 + \$4000 + \$22,000 = \$38,000$

Acid-test ratio $= \dfrac{\$38,000}{\$30,000} = 1.27$

9. Debt-to-equity ratio $= \dfrac{\$282,000}{\$118,000}$

$= 2.39$ or 239%

11. Total liabilities

$\$1,232,500 + \$650,000$
$= \$1,882,500$

Debt-to-equity ratio $= \dfrac{\$1,882,500}{\$1,280,000} = 1.47$ or 147%

13. Answers will vary.

15. Average owner's equity

$$\frac{\$845,000 + \$928,500}{2} = \$886,750$$

Ratio of net income after taxes to average owner's equity

$$= \frac{\$54,400}{\$886,750} = 0.0613 = 6.1\%$$

17. Average accounts receivable

$$= \frac{\$320,000 + \$450,000}{2} = \$385,000$$

Accounts receivable turnover rate

$$== \frac{\$6,500,000}{\$385,000}$$

$$= 16.9 \text{ times}$$

$$\text{Average age} = \frac{365}{16.9}$$

$$= 21.6 \text{ days}$$

19. Answers will vary.

Chapter 17 Review Exercises

1. Gross profit = net sales-cost of goods sold
$$= \$312,200 - \$124,800$$
$$= \$187,400$$

Net Income before taxes
   = gross profit − operating expenses
   $$= \$187,400 - \$89,200$$
   $$= \$98,200$$

2. Gross profit = net sales-cost of goods sold
$$= \$643,250 - \$379,520$$
$$= \$263,730$$

Net Income before taxes
   = gross profit − operating expenses
   $$= \$263,730 - \$124,800$$
   $$= \$138,930$$

3. Gross profit = net sales-cost of goods sold
$$= \$442,500 - \$300,900$$
$$= \$141,600$$

Net Income before taxes
   = gross profit − operating expenses
   $$= \$141,600 - \$98,400$$
   $$= \$43,200$$

4. Gross profit = net sales-cost of goods sold
$$= \$842,400 - \$606,520$$
$$= \$235,880$$

Net Income before taxes
   = gross profit − operating expenses
   $$= \$235,880 - \$212,300$$
   $$= \$23,580$$

Use the following formula for Exercises 5-8.

Cost of goods sold
   = Initial inventory + cost of goods purchased
   + freight − ending inventory

5. $\$215,400 + \$422,000 + \$26,300 - \$247,100$
   $$= \$416,600$$

6. $\$125,400 + \$94,300 + \$8200 - \$101,400$
   $$= \$126,500$$

7. $\$84,000 + \$52,400 + \$4300 - \$98,000$
   $$= \$42,700$$

8. $\$184,200 + \$245,000 + \$18,300 - \$165,400$
   $$= \$282,100$$

Use the following formulas for Exercises 9 and 10:

Net income after taxes = net sales −
− cost of goods sold
operating expenses − taxes

Ratio of net income after taxes to average owner's

$$\text{equity} = \frac{\text{Net income after taxes}}{\text{Average owner's equity}}$$

9. Net income after taxes

   $\$660,500 - \$0 - \$412,900 - \$58,800$
   $$= \$188,800$$

   Ratio of net income after taxes to average owner's equity
   $$\frac{\$188,800}{\$340,000} = 0.56$$

10. Net income after taxes

   $\$894,200 - \$462,800 - \$304,100 - \$36,700$
   $$= \$90,600$$

   Ratio of net income after taxes to average owner's equity
   $$\frac{\$90,600}{\$389,700} = 0.23$$

11.

LORI'S BOUTIQUE
INCOME STATEMENT
FOR THE YEAR ENDING DECEMBER 31

Gross Sales		$175,000	
Returns		− $8000	
Net Sales			$167,000
Inventory, January 1		$44,000	
Cost of Goods			
Purchased	$126,000		
Freight	+ $2000		
Total Cost of Goods Purchased		+ $128,000	
Total of Goods Available for Sale		$172,000	
Inventory, December 31		− $52,000	
Cost of Goods Sold			− $120,000
Gross Profit			$47,000
Expenses			
Salaries and Wages		$9000	
Rent		$4000	
Advertising		$1500	
Utilities		$1000	
Taxes on Inventory, Payroll		$2000	
Miscellaneous Expenses		+ $3000	
Total Expenses			$20,500
NET INCOME BEFORE TAXES			$26,500

12.

THE GUITAR WAREHOUSE
INCOME STATEMENT
FOR THE YEAR ENDING DECEMBER 31

Gross Sales		$2,215,000	
Returns			
		− $26,000	
Net Sales			$2,189,000
Inventory, January 1		$215,000	
Cost of Goods			
Purchased	$1,123,000		
Freight	+ $4000		
Total Cost of Goods Purchased		+ $1,127,000	
Total of Goods Available for Sale		$1,342,000	
Inventory, December 31		− $265,000	
Cost of Goods Sold			− $1,077,000
Gross Profit			$1,112,000
Expenses			
Salaries and Wages		$154,000	
Rent		$59,000	
Advertising		$11,000	
Utilities		$12,000	
Taxes on Inventory, Payroll		$10,000	
Miscellaneous Expenses		$9000	
Total Expenses			$255,000
NET INCOME BEFORE TAXES			$857,000
Income Taxes			− $242,300
NET INCOME			$614,700

13. $\dfrac{\text{Cost of goods sold}}{\text{Net sales}} = \dfrac{\$485,800}{\$812,200}$

$$= 0.5981 = 59.8\%$$

$\dfrac{\text{Operating expenses}}{\text{Net sales}} = \dfrac{\$104,300}{\$812,200}$

$$= 0.1284 = 12.8\%$$

14. $\dfrac{\text{Cost of goods sold}}{\text{Net sales}} = \dfrac{\$813,200}{\$1,329,400}$

$$= 0.6117 = 61.2\%$$

$\dfrac{\text{Operating expenses}}{\text{Net sales}} = \dfrac{\$387,100}{\$1,329,400}$

$$= 0.2911 = 29.1\%$$

15.

ANDY'S STEAK HOUSE			
	Amount	Percent	Percent from Table 17.1
Net Sales	$300,000	$\dfrac{\$300,000}{\$300,000} = 100\%$	100%
Cost of Goods Sold	$125,000	$\dfrac{\$125,000}{\$300,000} = \underline{41.7\%}$	<u>48.4%</u>
Gross Profit	$175,000	$\dfrac{\$175,000}{\$300,000} = \underline{58.3\%}$	<u>51.6%</u>
Wages	$72,000	$\dfrac{\$72,000}{\$300,000} = \underline{24.0\%}$	<u>26.4%</u>
Rent	$12,000	$\dfrac{\$12,000}{\$300,000} = \underline{4.0\%}$	<u>2.8%</u>
Advertising	$5700	$\dfrac{\$5700}{\$300,000} = \underline{1.9\%}$	<u>1.4%</u>
Total Expenses	$123,000	$\dfrac{\$123,000}{\$300,000} = \underline{41.0\%}$	<u>43.7%</u>
Net Income	$52,000	$\dfrac{\$52,000}{\$300,000} = \underline{17.3\%}$	<u>7.9%</u>

16.

| GASKETS, INC. |
| BALANCE SHEET |
| FOR DECEMBER 31 |

Assets

Current Assets			
Cash	$240,000		
Notes Receivabale	$180,000		
Accounts Receivabale	$460,000		
Inventory	$225,000		
Total Current Assets		$1,105,000	
Plant Assets			
Land	$180,000		
Buildings	$260,000		
Fixtures	$48,000		
Total Plant Liabilities		$488,000	
TOTAL ASSETS			$1,593,000

Liabilities

Current Liabilities			
Notes Payable	$410,000		
Accounts Payable	$882,000		
Total Current Liabilities		$1,292,000	
Long-term Liabilities			
Mortgages Payable	$220,000		
Long-term Notes Payable	$194,000		
Total Long-term Liabilities		$414,000	
Total Liabilities			$1,706,000

Owner's Equity

Owner's Equity	($113,000)	
TOTAL LIABILITIES AND OWNER'S EQUITY		$1,593,000

17.

Current Assets	Current Liabilities	Long-Term Liabilities	Owner's Equity	Liquid Assets
$342,000	$260,000	$140,000	$225,000	$120,000

$$\text{Current ratio} = \frac{\text{Current Assets}}{\text{Current Liabilities}} = \frac{\$342,000}{\$260,000} \approx 1.32$$

$$\text{Acid-test ratio} = \frac{\text{Liquid Assets}}{\text{Current Liabilities}} = \frac{\$120,000}{\$260,000} \approx 0.46$$

$$\text{Debt to equity ratio} = \frac{\text{Current Liabilities} + \text{Long Term Liabiities}}{\text{Owner's Equity}}$$

$$\frac{\$260,000 + \$140,000}{\$225,000} = \frac{\$400,000}{\$225,000} \approx 1.78 \text{ or } 178\%$$

18.

Current Assets	Current Liabilities	Long-Term Liabilities	Owner's Equity	Liquid Assets
$95,000	$115,000	$85,000	$48,000	$5000

$$\text{Current ratio} = \frac{\$95,000}{\$115,000} \approx 0.83$$

$$\text{Acid test ratio} = \frac{\$5000}{\$115,000} \approx 0.04$$

$$\text{Debt to equity ratio} = \frac{\$115,000 + \$85,000}{\$48,000} = \frac{\$200,000}{\$48,000} \approx 4.17 \text{ or } 417\%$$

19.

Current Assets	Current Liabilities	Long-Term Liabilities	Owner's Equity	Liquid Assets
$160,000	$205,000	$0	$185,000	$145,000

$$\text{Current ratio} = \frac{\$160,000}{\$205,000} \approx 0.78$$

$$\text{Acid test ratio} = \frac{\$145,000}{\$205,000} \approx 0.71$$

$$\text{Debt to equity ratio} = \frac{\$205,000 + \$0}{\$185,000} \approx 1.11 \text{ or } 111\%$$

20. Average accounts receivable $= \dfrac{\$875,400 + \$962,300}{2}$

$$= \$918,850$$

Accounts receivable turnover rate $= \dfrac{\$4,612,000}{\$918,850}$

$$= 5.0 \text{ times}$$

Average age $= \dfrac{365}{5.0} = 73 \text{ days}$

21. Average accounts receivable $= \dfrac{\$126,800 + \$92,400}{2}$

$$= \$109,600$$

Accounts receivable turnover rate $= \dfrac{\$942,500}{\$109,600}$

$$= 8.6 \text{ times}$$

Average age $= \dfrac{365}{8.6} \approx 42.4 \text{ days}$

22.

	Amount This Year	Percent This Year	Amount Last Year	Percent Last Year
Current assets				
Cash	$28,000	18.8%	$22,000	21.1%
Notes Receivable	$12,000	8.1%	$15,000	14.4%
Accounts Receivable	$39,000	26.2%	$31,500	30.1%
Inventory	$22,000	14.8%	$20,000	19.1%
Total Current Assets	$101,000	67.8%	$88,500	84.7%
Total Plant and Equipment	$48,000	32.2%	$16,000	15.3%
TOTAL ASSETS	$149,000	100.0%	$104,500	100.0%
Total Current Liabilities	$38,000	25.5%	$36,000	34.4%

Chapter 17 Summary Exercise

(a)

```
WALKER BICYCLE SHOP
INCOME STATEMENT
YEAR ENDING DECEMBER 31
```

Gross Sales		$212,000
Returns		$12,500
Net Sales		$199,500
Inventory, January 1		$44,000
Cost of Goods Purchased	$75,000	
Freight	+ $8000	
Total Cost of Goods Purchased		+ $83,000
Total of Goods Available for Sale		$127,000
Inventory, December 31		− $26,000
Cost of Goods Sold		
		− $101,000
Gross Profit		$98,500
Expenses		
Salaries and Wages	$37,000	
Rent	$12,000	
Advertising	$2000	
Utilities	$3000	
Taxes on Inventory, Payroll	$7000	
Miscellaneous Expenses	$4500	
Total Expenses		
		− $65,500
NET INCOME BEFORE TAXES		$33,000

(b)

$$\frac{\text{Gross sales}}{\text{Net sales}} = \frac{\$212,000}{\$199,500}$$

$$= 1.0626 = 106.3\%$$

$$\frac{\text{Returns}}{\text{Net sales}} = \frac{\$12,500}{\$199,500}$$

$$= 0.0626 = 6.3\%$$

$$\frac{\text{Cost of goods sold}}{\text{Net sales}} = \frac{\$101,000}{\$199,500}$$

$$= 0.5062 = 50.6\%$$

$$\frac{\text{Salaries and Wages}}{\text{Net sales}} = \frac{\$37,000}{\$199,500}$$

$$= 0.1854 = 18.5\%$$

$$\frac{\text{Rent}}{\text{Net sales}} = \frac{\$12,000}{\$199,500}$$

$$= 0.0601 = 6.0\%$$

$$\frac{\text{Utilities}}{\text{Net sales}} = \frac{\$3000}{\$199,500}$$

$$= 0.0150 = 1.5\%$$

(c)

WALKER BICYCLE SHOP BALANCE SHEET DECEMBER 31		
Assets		
Current Assets		
Cash	$62,000	
Notes Receivable	$2500	
Accounts Receivable	$8200	
Inventory	$26,000	
Total Current Assets		$98,700
Plant Assets		
Land	$7600	
Building	$28,000	
Fixtures	$13,500	
Toal Plant Assets		$49,100
TOTAL ASSETS		$147,800
Liabilities		
Current Liabilities		
Notes Payable	$4500	
Accounts Payable	$27,000	
Total Current Liabilities		$31,500
Long-term Liabilities		
Mortgages Payable	$15,000	
Long-term Notes Payable	$8000	
Total Long-term Liabilities		$23,000
Total Liabilities		$54,500
Owner's Equity		
Owner's Equity	$93,300	
TOTAL LIABILITITES AND OWNER'S EQUITY		$147,800

(d) Current ratio: $\dfrac{\$98,700}{\$31,500} = 3.13$

Acid-test ratio: $\dfrac{\$62,000 + \$2500 + \$8200}{\$31,500} = 2.31$

(e) Yes. Answers will vary.

SECURITIES AND DISTRIBUTION OF PROFIT AND OVERHEAD

18.1 Distribution of Profits in a Corporation

1. No dividends paid

3. $\$10 \times 0.03 = \$0.30$ per preferred share
$\$0.30 \times 50,000 = \$15,000$
$\$40,000 - \$15,000 = \$25,000$

$\dfrac{\$25,000}{175,000} = \$0.14$ per share of
common stock

5. $\$1000 \times 0.02 = \$20$ per preferred share
$\$20 \times 100,000 = \$2,000,000$
$\$2,000,000 - \$2,000,000 = \$0$ per share of
common stock

7. $\$320,000 - \$280,000 = \$40,000$

$\dfrac{\$40,000}{400,000} = \$0.10$ per share of
common stock

9. $\$100 \times 0.08 = \$8$ per preferred share
$\$8 \times 10,000 = \$80,000$
$\$592,000 - \$80,000 = \$512,000$

$\dfrac{\$512,000}{200,000} = \$2.56$ per share of
common stock

11. $\$850,000 \times 0.35 = \$297,500$
$\$40 \times 0.04 = \$1.60$ per preferred share
$\$1.60 \times 25,000 = \$40,000$
$\$297,500 - \$40,000 = \$257,500$

$\dfrac{\$257,500}{300,000} = \$0.86$ per share of
common stock

13. $\$100 \times 0.05 = \$5$
$\$5 \times 4 = \$20$ per preferred share
$\$20 \times 20,000 = \$400,000$
$\$2,675,000 - \$400,000 = \$2,275,000$

$\dfrac{\$2,275,000}{450,000} = \$5.06$ per share

15. $\dfrac{\$7,000,000}{1,200,000} = \$5.83$ per share

17. $\$150 \times 0.10 = \$15$
$\$15 \times 12,500 = \$187,500$
$\$1,500,000 - \$187,500 = \$1,312,500$

$\dfrac{\$1,312,500}{300,000} = \$4.38$ per share

(Notice that the 45% mentioned in the exercise is
not used in the solution. It should only be used
in finding the dividend per share.)

19. $\$50 \times 0.04 = \$2$ per preferred share each year
$\$2 \times 40,000 = \$80,000$

(a) Available last year

$\$680,000 \times 0.40 = \$272,000$
$\$272,000 - \$80,000 = \$192,000$

$\dfrac{\$192,000}{250,000} = \$0.77$ per share of
common stock

(b) Available this year

$\$765,000 \times 0.40 = \$306,000$
$\$306,000 - \$80,000 = \$226,000$

$\dfrac{\$226,000}{250,000} = \$0.90$ per share of
common stock

(c) Earnings per share

Last year: $\dfrac{\$680,000 - \$80,000}{250,000} = \$2.40$

This year: $\dfrac{\$765,000 - \$80,000}{250,000} = \$2.74$

Percent increase: $\dfrac{\$2.74 - \$2.40}{\$2.40} \approx 14.2\%$

21. Dividend per share of preferred stock

$\$100 \times 0.08 = \$8$

Dividend in 20,000 shares

$\$8 \times 20,000 = \$160,000$

Amount available for dividends

$\$620,000 \times 0.30 = \$186,000$

Dividend for each grandchild

$\dfrac{\$160,000}{6} = \$26,666.67$

23. Answers will vary.

18.2   Buying Stock

For Exercises 1-15, use Figure 18.2.

1. $57\frac{7}{16}$

3. $59\frac{5}{8}$

5. $30\frac{7}{8}$

7. 1.44

9. 808,300

11. $41\frac{9}{16}$ or 41.5625
$41.5625 \times 1000 = \$41,562.50$

13. $22\frac{1}{2}$ or 22.50
$22.50 \times 200 = \$4500$

15. $26\frac{7}{8}$ or 26.875
$26.875 \times 100 = \$2687.50$

For Exercises 17-27, use the broker's charges given in the text.

17. $27\frac{5}{8}$ or 27.625

$27.625 \times 200 = \$5525$

Find the broker's commission of 1.5%.

$5525 \times 0.015 = \$82.88$

Price

$5525 + \$82.88 = \$5607.88$

19. $37\frac{3}{4}$ or 37.75

$37.75 \times 1200 = \$45,300$

Find the broker's commission of 1.5%.

$45,300 \times 0.015 = \$679.50$

Price

$45,300 + \$679.50 = \$45,979.50$

21. $35\frac{5}{8}$ or 35.625

$35.625 \times 60 = \$2137.50$

Find the broker's commission of 1.5%.

$2137.50 \times 0.015 = \$32.06$

Now find the odd-lot differential of $0.125 per share.

$0.125 \times 60 = \$7.50$

Price

$2137.50 + \$32.06 + \$7.50 = \$2177.06$

23. $69\frac{1}{4}$ or 69.25
$69.25 \times 540 = \$37,395$

Find the broker's commission of 1.5%.

$37,395 \times 0.015 = \$560.93$

The odd-lot differential is charged only on 40 shares (since 500 shares make up a round lot).

$0.125 \times 40 = \$5$

Price

$37,395.00 + \$560.93 + \$5.00 = \$37,960.93$

For Exercises 25-29, use the broker's charges given in the text. Round to the nearest cent for SEC fee.

25. $30\frac{1}{2}$ or 30.50

$30.50 \times 300 = \$9150$

Find the SEC fee.

$\dfrac{\$9150}{300} = 30.5$ which gives a fee of $0.31.

Now find the sales commission.

$9150 \times 0.015 = \$137.25$

The seller receives

$9150 - \$0.31 - \$137.25 = \$9012.44.$

27. $40\frac{3}{4}$ or 40.75

$40.75 \times 100 = \$4075$

Find the SEC fee.

$\dfrac{\$4075}{300} = 13.58\overline{3}$ which gives a fee of $0.14.

Now find the sales commission.

$4075 \times 0.015 = \$61.13$

The seller receives

$4075 - \$0.14 - \$61.13 = \$4013.73$

29. $52\frac{3}{4}$ or 52.75

$\$52.75 \times 830 = \$43,782.50$

Find the SEC fee.

$\dfrac{\$43,782.50}{300} = 145.941\overline{6}$ which gives a fee of $1.46.

Now find the sales commission.

$\$43,782.50 \times 0.015 = \$656.74$

Find the odd-lot differential of $0.125 per share on 30 shares.

$\$0.125 \times 30 = \$3.75$

The seller receives

$\$43,782.50 - \$1.46 - \$656.74 - \$3.75$
$\quad = \$43,120.55.$

31. $52\frac{11}{16}$ or 52.6875

$\dfrac{\$0.48}{\$52.6875} = 0.0091 = 0.9\%$

33. $51\frac{11}{16}$ or 51.6875

$\dfrac{\$1.84}{\$51.6875} = 0.0355 = 3.6\%$

35. $36\frac{1}{4}$ or 36.25

$\dfrac{\$0.50}{\$36.25} = 0.0137 = 1.4\%$

37. $60\frac{5}{16}$ or 60.3125

$\dfrac{\$60.3125}{\$3.10} = 19.4 \approx 19$

39. $63\frac{1}{8}$ or 63.125

$\dfrac{\$63.125}{\$7.36} = 8.5 \approx 9$

41. $60\frac{3}{8}$ or 60.375
$\$60.375 \times 200 = \$12,075$
$120\frac{1}{2}$ or 120.50
$\$120.50 \times 300 = \$36,150$
$\$12,075 + \$36,150 = \$48,225$

Find the sales commission of 1%.

$\$48,225 \times 0.01 = \$482.25$

Price

$\$48,225 + \$482.25 = \$48,707.25$

43. Purchase price: $250 \times \$105 = \$26,250$
Commission: $\$26,250 \times 0.02 = \$525$
Odd-lot differential: $\$0.125 \times 50 = \$6.25$
Total price: $\$26,250 + \$525 + \$6.25 = \$26,781.25$

Sale of stock: $250 \times \$118.25 = \$29,562.50$

SEC fee: $\dfrac{\$29,562.50}{300} = 98.541\overline{6}$ which gives a fee of $0.99.
Commission: $\$29,562.50 \times 0.015 = \$443.44$
Odd-lot differential: $\$0.125 \times 50 = \$6.25$
The seller receives

$\$29,562.50 - \$0.99 - \$443.44 - \$6.25 = \$29,111.82$

Profit

$\$29,111.82 - \$26,781.25 = \$2330.57$

45. Answers will vary.

47. (a) Plan A: $500 payments semiannually at 8% compounded semiannually for 20 years
Amount of annuity: $S = R \cdot s_{\overline{n}|i}$

$R = \$500$ (the amount of each payment)
$n = 40$ (20×2, total number of payments)
$i = 0.04$ ($0.08 \div 2$, interest rate for each period)
Use column C of Appendix D.

$S = 500 \cdot s_{\overline{40}|.04} = 500(95.02551570) = \$47,512.76$

(b) Plan B: $500 payments semiannually at 12% compounded semiannually for 20 years.
Amount of an annuity: $S = R \cdot s_{\overline{n}|i}$

$R = \$500; \; n = 40; \; i = 0.06.$

Use column C of Appendix D.

$S = 500 \cdot s_{\overline{40}|0.06} = 500(154.76196562) = \$77,380.98$

(c) $\$77,380.98 - \$47,512.76 = \$29,868.22$

18.3   Bonds

For the exercises 1-25, use Figure 18.7. Remember that the numbers in the table are percents, and that most bonds have a face value of $1000.

1. $71\frac{1}{2}$

3. $9\frac{3}{4}\%$ or $\$1000 \times 0.0975 = \$97.50$

5. -2

7. The closing price is 133 or 133% = 1.33
The cost of one bond is

$1000 × 1.33 = $1330
$1330 × 30 = $39,900
$5 × 30 = $150

The total cost for 30 bonds is

$39,900 + $150 = $40,050

9. 94 = 94%

$1000 × 94% = $1000 × 0.94 = $940

$940 × 100 = $94,000
$5 × 100 = $500

$94,000 + $500 = $94,500

11. $90\frac{3}{8}$ = 90.375%

$1000 × 90.375% = $1000 × 0.90375
= $903.75

$903.75 × 50 = $45,187.50
$5 × 50 = $250

$45,187.50 + $250 = $45,437.50

13. $103\frac{1}{2}$ = 103.5%

$1000 × 103.5% = $1000 × 1.035
= $1035

$1035 × 25 = $25,875
$5 × 25 = $125

$25,875 + $125 = $26,000

15. Answers will vary.

17. $93\frac{1}{2}$ = 93.5%

$1000 × 93.5% = $1000 × 0.935
= $935

$935 × 20 = $18,700
$5 × 20 = $100

The total cost is

$18,700 + $100 = $18,800

19. I Pap: $81\frac{5}{8}$ = 81.625%

$1000 × 81.625% = $1000 × 0.81625
= $816.25

$816.25 × 10 = $8162.50
$5 × 10 = $50

$8162.50 + $50 = $8212.50

Kolmrg: $99\frac{3}{4}$ = 99.75%

$1000 × 99.75% = $1000 × 0.9975
= $997.50

$997.50 × 10 = $9975
$5 × 10 = $50

$9975 + $50 = $10,025

The total cost is

$8212.50 + $10,025 = $18,237.50

21. (a) ATT: $89\frac{1}{2}$ = 89.5%

$1000 × 89.5% = $1000 × 0.895 = $895

$895 × 10 = $8950
$5 × 10 = $50

$8950 + $50 = $9000

IBM: $90\frac{5}{8}$ = 90.375%

$1000 × 90.375% = $1000 × 0.90375
= $903.75

$903.75 × 10 = $9037.50
$5 × 10 = $50

$9037.50 + $50 = $9087.50

Total cost

$9000 + $9087.50 = $18,087.50

(b) ATT

$1000 × 6.5% = $1000 × 0.065 = $65
$65 × 10 = $650

IBM

$1000 × 5.375% = $1000 × 0.05375
= $53.75

$53.75 × 10 = $537.50

Annual interest

$650 + $537.50 = $1187.50

23. The tax free income is

$100,000 × $4\frac{1}{2}$% = $100,000 × 0.045
= $4500.

25. $P = \dfrac{I}{R}$

$= \dfrac{\$7000}{0.075}$

$= \$93,333.33$

18.4   Distribution of Profits and Losses in a Partnership

1. $\frac{1}{2} \times \$58,000 = \$29,000$ for each partner

3. Total amount contributed

   $\$80,000 + \$20,000 = \$100,000$

   Partner 1: $\dfrac{\$80,000}{\$100,000} = \dfrac{4}{5}$

   $\frac{4}{5} \times \$46,000 = \$36,800$

   Partner 2: $\dfrac{\$20,000}{\$100,000} = \dfrac{1}{5}$

   $\frac{1}{5} \times \$46,000 = \$9200$

5. $\$40,000 \times 12\% = \$40,000 \times 0.12 = \$4800$

   $\$25,000 - \$4800 = \$20,200$

   There are $3 + 2 = 5$ shares.

   Partner 1: $\frac{3}{5} \times \$20,200 = 0.60 \times \$20,200$
   $= \$12,120$

   $\$12,120 + \$4800 = \$16,920$

   Partner 2: $\frac{2}{5} \times \$20,200 = 0.40 \times \$20,200$
   $= \$8080$

7. Partner 1: $10,000 salary.
   Partner 2: $12,000 salary.
   $\$22,000 - (\$10,000 + \$12,000) = 0$
   There is no profit left after salaries are paid.

9. $\dfrac{\$84,500}{2} = \$42,250$ for each partner

11. Total investment

    $\$10,000 + \$20,000 + \$20,000 = \$50,000$

    Partner 1: $\dfrac{\$10,000}{\$50,000} = \dfrac{1}{5}$

    $\frac{1}{5} \times \$15,600 = \$3120$

    Partner 2: $\dfrac{\$20,000}{\$50,000} = \dfrac{2}{5}$

    $\frac{2}{5} \times \$15,600 = \$6240$

    Partner 3: $\dfrac{\$20,000}{\$50,000} = \dfrac{2}{5}$

    $\frac{2}{5} \times \$15,600 = \$6240$

13. $3 + 5 + 7 + 9 = 24$

    Partner 1: $\dfrac{3}{24} \times \$180,000 = \$22,500$

    Partner 2: $\dfrac{5}{24} \times \$180,000 = \$37,500$

    Partner 3: $\dfrac{7}{24} \times \$180,000 = \$52,500$

    Partner 4: $\dfrac{9}{24} \times \$180,000 = \$67,500$

15. $\$57,000 - \$15,000 = \$42,000$

    $1 + 4 = 5$ shares

    $\frac{1}{5} \times \$42,000 = \$8400$

    $\$8400 + \$15,000 = \$23,400$ for Finch

    $\frac{4}{5} \times \$42,000 = \$33,600$ for Renz

17. (a) $\$80,000 \times 10\% = \$80,000 \times 0.10 = \$8000$

    $\$60,000 - \$8000 = \$52,000$

    $1 + 3 = 4$ shares

    $\frac{1}{4} \times \$52,000 = \$13,000$

    $\$13,000 + \$8000 = \$21,000$ for Coker

    $\frac{3}{4} \times \$52,000 = \$39,000$ for Toms

    (b) $\$80,000 \times 10\% = \$8000$

    $\$6000 - \$8000 = (\$2000)$

    $1 + 3 = 4$ shares

    $\frac{3}{4} \times \$2000 = \$1500$ paid by Toms to Coker

    $\$6000 + \$1500 = \$7500$ to Coker

19. $\$15,000 \times 10\% = \$1500$

    $\$110,000 - (\$1500 + \$12,000) = \$96,500$

    $\$15,000 + \$25,000 + \$30,000 = \$70,000$

    Partner 1: $\dfrac{\$15,000}{\$70,000} = \dfrac{3}{14}$

    $\frac{3}{14} \times \$96,500 = \$20,679$

    $\$20,679 + \$1500 = \$22,179$

    Partner 2: $\dfrac{\$25,000}{\$70,000} = \dfrac{5}{14}$

    $\frac{5}{14} \times \$96,500 = \$34,464$

    $\$34,464 + \$12,000 = \$46,464$

    Partner 3: $\dfrac{\$30,000}{\$70,000} = \dfrac{3}{7}$

    $\frac{3}{7} \times \$96,500 = \$41,357$

21. Answers will vary.

18.5   Distribution of Overhead

1. Total square feet

$4000 + 8000 + 10,000 = 22,000$

Dept. 1: $\dfrac{4000}{22,000} = \dfrac{2}{11}$

$\dfrac{2}{11} \times \$330,000 = \$60,000$

Dept. 2: $\dfrac{8000}{22,000} = \dfrac{4}{11}$

$\dfrac{4}{11} \times \$330,000 = \$120,000$

Dept. 3: $\dfrac{10,000}{22,000} = \dfrac{5}{11}$

$\dfrac{5}{11} \times \$330,000 = \$150,000$

3. Total square feet

$2400 + 3600 + 4000 + 6000 = 16,000$

Department 1: $\dfrac{2400}{16,000} = \dfrac{3}{20}$

$\dfrac{3}{20} \times \$120,000 = \$18,000$

Department 2: $\dfrac{3600}{16,000} = \dfrac{9}{40}$

$\dfrac{9}{40} \times \$120,000 = \$27,000$

Department 3: $\dfrac{4000}{16,000} = \dfrac{1}{4}$

$\dfrac{1}{4} \times \$120,000 = \$30,000$

Department 4: $\dfrac{6000}{16,000} = \dfrac{3}{8}$

$\dfrac{3}{8} \times \$120,000 = \$45,000$

5.
$\begin{aligned}
15,000 \times \$8.00 &= \$120,000 \\
20,000 \times \$3.50 &= \$\ 70,000 \\
35,000 \times \$2.00 &= \$\ 70,000 \\
\text{Total value} &= \$260,000
\end{aligned}$

Product M: $\dfrac{\$120,000}{\$260,000} = \dfrac{12}{26}$

$\dfrac{12}{26} \times \$140,000 = \$64,615$

Product N: $\dfrac{\$70,000}{\$260,000} = \dfrac{7}{26}$

$\dfrac{7}{26} \times \$140,000 = \$37,692$

Product P: $\dfrac{\$70,000}{\$260,000} = \dfrac{7}{26}$

$\dfrac{7}{26} \times \$140,000 = \$37,692$

7.
$\begin{aligned}
140 \times \$100 &= \$14,000 \\
2000 \times \$15 &= \$30,000 \\
150 \times \$20 &= \$\ \ 3000 \\
1000 \times \$22 &\quad\ \ \$22,000 \\
\text{Total value} &= \$69,000
\end{aligned}$

Product 1: $\dfrac{\$14,000}{\$69,000} = \dfrac{14}{69}$

$\dfrac{14}{69} \times \$48,000 = \$9739$

Product 2: $\dfrac{\$30,000}{\$69,000} = \dfrac{10}{23}$

$\dfrac{10}{23} \times \$48,000 = \$20,870$

Product 3: $\dfrac{\$3000}{\$69,000} = \dfrac{1}{23}$

$\dfrac{1}{23} \times \$48,000 = \$2087$

Product 4: $\dfrac{\$22,000}{\$69,000} = \dfrac{22}{69}$

$\dfrac{22}{69} \times \$48,000 = \$15,304$

9. Total number of employees

$110 + 60 + 80 = 250$

Dept. X: $\dfrac{110}{250} = \dfrac{11}{25}$

$\dfrac{11}{25} \times \$650,000 = \$286,000$

Dept. Y: $\dfrac{60}{250} = \dfrac{6}{25}$

$\dfrac{6}{25} \times \$650,000 = \$156,000$

Dept. Z: $\dfrac{80}{250} = \dfrac{8}{25}$

$\dfrac{8}{25} \times \$650,000 = \$208,000$

11. Total number of employees

$100 + 120 + 140 + 60 = 420$

Dept. 1: $\dfrac{100}{420} = \dfrac{5}{21}$

$\dfrac{5}{21} \times \$800,000 = \$190,476$

Dept. 2: $\dfrac{120}{420} = \dfrac{2}{7}$

$\dfrac{2}{7} \times \$800,000 = \$228,571$

Dept. 3: $\dfrac{140}{420} = \dfrac{1}{3}$

$\dfrac{1}{3} \times \$800,000 = \$266,667$

Dept. 4: $\dfrac{60}{420} = \dfrac{1}{7}$

$\dfrac{1}{7} \times \$800,000 = \$114,286$

13. $2000 + 8000 + 6000 + 9000 + 1000 + 4000$
$= 30,000$ square feet

$\dfrac{2000}{30,000} = \dfrac{1}{15}$

$\dfrac{1}{15} \times \$360,000 = \$24,000$ to hoses

$\dfrac{8000}{30,000} = \dfrac{4}{15}$

$\dfrac{4}{15} \times \$360,000 = \$96,000$ to carburetors

$\dfrac{6000}{30,000} = \dfrac{1}{5}$

$\dfrac{1}{5} \times \$360,000 = \$72,000$ to water pumps

$\dfrac{9000}{30,000} = \dfrac{3}{10}$

$\dfrac{3}{10} \times \$360,000 = \$108,000$ to fuel pumps

$\dfrac{1000}{30,000} = \dfrac{1}{30}$

$\dfrac{1}{30} \times \$360,000 = \$12,000$ to gaskets

$\dfrac{4000}{30,000} = \dfrac{2}{15}$

$\dfrac{2}{15} \times \$360,000 = \$48,000$ to filters

15.
$$150 \times \$200 = \$\ 30,000$$
$$200 \times \$400 = \$\ 80,000$$
$$100 \times \$600 = \$\ 60,000$$
$$500 \times \ \$75 = \$\ 37,500$$
$$300 \times \$150 = \underline{\$\ 45,000}$$
Total value $= \$252,500$

$\dfrac{\$30,000}{\$252,500} \times \$68,000 = \$8079$ to construction

$\dfrac{\$80,000}{\$252,500} \times \$68,000 = \$21,545$ to plywood

$\dfrac{\$60,000}{\$252,500} \times \$68,000 = \$16,158$ to veneers

$\dfrac{\$37,500}{\$252,500} \times \$68,000 = \$10,099$ to wood chips

$\dfrac{\$45,000}{\$252,500} \times \$68,000 = \$12,119$ to furniture wood

17. $6 + 8 + 28 + 8 = 50$ employees

$\dfrac{6}{50} \times \$6000 = \$720$ to office

$\dfrac{8}{50} \times \$6000 = \$960$ to marketing

$\dfrac{28}{50} \times \$6000 = \$3360$ to distribution

$\dfrac{8}{50} \times \$6000 = \$960$ to accounting

19. Answers will vary.

Chapter 18 Review Exercises

1. (a) $\$50 \times 0.065 = \$3.25$ per preferred share

(b) $\$3.25 \times 8000 = \$26,000$
$\$460,000 - \$26,000 - \$317,000 = \$117,000$

$\dfrac{\$117,000}{180,000} = \$0.65$ per share of
common stock

2. (a) $\$150 \times 0.08 = \$12$ per preferred share

(b) $\$12 \times 15,000 = \$180,000$
$\$2,375,000 - \$180,000 - \$750,000 = \$1,445,000$

$\dfrac{\$1,445,000}{200,000} = \$7.23$ per share of
common stock

3. (a) $125 \times 0.07 = \$8.75$ per preferred share

(b) $8.75 \times 22{,}750 = \$199{,}062.50$

$2{,}640{,}000 - \$425{,}000 - \$199{,}062.50$
$= \$2{,}015{,}937.50$

$$\frac{\$2{,}015{,}937.50}{750{,}000} = \$2.69 \text{ per share of common}$$

4. $\dfrac{\$127{,}500}{82{,}000} = \$1.55$

5. $1{,}425{,}000 - \$675{,}000 = \$750{,}000$

$$\frac{\$750{,}000}{275{,}000} = \$2.73 \text{ per share}$$

6. $2{,}750{,}000 - \$900{,}000 = \$1{,}850{,}000$

$$\frac{\$1{,}850{,}000}{500{,}000} = \$3.70 \text{ per share}$$

7. $56\frac{1}{8}$

8. $34\frac{1}{2}$

9. $12\frac{5}{8}$

10. $+\dfrac{3}{8}$

11. 0.06

12. $4131 \times 100 = 413{,}100$

13. 0.5%

14. $21\frac{13}{16}$

15. $10\frac{3}{4}$

16. $7\frac{1}{2}$ or $\$7.50$
$7.50 \times 200 = \$1500$

17. $18\frac{3}{16}$ or $\$18.1875$
$18.1875 \times 300 = \$5456.25$

18. $69\frac{7}{16}$ or $\$69.4375$
$69.4375 \times 800 = \$55{,}550$

19. $41\frac{5}{8}$ or $\$41.625$

$41.625 \times 200 = \$8325$

Find the broker's commission of 1.5%

$8325 \times 0.015 = \$124.88$

Price

$8325 + \$124.88 = \$8449.88$

20. $73\frac{1}{8}$ or $\$73.125$

$73.125 \times 340 = \$24{,}862.50$

Find the broker's commission of 1.5%

$24{,}862.50 \times 0.015 = \$372.94$

Find the odd-lot differential of $0.125 per share on 40 shares.

$0.125 \times 40 = \$5$
Price

$24{,}862.50 + \$372.94 + \$5 = \$25{,}240.44$

21. $30\frac{5}{8}$ or $\$30.625$

$30.625 \times 100 = \$3062.50$

Find the SEC fee.
$\dfrac{\$3062.50}{300} = 10.208\overline{3}$ which gives a fee of $0.11.

Now find the sales commission.

$3062.50 \times 0.015 = \$45.94$

The seller receives

$3062.50 - \$0.11 - \$45.94 = \$3016.45$

22. $47\frac{1}{4}$ or $\$47.25$

$47.25 \times 180 = \$8505$

Find the SEC fee.
$\dfrac{\$8505}{300} = 28.35$ which gives a fee of $0.29.

Now find the sales-commission.

$8505 \times 0.015 = \$127.58$

Now find the odd-lot differential of $0.125 per share on 80 shares.

$0.125 \times 80 = \$10$

The seller receives

$8505 - \$0.29 - \$127.58 - \$10 = \$8367.13$

23. $104\frac{7}{8}$ or $\$104.875$

$$\frac{\$0.64}{\$104.875} = 0.0061 = 0.6\%$$

24. $19\frac{15}{16}$ or $\$19.9375$

$$\frac{\$0.04}{\$19.9375} = 0.0020 = 0.2\%$$

25. $83\frac{3}{8}$ or $\$83.375$

$$\frac{\$83.375}{\$1.42} = 58.7 \approx 59$$

26. $112\frac{13}{16}$ or $112.8125

$$\frac{\$112.8125}{\$2.99} = 37.7 \approx 38$$

27. $91\frac{3}{4}$

28. 1

29. 2033

30. No change

31. The closing price is $91\frac{3}{4}$
or $91\frac{3}{4}\% = 0.9175$

The cost of one bond is

$1000 \times 0.9175 = \$917.50$

$917.50 \times 30 = \$27,525$
$5 \times 30 = \$150$

The total cost for 30 bonds is

$27,525 + \$150 = \$27,675$

32. $6\frac{3}{4}\%$ or $1000 \times 0.0675 = \$67.50$

33. $90\frac{1}{2} = 90.5\%$

$1000 \times 90.5\% = \$1000 \times 0.905$
$= \$905$

$905 \times 50 = \$45,250$
$5 \times 50 = \$250$

$45,250 + \$250 = \$45,500$

34. $98\frac{1}{8} = 98.125\%$

$1000 \times 98.125\% = \$1000 \times 0.98125$
$= \$981.25$

$981.25 \times 100 = \$98,125$
$5 \times 100 = \$500$

$98,125 + \$500 = \$98,625$

35. $\dfrac{\$48,000}{3} = \$16,000$

Each partner receives $16,000.

36. $2 + 3 = 5$

Partner 1: $\dfrac{2}{5} \times \$120,000 = \$48,000$

Partner 2: $\dfrac{3}{5} \times \$120,000 = \$72,000$

37. $9000 + \$12,000 = \$21,000$

Partner 1: $\dfrac{\$9000}{\$21,000} = \dfrac{3}{7}$

$\dfrac{3}{7} \times \$90,000 = \$38,571$

Partner 2: $\dfrac{\$12,000}{\$21,000} = \dfrac{4}{7}$

$\dfrac{4}{7} \times \$90,000 = \$51,429$

38. Total square feet

$3000 + 5000 + 4000 = 12,000$

Dept. A: $\dfrac{3000}{12,000} = \dfrac{1}{4}$

$\dfrac{1}{4} \times \$100,000 = \$25,000$

Dept. B: $\dfrac{5000}{12,000} = \dfrac{5}{12}$

$\dfrac{5}{12} \times \$100,000 = \$41,667$

Dept. C: $\dfrac{4000}{12,000} = \dfrac{1}{3}$

$\dfrac{1}{3} \times \$100,000 = \$33,333$

39. $8000 \times \$12 = \$\ \ 96,000$
$40,000 \times \ \ \$6 = \$240,000$
$20,000 \times \ \ \$9 = \underline{\$180,000}$
Total value $= \$516,000$

Product: $\dfrac{\$96,000}{\$516,000} = \dfrac{8}{43}$

$\dfrac{8}{43} \times \$125,000 = \$23,256$

Product 2: $\dfrac{\$240,000}{\$516,000} = \dfrac{20}{43}$

$\dfrac{20}{43} \times \$125,000 = \$58,140$

Product 3: $\dfrac{\$180,000}{\$516,000} = \dfrac{15}{43}$

$\dfrac{15}{43} \times \$125,000 = \$43,605$

40. $70 + 55 + 45 + 60 = 230$ employees

Dept. A: $\dfrac{70}{230} = \dfrac{7}{23}$

$\dfrac{7}{23} \times \$85,000 = \$25,870$

Dept. B: $\dfrac{55}{230} = \dfrac{11}{46}$

$\dfrac{11}{46} \times \$85,000 = \$20,326$

Dept. C: $\dfrac{45}{230} = \dfrac{9}{46}$

$\dfrac{9}{46} \times \$85,000 = \$16,630$

Dept. D: $\dfrac{60}{230} = \dfrac{6}{23}$

$\dfrac{6}{23} \times \$85,000 = \$22,174$

41. Annual income

$8\frac{1}{4}\% = 0.0825$

$\$225,000 \times 0.0825 = \$18,562.50$

42. $\$320,000 \times 0.01 = \$3200$ *commission*
$\$320,000 - \$3200 = \$316,800$

Annual income

$\$316,800 \times 7\frac{5}{8}\% = \$316,800 \times 0.07625 = \$24,156$

(c) Shields:

$\$32,000 \text{ (salary)} + (\$0.73 \times 30,000) = \$53,900$

Abbot:

$\$40,000 \text{ (salary)} + (\$0.73 \times 20,000) = \$54,600$

Dougherty:

$\$48,000 \text{ (salary)} + (\$0.73 \times 50,000) = \$84,500$

(d) $\$250,000 \times 0.55 \times 0.25 = \$34,375$
$\$34,375 \times 0.12 = \$4125$

(e) $7 + 5 + 4 = 16$

Allocation

Math: $\dfrac{7}{16} \times \$142,000 = \$62,125$

English: $\dfrac{5}{16} \times \$142,000 = \$44,375$

Reading: $\dfrac{4}{16} \times \$142,000 = \$35,500$

(f) Yes: Probably not.

Chapter 18 Summary Exercise

(a) $\$15,000 \times 2 = $ 30,000 shares of common for Shields
$\$10,000 \times 2 = $ 20,000 shares of common for Abbot
$\$25,000 \times 2 = $ 50,000 shares of common for Dougherty
                Total $= 100,000$ shares

(b) Preferred stock dividend
$\$50 \times 0.08 = \$4$

Total paid to preferred

$10,000 \times \$4 = \$40,000$

Available for common stockholders

$250,000 \times 0.45 - \$40,000 = \$112,500 - \$40,000$
$= \$72,500$

Dividend per share of common stock

$\dfrac{\$72,500}{100,000} = \$0.73$

BUSINESS STATISTICS

19.1 Frequency Distributions and Graphs

For exercises 1-5, refer to the graphs in the text.

1. $65.08

3. $\dfrac{\$65.08 - \$2.25}{\$65.08} = 0.965 = 97\%$

5. 51.4%

	Number of Units	Frequency
7.	0-24	4
9.	50-74	6
11.	100-124	5

13. See the graph in the answer section of the textbook.

15. $6 + 3 + 5 + 9 = 23$

17. $4 + 3 = 7$

19. Answers will vary.

21. **(a)** Go across the graph from the working width (36 inches) to the $2\frac{1}{2}$ mph diagonal line, then down to the bottom to find almost 1 acre covered per hour.

(b) Go across the graph from the working width (8 ft. = 96 inches) to the 4 mph diagonal line, then down to the bottom to find about $3\frac{7}{8}$ acres covered per hour.

(c) Go across the graph from the working width (48 inches) to the vertical line for hourly acreage covered (1), then up to the top to find 2 mph.

(d) Go up the graph from the hourly acreage covered, $\left(4\frac{1}{2}\right)$ to the diagonal line, $\left(4\frac{1}{2}\right)$, then across and to the left to find about 99 inches.

23. Find the degrees of a circle for each category.

Category	Percent	Degrees
Interview	47%	$0.47 \times 360° = 169.2°$
Reference	26%	$0.26 \times 360° = 93.6°$
Computer test	8%	$0.08 \times 360° = 28.8°$
Other	19%	$0.19 \times 360° = 68.4°$

See the graph in the answer section of the textbook.

25. **(a)** The graph shows that the average annual yield in 1925 was about 3.5%.

(b) The graph shows that interest rates fluctuated from 2% to 5% between 1905 and 1965.

(c) Since 1965, the intersect rates fluctuated between 4% and 13%.

(d) Interest rate in 1941: 2%
Interest rate in 1980: 12%
Increase from 1941 to 1980: 10%

19.2 The Mean

1. $\text{Mean} = \dfrac{128 + 240 + 164 + 380}{4}$

$= \dfrac{912}{4} = 228$

3. $\text{Mean} = \dfrac{\begin{array}{c}(3800 + 3625 + 3904 + 3296 \\ + 3400 + 3650 + 3822 + 4020)\end{array}}{8}$

$= \dfrac{29,517}{8} = 3689.6 \text{ (rounded)}$

5. $\text{Mean} = \dfrac{4220 + 3840 + 3640 + 4080}{4}$

$= \dfrac{15,780}{4} = 3945 \text{ pounds (rounded)}$

7. $\text{Mean} = \dfrac{\$1280 + \$2650 + \$870 + \$940 + \$760}{5}$

$= \dfrac{\$6500}{5} = \$1300$

9. Mean value of cars

$$= \frac{\begin{array}{l}(\$385{,}000 + \$495{,}000 + \$873{,}000 + \$1{,}210{,}000 \\ + \$611{,}000 + \$802{,}000 + \$173{,}000 + \$708{,}000)\end{array}}{8}$$

$$= \frac{\$5{,}257{,}000}{8} = \$657{,}125$$

11.

Value	Frequency	Product
30	5	150
40	3	120
45	2	90
48	4	192
Total	14	552

$$\text{Mean} = \frac{552}{14} = 39.4 \text{ (rounded)}$$

13.

Value	Frequency	Product
125	6	750
130	4	520
150	5	750
190	3	570
220	2	440
230	5	1150
Totals	25	4180

$$\text{Mean} = \frac{4180}{25} = 167.2$$

15. Answers will vary.

17.

Salary	Number of Employees	Product
$ 18,000	8	$ 144,000
$ 21,000	10	$ 210,000
$ 28,000	8	$ 224,000
$ 29,000	6	$ 174,000
$ 38,000	4	$ 152,000
$ 41,000	3	$ 123,000
$ 53,000	2	$ 106,000
$162,000	1	$ 162,000
Totals	42	$1,295,000

Mean salary for employees

$$= \frac{\$1{,}295{,}000}{42} = \$31{,}000 \text{ (rounded)}$$

19.

Units	Grade	Units × Grade
4	D (1)	4
2	A (4)	8
3	C (2)	6
1	F (0)	0
3	D (1)	3
13		21

Grade point average $= \dfrac{21}{13} = 1.62$ (rounded)

21.

Interval	Freq.	Class Mark	Freq. × Class Mark
50-59	15	54.5*	817.5
60-69	20	64.5	1290
70-79	21	74.5	1564.5
80-89	27	84.5	2281.5
90-99	18	94.5	1701
100-109	2	104.5	209
	103		7863.5

$$\text{Mean} = \frac{7863.5}{103} = 76.3 \text{ (rounded)}$$

*Find the midpoint (or class mark) by averaging the lowest and highest numbers in the interval, or $\frac{50+59}{2} = 54.5$.

23.

Interval	Freq.	Class Mark	Freq. × Class Mark
25-49	18	37	666
50-74	15	62	930
75-99	30	87	2610
100-124	18	112	2016
125-149	32	137	4384
150-174	14	162	2268
175-199	7	187	1309
	134		14,183

$$\text{Mean} = \frac{14{,}183}{134} = 105.8 \text{ (rounded)}$$

25. Final average $= \dfrac{\text{Total of all scores}}{\text{Number of scores}}$

Let $x = $ the score needed on the third test.

$$70 = \frac{74 + 68 + x}{3}$$

$$70 \times 3 = 74 + 68 + x$$

$$210 = 142 + x$$

$$210 - 142 = 142 - 142 + x$$

$$68 = x$$

Wanda needs 68% on the third test.

19.3   The Median and the Mode

1. 12, 14, 32, 37, 65

   The median is the middle value when an odd number of values is arranged in numerical order.
   Median = 32

3. 75, 81, 95, 98

   The median is the mean of the two middle values when an even number of values is arranged in numerical order.
   $$\text{Median} = \frac{81 + 95}{2} = 88$$

5. 0.81, 0.82, 0.84, 0.86

   The median is the mean of the two middle values when an even number of values is arranged in numerical order.
   $$\text{Median} = \frac{0.82 + 0.84}{2} = 0.83$$

7. The mode is 60 because it appears most often in the list.

9. No value appears more often than any other. Hence, there is no mode.

11. The 6 and 4 appear most often. They are both modes. This is a bimodal data set.

13. Answers will vary.

15. Mean
   $$= \frac{68 + 90 + 56 + 82 + 110}{5} = 81.2 \text{ ft}$$
   56, 68, 82, 90, 110
   Median = 82 feet

17. Use the median.

19.

Interval	Class	Freq.	Freq. × Mark
100-109	104.5	10	1045
110-119	114.5	12	1374
120-129	124.5	8	996
130-139	134.5	2	269
Total		32	3684

Calculate the class mark by adding the endpoints and dividing by 2.
$$\text{Mean} = \frac{3684}{32} = 115.1 \text{ (rounded)}$$

There are 32 values. Divide 32 by 2 to get 16. The mean of the 16th and 17th number from smallest to largest is the median. (Use the class marks.) Use 114.5 as the 16th and 17th values.
$$\text{Median} = \frac{114.5 + 114.5}{2} = 114.5$$

The second class from the top has the most values.
Mode = 114.5.

21.

Interval	Class	Freq.	Freq. × Mark
2.76-2.85	2.805	0	0
2.86-2.95	2.905	3	8.715
2.96-3.05	3.005	48	144.24
3.06-3.15	3.105	5	15.525
3.16-3.25	3.205	0	0
Total		56	168.48

$$\text{Mean} = \frac{168.48}{56} = 3.01 \text{ (rounded)}$$

The company has a successful bagging operation.

19.4   Range and Standard Deviation

1. $$\text{Mean} = \frac{15 + 18 + 20 + 19}{4}$$
   Mean = 18

Data	Deviation	Square of Dev.
15	−3	9
18	0	0
20	2	4
19	1	1
Total		14

$$\frac{\text{Sum of Squares of Deviation}}{\text{Number}} = \frac{14}{4} = 3.5$$

$$\text{Standard deviation} = \sqrt{3.5}$$
$$= 1.9 \text{ (rounded)}$$

3. $$\text{Mean} = \frac{20 + 22 + 23 + 18 + 21 + 22}{6}$$
   $$\text{Mean} = \frac{126}{6} = 21$$

Data	Deviation	Square of Dev.
20	−1	1
22	1	1
23	2	4
18	−3	9
21	0	0
22	1	1
Total		16

$$\frac{\text{Sum of Squares of Deviation}}{\text{Number}} = \frac{16}{6}$$
$$= 2.67 \text{ (rounded)}$$

Standard deviation $= \sqrt{2.67} = 1.6$ (rounded)

5. Mean $= \dfrac{55 + 58 + 54 + 52 + 51 + 59 + 58 + 60}{8}$

Mean $= 55.9$ (round)

Data	Deviation	Square of Dev.
55	−0.9	0.81
58	2.1	4.41
54	−1.9	3.61
52	−3.9	15.21
51	−4.9	24.01
59	3.1	9.61
58	2.1	4.41
60	4.1	16.81
Total		78.88

$$\frac{\text{Sum of the Squares of Deviation}}{\text{Number of Data}} = \frac{78.88}{8}$$
$$= 9.86$$

Standard deviation $= \sqrt{9.86}$
$\qquad\qquad\qquad = 3.1$ (rounded)

7. The range is the distance between the largest and smallest value.
Range $= 61 - 18 = 43$

9. The range is the distance between the largest and smallest value.
Range $= 500 - 112 = 388$

11. Answers will vary.

13. Add the percentages to the right of the mean.
$34\% + 13.5\% + 2\% + 0.5\% = 50\%$

$0.50 \times 200 = 100$ students

15. Add the percentages between 79.5 and 80.5.
$34\% + 34\% = 68\%$

$0.68 \times 200 = 136$ students

17. Between 80.5 and 81, there are 13.5% of the 200 students.

$0.135 \times 200 = 27$ students

19. Add the percentages within one pound of the mean; two standard deviation.
$13.5\% + 34\% + 34\% + 13.5\% = 95\%$

$0.95 \times 200 = 190$ students

21. Less than 80 is outside two standard deviations from the mean. The chances of less than 80 sales occurring is 2.5%.

23. Between 80 and 160 is two standard deviations on either side of the mean. The chance of sales being between 80 and 160 is 95%.

25. Since 100 is the mean, about 50% of the scores are greater than 100.

27. The score of 115 is one standard deviation above the mean. Add the percentages above one standard deviation.

$13.5\% + 2\% + 0.5\% = 16\%$

About 16% of the scores are greater than 115.

29. A score of 70 is two standard deviations below the mean, and 130 is two standard deviations above the mean. Add the percentages in this area of the normal curve.

$13.5\% + 34\% + 34\% + 13.5\% = 95\%$

About 95% of the scores will be between 70 and 130.

31. A score of 55 falls three standard deviations below the mean. So, about 0.5% of the scores are less than 55.

33. The mean is 0.2 inches. Calculate one standard deviation from the mean.

$0.2 - 0.015 = 0.185$

Since 0.185 inch is less than the mean, 0.185 inch or more includes all the tire wear except the 16% less than one standard deviation. So, 84% of the wear is 0.185 inches or more.

35. The mean is 0.2 inch. Calculate two standard deviations from the mean.

$0.2 + 0.015 + 0.015 = 0.23$ inches.

Since 0.23 inch is two standard deviations above the mean, only 2.5% wear remains. So, 0.23 inch is 97.5% of the wear.

37. Calculate one standard deviation from the mean.

$47.6 - 2.7 = 44.9$ minutes

At least 44.9 minutes is from one standard deviation below the mean and beyond or 84%.

39. Calculate two standard deviations from the mean.

$47.6 + 2.7 + 2.7 = 53$ minutes.

No more than 53 is everything less than two standard deviations above the mean or 97.5%.

41. Calculate one standard deviation above and below the mean.

$47.6 + 2.7 = 50.3$ minutes.
$47.6 - 2.7 = 44.9$ minutes.

Between 44.9 minutes and 50.3 minutes is within one standard deviation above and below the mean or 68%.

43. Calculate three standard deviations above the mean.

$12.5 + 2.2 + 2.2 + 2.2 = 19.1$ claims.

This includes 99.5% of the claims. Only 0.5% should take more than 19.1 minutes.

$0.005 \times 200 = 1$ claim.

19.5   Index Numbers

1. Price relative

$= \dfrac{\text{Price this year}}{\text{Price in base year}} \times 100$

$= \dfrac{\$550}{\$225} \times 100$

$= 244.4$ (rounded)

3. Price relative

$= \dfrac{\text{Price this year}}{\text{Price in base year}} \times 100$

$= \dfrac{\$1200}{\$2000} \times 100$

$= 60$

5. Price relative

$= \dfrac{\text{Price this year}}{\text{Price in base year}} \times 100$

$= \dfrac{\$2.10}{\$1.25} \times 100$

$= 168$

7. Refer to Table 19.7. Move down the left column to Food and beverages and horizontally to the column headed Philadelphia. Find 154.3. A price index of 154.3 means that the goods which cost $100 in the base year would now cost $154.30.

9. Refer to the Table 19.7. Move down the left column to Medical and horizontally to the column headed New York. Find 244.5. A price index of 244.5 means that the goods which cost $100 in the base year would now cost $244.50.

11. Refer to the Table 19.7. Move down the left column to All items, and horizontally to the column headed Dallas. Find 151.4. A price index of 151.4 means that the goods which cost $100 during the base year would now cost $151.40.

13. Refer to the Table 19.7. Note that the price index for housing in New York is 171.7.
This means that a house costing $180,000 would now cost $180,000 $\times$ 1.717 = $309,000 (rounded).

15. Refer to the Table 19.7. Note that the price index for Food and beverages in Dallas is 157.0.
This means that food and beverages which cost $7200 in 1987 would now cost

$7200 $\times$ 1.57 = $11,300 (rounded).

17. To determine in which urban area housing costs have increased the fastest, scan the line of numbers to the right of each expense item in Table 19.7 in the textbook. The expense item with the highest number is Medical.

19. Answers will vary.

Chapter 19 Review Exercises

1.

Gallons of Gasoline	Number of weeks
10,000-10,999	1
11,000-11,999	3
12,000-12,999	10
13,000-13,999	3
14,000-14,999	2
15,000-15,999	1

2. $3 + 2 + 1 = 6$

3. See graph in the answer section of the textbook.

4.

Item	Dollar Amount	Percent of total	Degrees of a Circle
Newsprint	$12,000	20%	72°
Ink	$ 6000	10%	36°
Wire Service	$18,000	30%	108°
Salaries	$18,000	30%	108°
Other	$ 6000	10%	36°

5. See graph in the answer section of the textbook.

6. $20\% + 10\% + 30\% = 60\%$

7. Mean $= \dfrac{25 + 20 + 18 + 35 + 19}{5}$

$= 23.4$

Arrange in numerical order.

18, 19, 20, 25, 35

The median is the middle value when the number of values is odd.

Median $= 20$

There is no mode.

8. Mean $= \dfrac{85 + 80 + 82 + 82 + 88 + 90 + 92}{7}$

Mean $= 85.6$ (rounded)

Arrange in numerical order.

80, 82, 82, 85, 88, 90, 92

The median is the middle value when the number of values is odd

Median $= 85$

The mode is the value that appears most often.

Mode $= 82$

9. Mean

$= \dfrac{21 + 20 + 20 + 18 + 21 + 19 + 21 + 22}{8}$

$= 20.3$ (rounded)

Arrange in numerical order.

18, 19, 20, 20, 21, 21, 21, 22

The median is the mean of the two middle values when the number of values is even.

Median $= \dfrac{20 + 21}{2} = 20.5$

The mode is the value that appears most often.

Mode $= 21$

10. Mean $= \dfrac{42 + 44 + 41 + 44 + 45 + 44}{6}$

$= 43.3$ (rounded)

Arrange in numerical order.

41, 42, 44, 44, 44, 45

The median is the mean of the two middle values when the number of values is even.

Median $= \dfrac{44 + 44}{2} = 44$

The mode is the value that appears most often.

Mode $= 44$

11. Mean

$= \dfrac{8 + 7 + 6 + 6 + 7 + 7 + 5 + 9}{8}$

$= 6.9$ (rounded)

Arrange in numerical order.

5, 6, 6, 7, 7, 7, 8, 9

The median is the mean of the two middle values when the number of values is even.

Median $= \dfrac{7 + 7}{2} = 7$

The mode is the value that appears most often.

Mode $= 7$

12. Mean

$= \dfrac{2.5 + 2.4 + 2.4 + 2.3 + 2.4 + 2.6 + 2.0 + 2.2}{8}$

$= 2.4$ (rounded)

Arrange in numerical order.

2.0, 2.2, 2.3, 2.4, 2.4, 2.4, 2.5, 2.6

The median is the mean of the two middle values when the number of values is even.

Median $= \dfrac{2.4 + 2.4}{2} = 2.4$

The mode is the value that occurs most often.

Mode $= 2.4$

13.

Interval	Class Mark	Freq.	Freq. × Mark
10-14	12	6	72
15-19	17	3	51
20-24	22	5	110
25-29	27	7	189
30-34	32	5	160
35-39	37	9	333
		35	915

Mean $= \dfrac{915}{35} = 26.1$ (rounded)

14.

Interval	Class Mark	Freq.	Freq. × Mark
10-19	14.5	6	87
20-29	24.5	5	122.5
30-39	34.5	9	310.5
40-49	44.5	4	178
50-59	54.5	7	381.5
		31	1079.5

$$\text{Mean} = \frac{1079.5}{31} = 34.8 \text{ (rounded)}$$

15.

Interval	Class Mark	Freq.	Freq. × Mark
1-5	3	20	60
6-10	8	12	96
11-15	13	14	182
16-20	18	10	180
21-25	23	5	115
		61	633

The median is the middle score, which occurs in the 6-10 interval. Use the class mark.

Median = 8

The mode is the score that occurs most often, which is in the 1-5 interval.

Mode = 3

16.

Interval	Class Mark	Freq.	Freq. × Mark
50-59	54.5	3	163.5
60-69	64.5	5	322.5
70-79	74.5	18	1341
80-89	84.5	12	1041
90-99	94.5	4	378
		52	3246

The median is the middle score, which occurs in the 70-79 interval. Use the class mark.
Median = 74.5
The mode is the score that occurs most often, which is in the 70-79 interval.
Mode = 74.5

17. Range = $91 - 24 = 67$

$$\text{Mean} = \frac{62 + 24 + 38 + 91 + 56}{5} = 54.2$$

Data	Deviation	Square of Deviation
62	7.8	60.84
24	−30.2	912.04
38	−16.2	262.44
91	36.8	1354.24
56	1.8	3.24
Total		2592.8

$$\frac{\text{Sum of Squares of Deviation}}{\text{Number}} = \frac{2592.8}{5}$$
$$= 518.56 \text{ (rounded)}$$

Standard deviation $= \sqrt{518.56} = 22.77$ (rounded)

18. Range = $18 - 5 = 13$

$$\text{Mean} = \frac{5 + 7 + 12 + 10 + 7 + 12 + 18}{7}$$
$$= 10.14 \text{ (rounded)}$$

Data	Deviation	Square of Deviation
5	−5.14	26.4196
7	−3.14	9.8596
12	1.86	3.4596
10	−0.14	0.0196
7	−3.14	9.8596
12	1.86	3.4596
18	7.86	61.7796
Total		114.8572

$$\frac{\text{Sum of Squares of Deviation}}{\text{Number}} = \frac{114.8572}{7}$$
$$= 16.41 \text{ (rounded)}$$

Standard deviation $= \sqrt{16.41} = 4.05$ (rounded)

19. Range = $86 - 65 = 21$

$$\text{Mean} = \frac{82 + 86 + 78 + 74 + 65}{5} = 77$$

Data	Deviation	Square of Deviation
82	5	25
86	9	81
78	1	1
74	−3	9
65	−12	144
Total		260

$$\frac{\text{Sum of Squares of Deviation}}{\text{Number}} = \frac{260}{5}$$

$$= 52$$

Standard deviation $= \sqrt{52} = 7.21$ (rounded)

20. Range $= 162 - 120 = 42$

$$\text{Mean} = \frac{150 + 145 + 130 + 120 + 162 + 158}{6}$$

$$= 144.17 \text{ (rounded)}$$

Data	Deviation	Square of Deviation
150	5.83	33.9889
145	0.83	0.6889
130	−14.17	200.7889
120	−24.17	584.1889
162	17.83	317.9089
158	13.83	191.2689
Total		1328.8334

$$\frac{\text{Sum of Squares of Deviation}}{\text{Number}} = \frac{1328.8334}{6}$$

$$= 221.47 \text{ (rounded)}$$

Standard deviation $= \sqrt{221.47} = 14.88$ (rounded)

21. Price relative $= \dfrac{\$32,300}{\$22,500} \times 100$

$$= 143.6 \text{ (rounded)}$$

22. Price relative $= \dfrac{\$410}{\$250} \times 100$

$$= 164.0$$

23. Price relative $= \dfrac{\$950}{\$1800} \times 100$

$$= 52.8 \text{ (rounded)}$$

24. Price relative $= \dfrac{\$300}{\$185} \times 100$

$$= 162.2 \text{ (rounded)}$$

25. Refer to the Table 19.7. Note that the price index for All items in Dallas is 151.4
This means that a family would need $33,500 × 1.514 = $50,700 (rounded) today.

26. Calculate one standard deviation below the mean.

$7.2 - 0.08 = 7.12$

Find the sum of the percentages.

$34\% + 34\% + 13.5\% + 2\% + 0.5\% = 84\%$

About 84% of the items can be expected to have a coating thicker than 7.12 millimeters.

27. The graph does not clearly show an increase or decrease in sales. See the graph in the answer section of textbook.

28.

Hours	Grade	Hours × Grade
3	A(4)	12
3	C(2)	6
4	B(3)	12
3	D(1)	3
13		33

Grade point average $= \dfrac{33}{13} = 2.54$ (rounded)

Hours	Grade	Hours × Grade
3	A(4)	12
3	C(2)	6
4	B(3)	12
3	F(0)	0
13		30

Grade point average $= \dfrac{30}{13} = 2.31$ (rounded)

29. Price relative $= \dfrac{\$195,000}{\$110,000} \times 100$

$$= 177.3 \text{ (rounded)}$$

Cost estimate $= \$90,000 \times 1.773$
$$= \$159,570$$

30. Yes. His personal problems seemed to influence his work performance.

Chapter 19 Summary Exercise

(a) Store 1

Median
6.5, 6.8, 6.9, 7.0, 7.5, 7.6, 7.8, 8.0, 8.2
Median = $7.5

Mean

$$\frac{6.5 + 6.8 + 7.0 + 6.9 + 7.5 + 7.8 + 8.0 + 7.6 + 8.2}{9}$$

Mean = $7.4
There is no mode.

Store 2

Median

6.2, 8.2, 8.2, 8.7, 9.6
Median = $8.2

Mean

$$\frac{8.2 + 6.2 + 8.2 + 8.7 + 9.6}{5}$$

Mean = $8.2
Mode = $8.2

(b) See the graph in the answer section of the textbook.

(c) Sales at Store 2 seem to be growing faster than at Store 1.

Cumulative Review Exercises (Chapters 16-19)

1. $c = \$4840$; $n = 4$; $s = \$200$

(a) $\frac{1}{4} = 0.25 = 25\%$

(b) $d = \frac{c-s}{n} = \frac{\$4840 - \$200}{4}$
$$= \$1160$$

(c) $b = c - 2d = \$4840 - (2 \times \$1160)$
$$= \$2520$$

2. $c = \$16,400$; $n = 10$; $s = \$0$

rate $= \frac{2}{n} = \frac{2}{10} = 20\%$

(a) $d = r \times b = 0.20 \times \$16,400$
$$= \$3280$$

(b) $b = c - d = \$16,400 - \$3280$
$$= \$13,120$$

$d = r \times b = 0.20 \times \$13,120$
$$= \$2624$$

3. $c = \$12,820$; $s = \$400$; $n = 8$

$$\frac{n(n+1)}{2} = \frac{8 \times 9}{2} = 36$$

Year	Computation	Amt. of Deprec.	Accum. Deprec.	Book Value
0	----	----	----	$12,820
1	($\frac{8}{36} \times \$12,420$)	$2760	$2760	$10,060
2	($\frac{7}{36} \times \$12,420$)	$2415	$5175	$7645

4. $c = \$42,250$; $n = 11,500$; $s = \$2000$

Depreciation amount

$c - s = \$42,250 - \$2000 = \$40,250$

Depreciation per unit

$$\frac{c-s}{n} = \frac{\$42,250 - \$2000}{11,500} = \$3.50$$

Depreciation for the first year

$12 \times 365 = 4380$ hours
$4380 \times \$3.50 = \$15,330$

Book value

$\$42,250 - \$15,330 = \$26,920$

5. $c = \$115,800$; $n = 5$; $s = \$3800$

April to Dec. 31 is 9 months

(a) Straight-line method

$$d = \frac{c-s}{n} = \frac{\$115,800 - \$3800}{5}$$
$$= \$22,400$$

$$\$22,400 \times \frac{9}{12} = \$16,800$$

(b) Doubling-decling-balance method

rate $= \frac{2}{n} = \frac{2}{5} = 40\%$

$d = r \times b = 0.40 \times \$115,800$
$$= \$46,320$$

$$\$46,320 \times \frac{9}{12} = \$34,740$$

(c) Sum-of-the-year's-digits method

$$\frac{n(n+1)}{2} = \frac{5 \times 6}{2} = 15; \text{ rate} = \frac{5}{15}$$

$d = r \times (c - s) = \frac{5}{15} \times (\$115,800 - \$3800)$
$$= \$37,333$$

$$\$37,333 \times \frac{9}{12} = \$28,000$$

6. (a) Tugboat-10 years

(b) Delivery Van-5 years

(c) 4-year-old race horse-3 years

(d) 32-unit apartment house-27.5 years

(e) couch for dentist office-7 years

7.

Year	Computation	Amt. of Deprec.	Accum. Deprec.	Book Value
0	----	----	----	$96,000
1	$(33.33\% \times \$96,000)$	$32,000	$32,000	$64,000
2	$(44.45\% \times \$96,000)$	$42,672	$74,672	$21,328
3	$(14.81\% \times \$96,000)$	$14,218	$88,890	$7110
	$(7.41\% \times \$96,000)$	$7110*	$96,000	0
	*to depreciate to 0			

8.

THE FASHION SHOPPE
INCOME STATEMENT
YEAR ENDING DECEMBER 31

Gross Sales		$240,800
Returns		− $4300
Net Sales		$236,500
Inventory, January 1	$48,300	
Cost of GoodsP urchased		
$102,000		
Freight + $2900		
Total Cost of Goods Purchased	+ $153,200	
Total of Goods Available for Sale	$104,900	
Inventory, December 31	− $41,500	
Cost of Goods Sold		− $111,700
Gross Profit		$124,800
Expenses		
Salaries and Wages	$32,400	
Rent	$15,000	
Advertising	$2200	
Utilities	$3100	
Taxes on Inventory, Payroll	$6100	
Miscellaneous Expenses	+ $8900	
Total Expenses		− $67,700
NET INCOME BEFORE TAXES		$57,100
Income Taxes		− $11,400
NET INCOME		$45,700

9.   Cost of goods sold   Net income before taxes

$$\frac{\$111,700}{\$236,500} = 0.4723 = 47.2\% \qquad \frac{\$57,100}{\$236,500} = 0.2414 = 24.1\%$$

Gross profit   Net income

$$\frac{\$124,800}{\$236,500} = 0.5276 = 52.8\% \qquad \frac{\$45,700}{\$236,500} = 0.1932 = 19.3\%$$

10. THE FASHION SHOPPE
    BALANCE SHEET FOR DECEMBER 31

	Assets		
Currents Assets	$28,400		
Cash	$8400		
Notes Receivable	$3800		
Accounts Receivable	$41,500		
Inventory			
Total Current Assets		$82,100	
Plant Assets			
Land	$0		
Buildings	$0		
Fixtures	$12,200		
Total Plant Assets		$12,200	
TOTAL ASSETS			$94,300

	Liablities		
Current Liabilities			
Notes Payable	$4800		
Accounts Payable	$32,500		
Total Current Liabilities		$37,300	
Long-term Liabilities			
Mortgages Payable	$0		
Long-term Notes Payable	$0		
Total Long-term Liabilities		$0	
Total Liabilities		$37,300	

	Owner's Equity	
Owner's Equity		$57,000
Total Liabilities and Oqner's Equity		$94,300

11. Current ratio $= \dfrac{\text{Current assets}}{\text{Current liabilities}}$

$\qquad = \dfrac{\$82,100}{\$37,300} = 2.20$

Liquid assets $= \$28,400 + \$8400 + \$3800 = \$40,600$

Acid-test ratio $= \dfrac{\text{Liquid assets}}{\text{Current liabilities}}$

$\qquad = \dfrac{\$40,600}{\$37,300} = 1.09$

Average owner's equity $= \dfrac{\$42,800 + \$57,000}{2}$

$\qquad = \$49,900$

$\dfrac{\text{Net income after taxes}}{\text{Average owner's equity}} = \dfrac{\$45,700}{\$49,900}$

$\qquad = 0.92$

12. $80 \times 0.05 = \$4$ per preferred share

$4 \times 50 = \$200$

13. $60 \times 0.06 = \$3.60$ per preferred share each year

$3.60 \times 10,000 \times 2 = \$72,000$

$80,750 - \$72,000 = \$8750$

$$\frac{\$8750}{25,000} = \$0.35 \text{ per share of common stock}$$

14. Earnings per share of common stock

$$\frac{\$85,000\text{-}\$8000}{200,000} = \$0.385$$

15. $107\frac{1}{4}$ or 107.25

$107.25 \times 80 = \$8580$

Find the broker's commission of 1%

$8580 \times 0.01 = \$85.80$

Now find the odd-lot differential of $0.125 per share.

$0.125 \times 80 = \$10$

Price

$8580 + \$85.80 + \$10 = \$8675.80$

16. $62\frac{1}{2}$ or 62.50

   (a) Current yield

$$\frac{\$0.40}{\$62.50} = 0.0064$$
$$= 0.6\% \text{ (rounded)}$$

   (b) PE ratio

$$\frac{\$62.50}{\$2.06} = 30.3 \approx 30$$

17. $85\frac{1}{4}$ or 85.25

$85.25 \times 100 = \$8525$

Find the SEC fee.

$$\frac{\$8525}{300} = 28.41\overline{6}, \text{ which gives}$$

a fee of $0.29.

Sales commission = $15

The seller receives

$8525 - \$0.29 - \$15 = \$8509.71$

18. Low for the year: $20

Yield: 0.3%

High for the day: $24\frac{1}{8}$ or 24.125

19. Current yield: 6.0%

Volume: 5 bonds

Close: $90\frac{3}{8}$ or $1000 \times 0.90375 = \$903.75$

20. Net asset value: $30.82

Return on investment; 27.8%

21. $97\frac{1}{4} = 97.25\%$

$1000 \times 97.25\% = \$1000 \times 0.9725$
$= \$972.50$

$972.50 \times 25 = \$24,312.50$
$5 \times 25 = \$125$

The total cost for 25 bonds is

$24,312.50 + \$125 = \$24,437.50$

22. $92\frac{3}{8} = 92.375\%$

$1000 \times 92.375\% = \$1000 \times 0.92375$
$= \$923.75$

$923.75 \times 40 = \$36,950$

Commission

$5 \times 40 = \$200$

Net proceeds

$36,950 - \$200 = \$36,750$

23. **(a)** Equal shares

$$\frac{1}{2} \times \$48,000 = \$24,000 \text{ to each partner}$$

   (b) Ratio of 3:1

There are $3 + 1 = 4$ shares

Partner 1: $\frac{3}{4} \times \$48,000 = \$36,000$

Partner 2: $\frac{1}{4} \times \$48,000 = \$12,000$

   (c) Original investment

$40,000 + \$60,000 = \$100,000$

Partner 1: $\frac{\$40,000}{\$100,000} = \frac{2}{5}$

$\frac{2}{5} \times \$48,000 = \$19,200$

Partner 2: $\frac{\$60,000}{\$100,000} = \frac{3}{5}$

$\frac{3}{5} \times \$48,000 = \$28,800$

24. Original investment

$60,000 + $20,000 = $80,000
$85,000 − $28,000 = $57,000

Padgett: $\dfrac{\$60,000}{\$80,000} = \dfrac{3}{4}$

$\dfrac{3}{4} \times \$57,000 = \$42,750$

Harden: $\dfrac{\$20,000}{\$80,000} = \dfrac{1}{4}$

$\dfrac{1}{4} \times \$57,000 = \$14,250$

$14,250 + $28,000 = $42,250

25. Machining: $\dfrac{30,000}{80,000} = \dfrac{3}{8}$

$\dfrac{3}{8} \times \$340,000 = \$127,500$

Stamping: $\dfrac{15,000}{80,000} = \dfrac{3}{16}$

$\dfrac{3}{16} \times \$340,000 = \$63,750$

Assembly: $\dfrac{35,000}{80,000} = \dfrac{7}{16}$

$\dfrac{7}{16} \times \$340,000 = \$148,750$

26. Dept. A: $\dfrac{20}{50} = \dfrac{2}{5}$

$\dfrac{2}{5} \times \$180,000 = \$72,000$

Dept. B: $\dfrac{12}{50} = \dfrac{6}{25}$

$\dfrac{6}{25} \times \$180,000 = \$43,200$

Dept. C: $\dfrac{18}{50} = \dfrac{9}{25}$

$\dfrac{9}{25} \times \$180,000 = \$64,800$

27. $12,800 \times \$25 = \$320,000$
$14,200 \times \$15 = \$213,000$
$180 \times \$200 = \$\ 36,000$
Total Value = $569,000

Product 1

$\dfrac{\$320,000}{\$569,000} \times \$140,000 = \$78,734.62$

Product 2

$\dfrac{\$213,000}{\$569,000} \times \$140,000 = \$52,407.73$

Product 3

$\dfrac{\$36,000}{\$569,000} \times \$140,000 = \$8857.64$

28. 1998: $560
1999: $580

$\dfrac{\$580 - \$560}{\$560} = 3.6\%$

29. 1991-92: $300
1998-99: $150

$\dfrac{\$300 - \$150}{\$300} = 0.50 = 50\%$

30. See the chart and graph in the answer section of the textbook.

31. (a) Mean

$\dfrac{18 + 14 + 16 + 17 + 16 + 19}{6} = 16.7$

Median
Arrange in numerical order.

14, 16, 16, 17, 18, 19

The median is the mean of the two middle values when the number of values is even.

Median $= \dfrac{16 + 17}{2} = 16.5$

The mode is the value that appears most often.
Mode $= 16$

(b) Mean

$\dfrac{24 + 32 + 25 + 24 + 31 + 28 + 27}{7} = 27.3$

Mean
Arrange in numerical order.

24, 24, 25, 27, 28, 31, 32

The median is the middle value when the number of values is odd.

Median $= 27$

The mode is the value that appears most often.
Mode $= 24$

32.

Hours	Grade	Hours × Grade
3	B(3)	9
4	C(2)	8
3	C(2)	6
3	A(4)	12
13		35

Grade point average $= \dfrac{35}{13} = 2.69$ (rounded)

33. Mean

$$\frac{78 + 82 + 71 + 69}{4} = 75$$

Range

$$82 - 69 = 13$$

Standard deviation

Data	Deviation	Square of Deviation
78	3	9
82	7	49
71	−4	16
69	−6	36
Total		110

$$\frac{\text{Sum of the squares of Deviation}}{\text{Number}} = \frac{110}{4} = 27.5$$

Standard deviation $= \sqrt{27.5} = 5.2$ (rounded)

34. The category between 0.625 inches and 0.715 inches is everything between the mean and two standard deviations above the mean. Find the sum of the percentages

$$34\% + 13.5\% = 47.5\%$$

35. Price relative

$$= \frac{\text{Price this year}}{\text{Price last year}} \times 100$$

$$= \frac{\$680}{\$1200} \times 100$$

$$= 57 \text{ (rounded)}$$

36. Refer to Table 19.7. Note that the price index for Housing in Chicago is 160.2. This means that a house that cost $68,000 in 1987 would not cost about $68,000 × 1.602 = $109,000 (rounded to the nearest thousand).

CALCULATOR BASICS

A.1 Scientific Calculators

A TI–30Xa calculator was used for these exercises.

1. 384.92 [+] 407.61 [+] 351.14 [+] 27.93 [=] 1171.6

3. 6850 [+] 321 [+] 4207 [=] 11,378

5. 4270.41 [−] 365.09 [=] 3905.32

7. 384.96 [−] 129.72 [=] 255.24

9. 365 [×] 43 [=] 15,695

11. 3.7 [×] 8.4 [=] 31.08

13. 375.4 [÷] 10.6 [=] 35.41509434

which rounds to 35.42.

15. 96.7 [÷] 3.5 [=] 27.62857143

which rounds to 27.63.

17. [(] 9 [×] 9 [)] [÷] [(] 2 [×] 5 [)] [=] 8.1

19. [(] 87 [×] 24 [×] 47.2 [)] [÷] [(] 13.6 [×] 12.8 [)] [=] 566.1397059

which rounds to 566.14.

21. [(] 2 [×] 3 [+] 4 [)] [÷] [(] 6 [+] 10 [)] [=] 0.625

which rounds to 0.63.

23. [(] 640 [−] 0.6 [×] 12 [)] [÷] [(] 17.5 [+] 3.2 [)] [=] 30.57004831

which rounds to 30.57.

25. [(] 14 [y^x] 2 [−] 3.6 [×] 6 [)] [÷] [(] 95.2 [÷] 0.5 [)] [=] 0.9159663866

which rounds to 0.92.

27. 7 [$a^{b/c}$] 5 [$a^{b/c}$] 8 [÷] [(] 1 [+] [$a^{b/c}$] 3 [$a^{b/c}$] 8 [)] [=] $5\frac{6}{11}$

29. [(] [$a^{b/c}$] 3 [$a^{b/c}$] 4 [÷] [$a^{b/c}$] 5 [$a^{b/c}$] 8 [)] [y^x] 3 [÷] 3 [$a^{b/c}$] 1 [$a^{b/c}$] 2 [=] 0.493714286

which rounds to 0.49.

31. Answers will vary.

33. 397 [×] 23.86 [+] 125 [×] 28.74 [+] 740 [×] 21.76 [=] 29,167.32

The total paid by the bookstore is $29,167.32.

35. 32 [×] [(] 2 [×] [(] 9.25 [+] 2 [×] 6.80 [)] [)] [+] 40 [×] [(] 9.25 [+] 3 [×] 6.80 [)] [=] 2648.4

The payroll is $2648.40.

37. (a) 0.065 [×] 17,908.43 [=] 1164.04795

The tax on the new car is $1164.05.

(b) 0.065 [×] 1463.58 [=] 95.1327

The tax on the office word processor is $95.13.

39. 8000 [+] [(] 528.31 [×] 12 [×] 30 [)] [−] 80,000 [=] 118,191.6

The sum of her down payment and all monthly payments exceeds the purchased price by $118,191.60.

41. 15,000 [+] 2800 [+] 28,000 [−] 32,400 [=] 13,400

Ben needs an additional $13,400.

A.2 Financial Calculators

An HP-12C calculator was used for these exercises.

1. 10 [n] 8 [i] 3500 [PV] [FV] −7,556.24

3. 10 [n] 3 [i] 12000 [FV] [PV] −8,929.13

5. 7 [n] 8 [i] 300 [PMT] [FV] −2,676.84

7. 30 [n] 319.67 [PMT] 12000 [CHS] [FV] [i] 1.50 (1.5%)

9. 360 [n] 1 [i] 83500 [PV] [PMT] −858.89

11. 4 [i] 85383 [PV] 5600 [CHS] [PMT] [n]
    24.00 (24)

13. 16 [n] 1.5 [i] 2000 [PV] [FV] −2,537.97

15. 360 [n] 0.75 [i] 86500 [PV] [PMT] − 696.00

17. 0.8 [i] 12000 [CHS] [PMT] 340000 [FV] [n]
    26.00 (26)

19. 345000 [Enter] 0.75 [×] 258,750.00

    60 [n] 1 [i] 258750 [PV] [PMT] −5,755.75

THE METRIC SYSTEM

1. $68 \text{ cm} = \dfrac{68}{100} = 0.68 \text{ m}$

3. $4.7 \text{ m} = 4.7 \times 1000 = 4700 \text{ mm}$

5. $8.9 \text{ kg} = 8.9 \times 1000 = 8900 \text{ g}$

7. $39 \text{ cL} = \dfrac{39}{100} = 0.39 \text{ L}$

9. $46{,}000 \text{ g} = \dfrac{46{,}000}{1000} = 46 \text{ kg}$

11. $0.976 \text{ kg} = 0.976 \times 1000 = 976 \text{ g}$

13. $36 \text{ m} = 36 \times 1.09 = 39.24 \text{ yards}$

15. $55 \text{ yards} = 55 \times 0.914 = 50.27 \text{ m}$

17. $4.7 \text{ m} = 4.7 \times 3.28 = 15.42 \text{ feet}$

19. $3.6 \text{ feet} = 3.6 \times 0.305 = 1.10 \text{ m}$

21. $496 \text{ km} = 496 \times 0.62 = 307.52 \text{ miles}$

23. $768 \text{ miles} = 768 \times 1.609 = 1235.71 \text{ km}$

25. $683 \text{ g} = 683 \times 0.00220 = 1.50 \text{ pounds}$

27. $4.1 \text{ pounds} = 4.1 \times 454 = 1861.4 \text{ g}$

29. $38.9 \text{ kg} = 38.9 \times 2.20 = 85.58 \text{ pounds}$

31. $1 \text{ kg} = 1000 \text{ g}$

$\dfrac{1000}{5} = 200$

There are 200 nickels in 1 kilogram of nickels.

33. $3\text{L} = 3 \times 1000 = 3000 \text{ mL}$

$3000 \text{ mL} \times \dfrac{0.0002 \text{ g}}{1 \text{ mL}} = 0.6 \text{ g}$

The helium weights 0.6 g.

35. Answers will vary.

37. $C = \dfrac{5(104 - 32)}{9} = 40° \text{ C}$

39. $C = \dfrac{5(536 - 32)}{9} = 280° \text{ C}$

41. $C = \dfrac{5(98 - 32)}{9} = 37° \text{ C (rounded)}$

43. $F = \dfrac{9 \times 35}{5} + 32 = 95° \text{ F}$

45. $F = \dfrac{9 \times 10}{5} + 32 = 50° \text{ F}$

47. $F = \dfrac{9 \times 135}{5} + 32 = 275° \text{ F}$

49. Unreasonable

51. Reasonable

53. Reasonable